STERLING
Test Prep

ACS
Physical Chemistry

Comprehensive Content

3rd edition

Copyright © Sterling Test Prep

Customer Satisfaction Guarantee

Your feedback is important because we strive to provide the highest quality educational materials. Email us comments or suggestions.

info@sterling-prep.com

We reply to emails – check your spam folder

All rights reserved. This publication's content, including the text and graphic images or part thereof, may not be reproduced, downloaded, disseminated, published, converted to electronic media, or distributed by any means without prior written consent from the publisher. Copyright infringement violates federal law and is subject to criminal and civil penalties.

3   2   1

ISBN-13: 979-8-8855732-2-1

Sterling Test Prep materials are available at quantity discounts.

Contact info@sterling–prep.com

Sterling Test Prep
6 Liberty Square #11
Boston, MA 02109

©2025 Sterling Test Prep

Published by Sterling Education

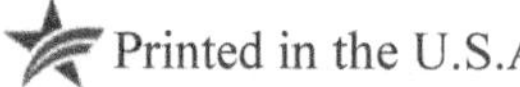 Printed in the U.S.A.

 Copyright © Sterling Test Prep.

STERLING
Test Prep

Thousands of students use our study aids to achieve high test scores!

Scoring well on ACS Physical Chemistry exam is a challenging task. This book provides a thorough review of physical chemistry topics tested on ACS, covering the knowledge areas necessary to answer test questions. Understanding key concepts, extracting and analyzing information, and distinguishing between similar answer choices are more effective than mere memorization. This book helps you master physical chemistry content and develop the ability to apply your knowledge on the exam.

This text is clearly presented and systematically organized to provide targeted ACS Physical Chemistry preparation. Learn the scientific foundations and essential concepts needed to master the material and answer exam questions. These review chapters teach important physical chemistry principles and relationships and how they apply to questions.

From the foundations of physical and chemical properties of matter to complex mechanisms of atomic particles, you will develop a better understanding of chemical phenomena on microscopic and macroscopic levels. This book is an essential overview of chemical kinetics, reaction mechanisms, thermodynamics, and quantum chemistry concepts. Learn about intra and intermolecular interactions, properties of gases, liquids and solids, chemical and phase equilibria, atomic theory and electronic structure.

Experienced chemistry instructors analyzed the test content and developed this study material that builds knowledge and skills crucial for success. Our test preparation experts systematized the content to match the current ACS exam for targeted and effective learning.

With this study material, you will significantly improve your test score.

250426akp

Featured on

Any duplication (copies, uploads, PDFs) is illegal.

Copyright © Sterling Test Prep.

ACS Physical Chemistry Practice Questions provides high-yield practice questions covering topics tested on ACS. Develop the ability to apply your knowledge and quickly choose the correct answer to increase your test score.

ACS Organic Chemistry Practice Questions provides high-yield practice questions covering topics tested on ACS Organic Chemistry exam. Develop the ability to apply your knowledge and quickly choose the correct answer to increase your test score.

Visit our Amazon

Any duplication (copies, uploads, PDFs) is illegal.

If you benefited from this book, please leave a review on Amazon so others can learn from your input. Reviews help us understand our customers' needs and experiences while keeping our commitment to quality.

Copyright © Sterling Test Prep.

Table of Contents

Table of Contents (*continued*)

CHAPTER 1: PHASES & PHASE EQUILIBRIA (*continued*)

Copyright © Sterling Test Prep.

Table of Contents (*continued*)

Any duplication (copies, uploads, PDFs) is illegal.

Table of Contents (*continued*)

CHAPTER 2: THERMODYNAMICS (*continued*)

Copyright © Sterling Test Prep.

Table of Contents (*continued*)

CHAPTER 2: THERMODYNAMICS (*continued*)

Any duplication (copies, uploads, PDFs) is illegal.

Table of Contents (*continued*)

CHAPTER 2: THERMODYNAMICS (*continued*)

Copyright © Sterling Test Prep.

Table of Contents (*continued*)

CHAPTER 2: THERMODYNAMICS (*continued*)

Any duplication (copies, uploads, PDFs) is illegal.

Table of Contents (*continued*)

CHAPTER 2: THERMODYNAMICS (*continued*)

Copyright © Sterling Test Prep.

Table of Contents (*continued*)

Table of Contents (*continued*)

Copyright © Sterling Test Prep.

Table of Contents (*continued*)

Any duplication (copies, uploads, PDFs) is illegal.

Table of Contents (*continued*)

Copyright © Sterling Test Prep.

Table of Contents (*continued*)

Table of Contents (*continued*)

Copyright © Sterling Test Prep.

Table of Contents (*continued*)

CHAPTER 5: STOICHIOMETRY (*continued*)

Any duplication (copies, uploads, PDFs) is illegal.

Table of Contents (*continued*)

CHAPTER 6: CHEMICAL KINETICS & EQUILIBRIA (*continued*)

Copyright © Sterling Test Prep.

Table of Contents (*continued*)

Table of Contents (*continued*)

 Copyright © Sterling Test Prep.

Table of Contents (*continued*)

Table of Contents (*continued*)

Copyright © Sterling Test Prep.

Table of Contents (*continued*)

Any duplication (copies, uploads, PDFs) is illegal.

Table of Contents (*continued*)

Copyright © Sterling Test Prep.

Table of Contents (*continued*)

CHAPTER 9: ELECTROCHEMISTRY (*continued*)

Table of Contents (*continued*)

Copyright © Sterling Test Prep.

Table of Contents (*continued*)

Table of Contents (*continued*)

Copyright © Sterling Test Prep.

Table of Contents (*continued*)

Copyright © Sterling Test Prep.

CHAPTER 1

Phases
&
Phase Equilibria

- Phase Equilibria

- Phase Diagrams

- Molality

- Colligative Properties

- Gas Phase

- Gas Pressure

- Kinetic Molecular Theory of Gases

- Gas Laws

- Partial Pressure

- Practice Questions & Detailed Explanations

Copyright © Sterling Test Prep.

Phase Equilibria

Phases of substances

Most substances exist in solid, liquid, or gas form.

A *solid* has a definite shape and volume (e.g., ice). The molecules in a solid vibrate about a fixed position, and a solid *cannot be compressed*.

A *liquid* has a definite volume but takes its container shape (e.g., water). The molecules move about but are close and *bound by intermolecular forces*.

A *gas* has neither a definite volume nor a definite shape; it takes the container's volume and shape (e.g., steam or water vapor). The molecules *move independently* and are not held together by intermolecular forces, and gas is *easily compressible*.

Freezing and melting point

The *freezing point* is the temperature at which a liquid changes state to become a solid (i.e., solidification).

The *melting point* is when a solid changes its state and becomes a liquid.

The melting point of a substance is the temperature at which it changes from a solid to a liquid, and it has the same temperature range as the *freezing point*; the reverse reaction occurs at the same temperature.

Theoretically, a substance's melting point and freezing point should be identical.

Experimentally, a difference in the respective processes' quantities may be observed.

Vaporization

Vaporization is the phase transition at which a liquid becomes a gas.

There are two types of vaporization: *boiling* and *evaporation*.

The *boiling point* is when the liquid is heated to a temperature (i.e., the kinetic energy of the molecules) where the vapor pressure is high enough for bubbles to form inside the liquid. It is in the same temperature range as the *condensation point*.

As a liquid is boiling, the temperature remains constant until the liquid has converted into gas.

Since the boiling point is based on pressure, the precise value depends on the environment (e.g., elevation relative to sea level).

For example, at high-altitude and low-pressure, a substance's boiling point is lower, requiring less heat energy to increase the vaporizing molecule's kinetic energy to break intermolecular attraction to adjacent molecules.

Any duplication (copies, uploads, PDFs) is illegal.

Evaporation occurs at temperatures below the boiling point. It occurs at the liquid's surface when the gaseous phase above the liquid is not saturated with the evaporating substance.

Boiling occurs from the bulk of the liquid, not just the surface.

Evaporation tends to occur more quickly in liquids with higher vapor pressure.

Condensation and sublimation

Condensation (reverse of evaporation) is when the substance changes from the gas phase to the liquid phase. Condensation commonly occurs when a vapor is cooled or compressed to its saturation limit.

The *condensation po*int is the temperature at which condensation occurs.

For example, condensation is the formation of water droplets in atmospheric clouds.

Sublimation (reverse of deposition) is the phase transition from a solid to a gas, essentially evaporation that occurs directly from the solid phase below the melting point.

Deposition (reverse of sublimation) is the phase transition from a solid directly into a gas (i.e., the matter is not present as an intermediate liquid). This process is observed in the winter when the temperature rises above freezing, and the snow does not thaw to liquid, yet the snow melts.

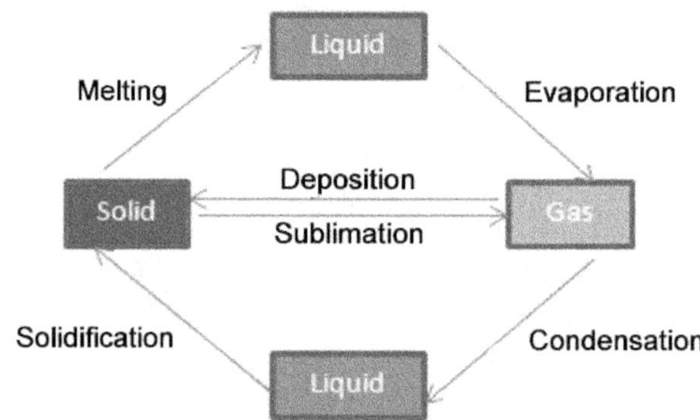

Interconversion of states of matter

Copyright © Sterling Test Prep.

Phase Diagrams

Phase change diagram

When an ice block is placed in the Sun, it slowly absorbs the thermal energy through radiation and undergoes a series of phase changes. The ice cube changes from a solid to a liquid to a gas.

The *melting point* of a substance is the temperature at which it changes from a solid to a liquid, and it has the same temperature range as the *freezing point*; the reverse reaction occurs at the same temperature.

The *boiling point* (vaporization) of a substance is the temperature at which it changes phase from a liquid to a gas, and it is the same temperature range as the *condensation point*.

From the graph below, once the ice cube absorbs enough heat to melt (i.e., reaches 0 °C), its temperature remains constant.

Melting is points B to C on the graph where the molecules have reached a temperature of 0 °C. The ice melts only when this occurs, as illustrated by the graph's horizontal plateau from B to C.

When the liquid water changes to a gas, the liquid absorbs enough heat to vaporize (i.e., at 100 °C). The liquid temperature remains constant from point D to E until all molecules reach a temperature of 100 °C. The horizontal plateau from D to E signifies the phase transition of vaporization.

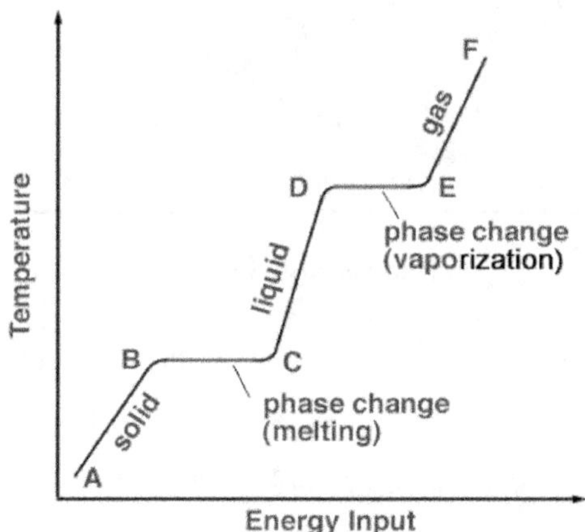

Any duplication (copies, uploads, PDFs) is illegal.

Heat during phase changes

Although the energy (heat) input increases with temperature, it does not remain constant as the temperature does during a phase change.

Technically, heat converts the potential energy (PE) stored within each atom to kinetic energy (KE) before changing phase.

As energy is continuously added to the process, the substance uses the added heat to transition each molecule before increasing the temperature.

Pressure *vs.* temperature relationship

A *phase diagram* displays the phases of a substance. A phase diagram is a graph plotted as pressure *vs.* temperature.

The graph is split into three disproportionate parts, representing the substance's three phases.

Each point on the graph represents a possible combination of pressure and temperature for the system.

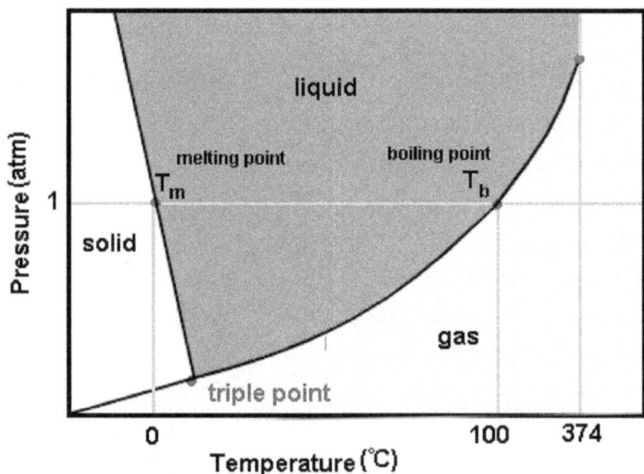

The horizontal line intersects on the phase diagram the temperature where the substance changes phases.

The *melting point* and *boiling point* are found by drawing a horizontal line at 1 atm of pressure, atmospheric pressure.

The *triple point* is shown when all three lines intersect and three phases exist simultaneously.

The *critical point* is the temperature above which the substance is a gas, regardless of pressure.

 Copyright © Sterling Test Prep.

Molality

Molal concentration

Molality (or *molal concentration*) measures the moles (i.e., the concentration of solutes) in a solvent in terms of the mass (on kg) of the solvent.

The SI unit for molality is mol/kg.

Molality m is different from the uppercase M for molarity (i.e., M = moles/liter).

> *Molality* = (moles of solute) / (mass in kg of solvent)

Calculating molality

For example, if 3.00 moles of sucrose are mixed into 1.00 L water until the sucrose is entirely dissolved (i.e., saturated solution). What is the molality of the solution? (Use the density of water = 1.00 g/mL or 1.00 kg/L).

Solution, the mass of the liter of water is 1.00 kg.

Use the solvent mass of 1. Kg/L and solve for molality.

> m = 3.00 mol / 1.00 kg

> m = 3.00 mol/kg of sucrose

> m = 3.00

The molality of the solution is 3.00 m.

Instead of expressing the number of moles, the solute's mass is given in grams.

When calculating molality, the solvent's mass is the pure solvent, excluding mass of the solute.

Molarity is the volume of solution, including the mass of the solute *and* the solvent.

> *Molality m* is expressed as mol solute/kg solvent.

> *Molarity M* is expressed as mol solute/L solution

For example, suppose 16.88 grams of sodium chloride (NaCl) are dissolved into 4.00 kg of water. What is the molality of the solution?

Convert grams of NaCl to moles (molar mass of NaCl = 58.44 g/mol).

> moles = (mass) / (molar mass)

> moles = (116.88 g) / (58.44 g/mol)

> moles = 2.00 mol NaCl

Solve for molality:

Any duplication (copies, uploads, PDFs) is illegal.

m = (moles of solute) / (mass in kg of solvent)

m = (2.00 mol) / (4.00 kg of solvent)

m = 0.500 mol/kg of the solution

m = 0.500

The molality of the solution is 0.500 mol/kg.

Copyright © Sterling Test Prep.

Colligative Properties

Solute particles

Colligative properties depend on the ratio of solute particles to solvent but *not* on the type of chemical species.

Solute particles in mixtures increase the strength of intermolecular (e.g., dipole-dipole) bonds. It is more difficult to boil the solution (i.e., boiling point elevation) or freeze the solution (i.e., freezing point depression).

Colligative properties include:

1) the relative lowering of vapor pressure,

2) the elevation of boiling point,

3) the depression of freezing point, and

4) osmotic pressure.

Van 't Hoff factor

The colligative properties calculations incorporate the van 't Hoff factor (i), which measures a solute's effect on the solution's colligative properties.

The van 't Hoff factor is:

$$\frac{[\text{concentration of particles produced when a substance dissolves}]}{[\text{concentration of a substance calculated from its mass}]}$$

Concentration (represented in [ ]) is converted to reflect the total number of particles in the solution. (e.g., 5 g of NaCl dissociates into Na^+ and Cl^-).

For example, glucose ($C_6H_{12}O_6$) has an i value of 1 because it does not dissociate in solution.

Nonelectrolytes do not dissociate in solution (e.g., CH_4 and O_2).

The salt NaCl has an i value of 2 because it dissociates into 1 Na^+ and 1 Cl^- (2 resulting ions per one reactant molecule).

Vapor pressure depression (Raoult's Law)

In 1888, French chemist Francois-Marie Raoult (1830-1901) proved that the vapor pressure of a solution equals the mole fraction of the solvent times the vapor pressure of the pure solvent.

Any duplication (copies, uploads, PDFs) is illegal.

Raoult's Law is expressed by:

$$P = \chi_{solvent} \cdot P^{\circ}_{solvent}$$

where P is the vapor pressure, $\chi_{solvent}$ is the mole fraction of solvent (moles of solvent / total moles of solute *and* solvent), and $P^{\circ}_{solvent}$ is the vapor pressure of the pure solvent

Calculate the χ_{solute} using the van 't Hoff factor (i.e., 1 mol of NaCl yields 2 mol of ions in the solution).

For example, suppose 2.00 moles of sucrose is added to a pitcher containing 2.00 liters of water. What is the vapor pressure of the sucrose solution? (Use the value 1.00 L water = 1.00 kg, the vapor pressure of pure water = 23.8 mmHg and the molecular mass of H_2O = 18.02 g/mol)

Convert the liters of water into mass:

2.00 L water = 2 Kg

2 Kg = 2,000 g

Convert the mass of the water into moles:

molecular mass = mass / moles

moles = mass / molecular mass

moles = (2,000 g) / (18.02 g/mol)

moles = 110.9 moles H_2O

Solve for the mole fraction, $\chi_{solvent}$:

$\chi_{solvent}$ = (moles of solvent) / (moles of the solute *and* solvent)

$\chi_{solvent}$ = (110.9 moles H_2O) / (moles of the solute *and* solvent)

$\chi_{solvent}$ = 110.9 moles H_2O / (110.9 moles solvent + 2 moles solute)

$\chi_{solvent}$ = 0.98

Use Raoult's Law to determine the pressure:

$P = \chi_{solvent} \cdot P^{\circ}_{solvent}$

$P = (0.98) \cdot (23.8$ mmHg$)$

$P = 23.4$ mmHg

The addition of a solute caused the vapor pressure to decrease.

Copyright © Sterling Test Prep.

Boiling point elevation ($\Delta T_b = K_b m$)

Adding a solute to a solvent stabilizes the solvent in the liquid phase, thus lowering the solvent molecules' tendency to move to the gas or solid phases. Therefore, the boiling point increases.

The *boiling point elevation* is proportional to the decrease of vapor pressure in a dilute solution:

$$\Delta T_b = k_b \cdot m \cdot i$$

where ΔT_b is the increase in boiling point, k_b is the molal boiling point constant (a given value determined experimentally), m is the molality (mol solute/kg solvent), and i is the van 't Hoff factor

For example, what is the boiling point elevation if 6.4 g of ammonia is dissolved in 0.3 kg of water? (Use the k_b for water = 0.52 °C/m and the molar mass of ammonia = 17.031 g/mol)

Convert the mass of ammonia (NH_3) into moles:

molecular mass = mass / moles

moles = mass / molecular mass

moles = (6.4 g) / (17.031 g/mol)

moles = 0.38 mol NH_3

Calculate the molality of the solution:

molality = (moles of solute) / (mass (in kg) of solvent)

m = (0.38 mol NH_3) / (0.3 kg of water)

m = 1.27 m

Calculate the boiling point elevation:

$$\Delta T_b = k_b \cdot m \cdot i$$

$$\Delta T_b = (0.52 \text{ °C/m}) \cdot (1.27 \text{ m}) \cdot (1)$$

$$\Delta T_b = 0.66 \text{ °C}$$

The boiling point increased by 0.66 °C due to the addition of 6.4 g of NH_3.

Any duplication (copies, uploads, PDFs) is illegal.

Freezing point depression ($\Delta T_f = K_f m$)

The *freezing point depression* (i.e., lowering the freezing point) is calculated by the following equation: (note the negative sign indicates that the change is decreased).

The freezing point depression is expressed as:

$$\Delta T_f = -k_f \cdot m \cdot i$$

where ΔT_f is the decrease in freezing point, k_f is the molal freezing point constant (a given value determined experimentally), m is the molality (mol solute/kg solvent), and i is the van 't Hoff factor (given since it is determined experimentally)

For example, a 48.0 g sample of a nonelectrolyte (does not dissociate in solution) is dissolved in 500.0 g of water to produce a solution with a freezing point of –3.5 °C. What is the molar mass of the compound? (Use k_f of water = 1.86 °C/m, the freezing point of pure water = 0 °C)

The freezing point of pure water = 0 °C, and the freezing point depression, ΔT_f = 3.5 °C.

Calculate the moles using the freezing point depression expression:

$$\Delta T_f = k_f \cdot m \cdot i$$

$$3.5 \ °C = (1.86 \ °C/m) \cdot (x \ / \ 0.5 \ kg) \cdot (1)$$

$$3.5 \ °C = (3.72 \ °C/m) \cdot (x)$$

$$3.5 \ °C \ / \ (3.72 \ °C/m) = x$$

$$x = 0.94 \ mol \ nonelectrolyte$$

Calculate molar mass by dividing the sample's mass by moles in the sample.

Molar mass = mass / moles

Molar mass = 48.0 g / 0.94 mol

Molar mass = 51.1 g/mol

The molar mass of the unknown nonelectrolyte is 51.1 g/mol.

Copyright © Sterling Test Prep.

Osmotic pressure

Osmotic pressure is the *minimum pressure that needs to be applied to a solution to prevent the inward flow of water across a semipermeable membrane.*

The semipermeable membrane allows solvent particles to pass, not solute particles.

The *osmotic pressure* (Π) is:

$$\Pi = MRT \cdot i$$

where Π is the osmotic pressure, M is the molarity (mol/L), R is the ideal gas constant (0.08206 L·atm/mol·K), and T is the temperature (K)

Osmotic pressure determines whether and in what direction osmosis occurs.

Osmosis is a solvent's movement across a semi-permeable membrane from a low solute concentration (i.e., high solvent concentration) to an area of high solute concentration (i.e., low solvent concentration).

The solvent moves from a low Π value to an area across the semipermeable membrane with a high Π value.

For example, what is the osmotic pressure of a solution prepared by adding 10.5 g of sucrose ($C_{12}H_{22}O_{11}$) to enough water to make 300.0 mL of solution at 25 °C? (Use the molecular mass of sucrose = 342.0 g/mol and the conversion of 300.0 mL = 0.30 L)

Calculate the number of moles of $C_{12}H_{22}O_{11}$ by dividing the mass of the sample by the molar mass of the sample.

$$\text{moles} = (\text{mass}) / (\text{molar mass})$$

$$\text{moles} = (10.5 \text{ g } C_{12}H_{22}O_{11}) / (342.0 \text{ g/mol } C_{12}H_{22}O_{11})$$

$$\text{moles} = 0.03 \text{ mol } C_{12}H_{22}O_{11}$$

Calculate the molarity, M:

$$\text{Molarity} = (\text{moles solute}) / (\text{liters of solution})$$

$$\text{Molarity} = (0.03 \text{ mol } C_{12}H_{22}O_{11}) / (0.30 \text{ L solution})$$

$$\text{Molarity} = 0.10 \text{ mol/L}$$

Convert temperature to Kelvin:

$$T = {}^\circ C + 273$$

$$T = 25 + 273$$

$$T = 298 \text{ K}$$

Any duplication (copies, uploads, PDFs) is illegal.

Since sucrose does not dissociate, calculate osmotic pressure (Π) using $i = 1$.

$$\Pi = MRT{\cdot}i$$

$$\Pi = (0.10 \text{ mol/L}){\cdot}(0.08206 \text{ L}{\cdot}\text{atm/mol}{\cdot}\text{K}){\cdot}(298 \text{ K}){\cdot}(1)$$

$$\Pi = 2.45 \text{ atm}$$

The osmotic pressure for the sucrose solution is 2.45 atm.

Colloids

A *solution* is a homogeneous mixture that consists of one phase (i.e., the solution stays mixed).

A *colloid* is a solution with microscopic insoluble particles dispersed throughout the solution.

A colloid has a dispersed phase (i.e., suspended particles) and a continuous phase (i.e., a medium of suspension that holds the particles).

A colloid stays mixed (i.e., the particles will not settle) unless centrifuged under the influence of gravity.

A typical colloid le is homogenized milk, consisting of butterfat globules dispersed within a water-based solution.

For example, a colloid forms when water and oil are shaken together.

Henry's Law

In 1803, English chemist William Henry (1774-1836) showed that at a constant temperature, the solubility of a gas in a liquid is directly proportional to the gas's partial pressure above the liquid (when the gas is in equilibrium with the liquid).

Henry's Law is described by :

$$P_{\text{solute}} = k_{\text{H}}{\cdot}c$$

where P_{solute} is the partial pressure of the solute at the solution's surface, k_{H} is Henry's Law constant (unique for each solute-solvent pair), and c is the concentration of the dissolved gas

For example, how many grams of carbon dioxide (CO_2) gas is dissolved in a 0.5 L bottle of carbonated water if the manufacturer uses a pressure of 2.5 atm in the bottling process? (Use the K_{H} of CO_2 in water = 29.76 atm/(mol/L) and molar mass of CO_2 = 44 g/mol)).

Determine the concentration of CO_2 using Henry's Law:

$$P_{\text{solute}} = k_{\text{H}}{\cdot}c$$

$$2.7 \text{ atm} = [29.76 \text{ atm/(mol/L)}]{\cdot}c$$

$$c = 0.09 \text{ mol/L}$$

  Copyright © Sterling Test Prep.

Calculate the moles of CO_2 in 0.5 L by dividing the mol/L by 2.

      moles / grams = molar mass

      0.09 / 2 = 0.045 mol

Convert moles to grams using the molar mass of CO_2, 44 g/mol (from the periodic table).

      0.045 mol $\times$ (44 g/mol) = 1.98 g

There are 1.98 grams of CO_2 in a 0.5 L bottle of carbonated water.

Notes for active learning

Copyright © Sterling Test Prep.

Gas Phase

Gaseous state

Gas is a fundamental state of matter, distinguished by the vast separation of the individual gas particles.

Unlike solids and liquids, gases have an indefinite shape and volume because they expand to fill their containers.

As a result, gases are subject to pressure, volume, and temperature changes.

Some common elements and compounds exist in the gaseous state under normal conditions of pressure and temperature, such as oxygen (O_2), nitrogen (N_2), and carbon dioxide (CO_2).

A vapor is a gaseous form usually existing as a liquid or solid at ordinary pressures and temperatures (e.g., water vapor).

Four variables are needed to describe the gaseous state:

1. Quantity of gas, n (in moles)

2. The temperature of the gas, T (in Kelvin)

3. The volume of gas, V (in liters)

4. The pressure of gas, P (in kPa)

Absolute temperature

Temperature is an objective measure of particles' kinetic energy, and the volume of a gas (i.e., the space that it occupies) is proportional to its temperature.

In the 19th century, experiments revealed that gases expand as temperature rises and compress when temperature drops, which led to the determination of the expansion coefficient of gases per degree Celsius.

This posed an interesting question: what happens if the temperature is low enough to reach a point where the gas's calculated volume is zero?

This *absolute zero* temperature was calculated to be –273.15 °C.

Gases do not reach zero volume because gases liquefy or solidify before reaching this temperature.

Additionally, gas cannot have a zero volume because the gas molecules' atoms occupy a certain amount of space.

Kelvin scale

In 1848, British physicist Lord William Thomson Kelvin (1824-1907) proposed a temperature scale based on this absolute zero value.

The Kelvin (K) scale starts with absolute zero as 0 K, and the increments are identical to the Celsius scale.

To convert from Kelvin to Celsius, add 273.15; and for the opposite conversion, subtract 273.15.

Kelvin should be converted to Celsius before converting it into Fahrenheit.

Kelvin temperatures are not written with a degree symbol (273 K *vs.* 273 °C or 273 °F).

Copyright © Sterling Test Prep.

Gas Pressure

Pressure as force

The *pressure* is the average force applied to or experienced by a given area.

Gas molecules are in motion, so in a container, the gas molecules continuously move around and collide with the container's walls. Since the force is applied over a specific area, this force is the container's gas pressure.

Newton's Second Law, $F = ma$, can be used to determine the units of pressure.

The SI units of mass (m) and acceleration (a) are the kilogram (kg) and the meter per second squared (m/s^2).

The unit of force (F) is derived from the equation:

$$F = ma, \text{ is kg·m/s}^2$$

This derived unit is Newton (N).

The pressure is the force exerted over a specific area.

The SI unit of length is a meter (m), so the SI unit for area is m^2.

Therefore, the unit for pressure is N/m^2, known as the Pascal (Pa).

Simple mercury barometer

A *barometer* measures Earth's atmospheric pressure. 30 inHg is considered normal.

Dry air creates more pressure than wet air, so dry air is more massive than the same amount of wet air.

Consequently, *high barometric readings* (i.e., 30.70 inHg) indicate dry air and fair weather.

Low barometric readings (i.e., 29.80 inHg) indicate an increased chance of rain or showers.

Hurricane conditions are indicated by barometric readings of 27.30 inHg.

A *mercury barometer* (shown below) is made by inverting a column filled with mercury and placing it in a mercury dish without allowing air to enter the tube. Some mercury flows out when the tube is inverted, leaving a space in the tube above the mercury. This space is nearly a vacuum. Therefore, there is no pressure pushing down on the column.

The pressure exerted by the mercury column balances the pressure of the atmosphere.

The mercury column's height is about 760 mm above the surface of the mercury in the dish at sea level.

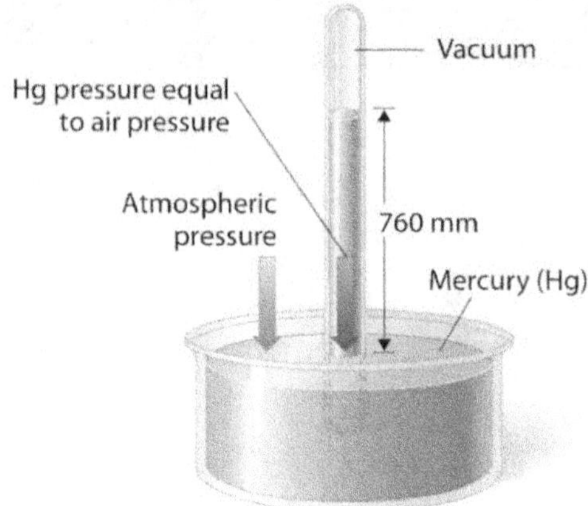

Expressing pressure

Any liquid may be used to measure atmospheric pressure; the height of the column depends on the density of the liquid. A comparison of mercury (density = 13.6 g/cm^3) and water (density = 1 g/cm^3) indicates that the column would be almost 34 feet high if a barometer were filled with water.

The pressure exerted by the mercury barometer is recorded in units of millimeters of mercury (mmHg), a unit sometimes called the torr in honor of the Italian physicist Evangelista Torricelli (1608-1647), who invented the mercury barometer in 1643.

The standard atmosphere (atm) is another unit of pressure, and its conversion factors are listed:

1 atm = 101,325 Pa

1 atm = 760 mm Hg = 760 torr = 76 cm Hg

1 atm = 14.7 lb/in^2 (psi, or pound per square inch)

A bar is a metric unit of pressure equal to 100,000 Pa; it is about equal to atmospheric pressure (101,325 Pa).

Closed-tube and open-tube manometers

A *manometer* is an instrument that measures the pressure of gases. It is simply a bent piece of tubing and, in principle, operates like a barometer.

There are two major types of manometers.

A *closed-tube manometer* (shown on the left below) measures pressures below atmospheric pressures, usually the container's gas pressure. Since the end is sealed, it contains a vacuum.

The *open-tube manometer* (shown on the right below) measures gas pressures near atmospheric pressures.

　　　　　　　　　　　　　　　　　　　　　　　　　Copyright © Sterling Test Prep.

The *pressure* is the difference in the mercury levels' heights in the two arms. The mercury levels in the manometer's two arms relate the gas pressure to the atmospheric pressure.

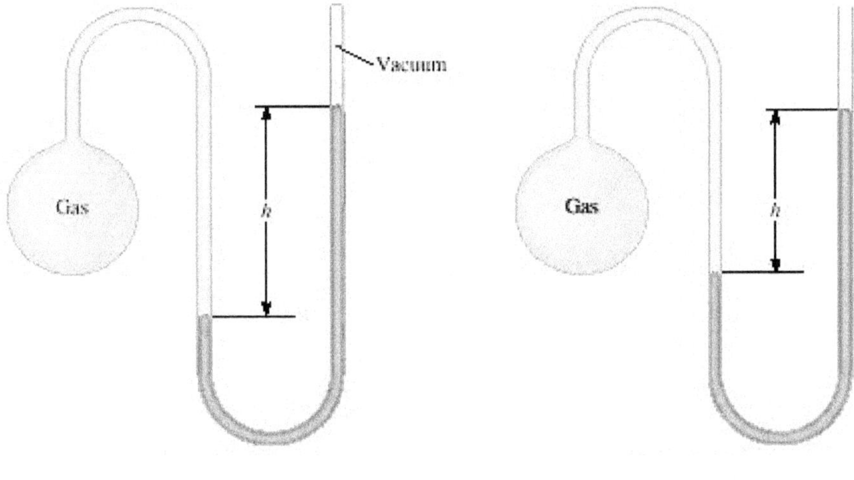

$$P_{gas} = Ph \qquad\qquad P_{gas} = P_h + P_{atm}$$

Left: gas pressure is less than atmospheric pressure (closed tube)

Right: gas pressure is higher than atmospheric pressure (open tube)

Molar volume at STP

In 1811, Italian scientist Amedeo Avogadro suggested that *"equal volumes of gases measured at the same temperature and pressure conditions contain the same number of molecules."*

In chemistry, *molar volume* is usually measured at "standard temperature and pressure" (i.e., STP), which is a temperature of 273.15 K (0 °C, 32 °F) and absolute pressure of exactly 10^5 Pa (100 kPa, 1 bar).

At these standard conditions, a mole of gas occupies 22.7 L of space.

Before 1982, when the International Union of Pure and Applied Chemistry (IUPAC, the organization that governs chemistry standards) changed the definition of STP, the standard pressure was defined as 101,325 Pa (1 atm).

In these conditions, the volume of 1 mol of gas occupies 22.4 L.

For problems, use the volume of one mole of gas as 22.7 L unless indicated otherwise.

Any duplication (copies, uploads, PDFs) is illegal.

Notes for active learning

Copyright © Sterling Test Prep.

Kinetic Molecular Theory of Gases

Macroscopic properties of gas

The *kinetic molecular theory of gases* relates the microscopic values of atomic kinetic energy to quantities like temperature and pressure.

The *kinetic molecular theory* (KMT) uses atomic theory to describe gases' behavior. This theory developed over 100 years, culminating in an 1857 paper by German physicist Rudolf Clausius (1822-1888).

KMT explains the macroscopic properties of a gas (e.g., pressure and temperature) in terms of its microscopic components (e.g., molecules or atoms).

The KMT of gases originated in the ancient idea that matter consists of tiny invisible atoms in rapid motion.

The idea was developed between the 17th and 19th centuries to explain gases' properties.

The kinetic theory of gases requires *five critical assumptions.*

1.   Gas molecules move in random molecular motion and continue straight until they collide with something.

2.   The gas molecule is a "point mass," a particle so small that mass is nearly zero.

     An ideal gas particle has a *negligible volume.*

3.   No attractive intermolecular molecular forces for gases participate.

4.   Collisions between gas molecules are "perfectly elastic."

     Therefore, the gas molecules' total kinetic energy (KE) remains *constant before and after* the collision since intermolecular attractive and repulsive forces are nonexistent.

5.   The average kinetic energy (*KE*) for gas particles is proportional to the absolute temperature, regardless of the chemical identity or atomic mass.

     At 0 K (absolute zero or -273 °C), the molecules (and orbiting electrons) are not moving and have no volume.

These rules display gases as ideal substances and only approximate their actual behavior.

However, their description is remarkably accurate.

From these assumptions, laws are derived to describe the behavior of gases.

Any duplication (copies, uploads, PDFs) is illegal.

Boltzmann's constant

The final assumption (#5) for KE is expressed by:

$$KE = \tfrac{1}{2}\, mv^2$$

$$\tfrac{1}{2}\, mv^2 = (3/2)\, k_B T$$

so,

$$KE = (3/2)\, k_B T$$

where *KE* is kinetic energy, *m* is mass, *v* is velocity, *T* is the temperature (Kelvin) and k_B is *Boltzmann's constant* ($k_B = 1.38 \times 10^{-23}$ J/K)

Boltzmann's constant relates the macroscopic and microscopic behavior of gases to their particles

The $KE = (3/2)\, k_B T$ equation is significant because it states that a gas particle's average kinetic energy is proportional to its absolute temperature.

Increasing the temperature of the gas particles increases the total speed and overall energy.

Gas atoms are infinitesimally small; nearly impossible to accurately measure a particle's speed.

The speed of gas particles is defined by the root-mean-square speed (u_{rms}).

The *root-mean-square speed* equation:

$$u_{rms} = \sqrt{\frac{3RT}{MM}}$$

where *R* is the ideal gas constant, *T* is the absolute temperature (K), and *MM* is the molar mass of the gas sample

Root-mean-square speed written with the *Boltzmann constant*, k_B:

$$u_{rms} = \sqrt{\frac{3kT}{m}}$$

where *k* is the Boltzmann constant, *T* is the absolute temperature (K), and *m* is the mass of one gas molecule

Copyright © Sterling Test Prep.

Boltzmann distribution plot

A plot of each gas particle's distribution of speeds at a given temperature shows a slightly asymmetric curve.

This speed-distribution curve is the *Maxwell-Boltzmann distribution curve.*

The *Boltzmann distribution plot* shows that the curve's peak corresponds to the most probable speed.

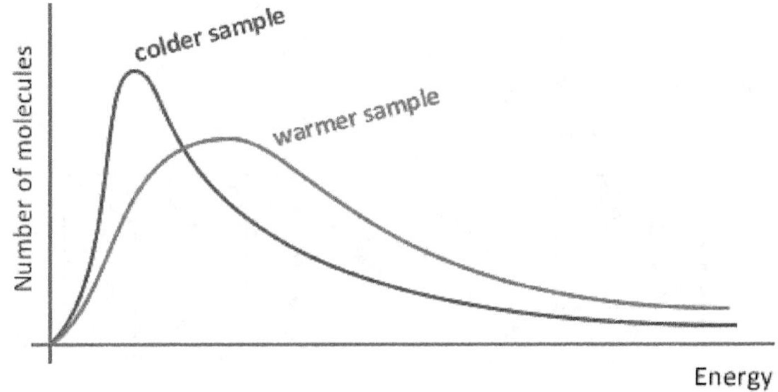

Maxwell-Boltzmann Distribution curves of molecular speeds at two temperatures

If the temperature (K) remains constant, the total kinetic energy remains unchanged.

However, the energy is distributed in several ways, and the gas particles travel at many speeds.

This distribution changes continually as the gas atoms collide with each other and with the container walls.

The curve flattens (i.e., net area under the curve is greater) and shifts to the right at higher temperatures, indicating that more gas molecules move at higher speeds and therefore possess more kinetic energy.

Effusion and diffusion

Effusion is the flow of gas particles *under pressure* from one compartment through a small opening, as shown.

Diffusion, the process whereby a substance (solute or particle) spreads from a region of high concentration to one of lower concentration.

Diffusion is explained using the kinetic theory of motion.

Any duplication (copies, uploads, PDFs) is illegal.

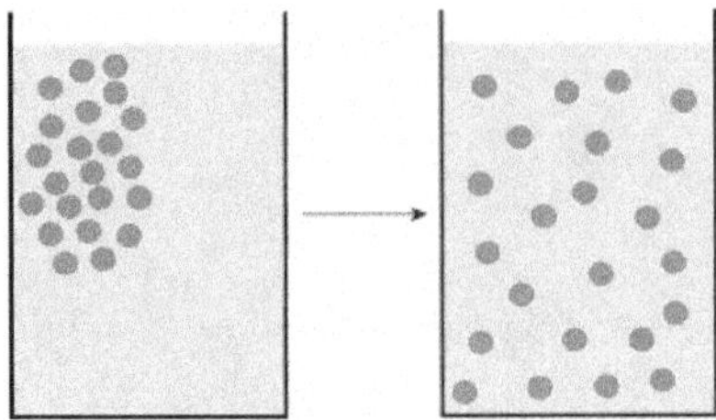

Diffusion of solutes in a solvent

According to the kinetic theory, heavier gases diffuse more slowly than lighter gases because of the difference in their travel speed.

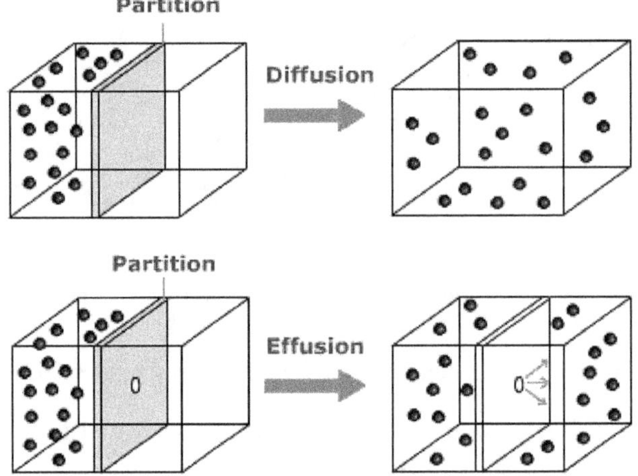

Diffusion and effusion of gas particles

Graham's Law

Scottish chemist Thomas Graham (1805-1869) developed *Graham's Law*, which states that *at constant temperature and pressure conditions, the rates at which two gases diffuse are inversely proportional to the square root of their molar masses.*

Graham's Law is expressed by:

$$\frac{r_1}{r_2} = \sqrt{\frac{MM_2}{MM_1}}$$

where r_1 and r_2 are the diffusion rates of gas 1 and gas 2, respectively, and MM_1 and MM_2 are the molar masses of the gases

Graham's Law applies to the effusion of gas particles.

The equation for effusion is the same as for diffusion.

Copyright © Sterling Test Prep.

Gas Laws

Ideal gas law

The *ideal gas law* was proposed in 1834 by French physicist Émile Clapeyron (1799-1864). It is a combination of the gas laws discussed, and it is the equation regarding the state of a hypothetical gas.

The *ideal gas law* explains the relationship between pressure (P), volume (V), and temperature (T).

The gas law is expressed as:

$$PV = nRT$$

where P is the pressure, V is the volume, T is the temperature, n is the number of moles, and R is the gas constant (use $R = 0.082$ L·atm/K·mol)

One mole $= 6.023 \times 10^{23}$ molecules; technically, the number of H atoms in one gram of H.

Because atoms are so small, it is more practical to count atoms in moles than to count them individually.

The ideal gas law applies to the pressure exerted by a gas on a cylinder with a moving wall.

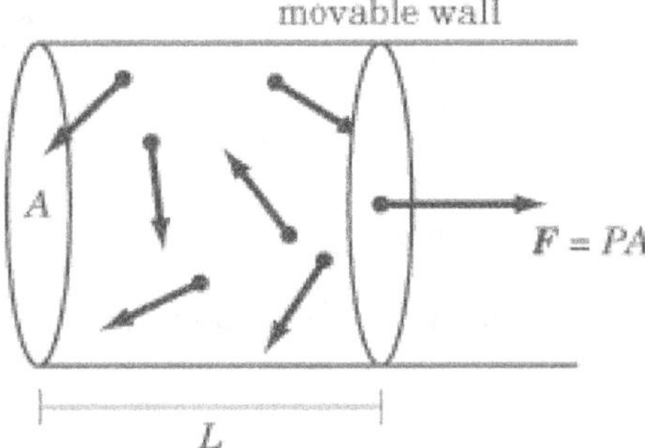

Since pressure is equal to $P = F/A$, the force that the gas exerts on the wall is equal to $F = PA$.

If this force moves the wall back a length of L, the volume of the cylinder increases by $\Delta V = LA$.

By solving for A and substituting it into the equation for force, the result is $F = P\Delta VL$, equal to $P\Delta V = FL$.

For example, suppose 2.5 moles of gas are at standard temperature and pressure. What is the volume occupied by the gas? (Use standard pressure $= 1.00$ atm and standard temperature $= 273.0$ K)

Use the ideal gas law:

$$PV = nRT$$

Solve for V

$$V = (nRT) / P$$

$$V = [(2.5 \text{ mol})\cdot(0.082 \text{ L·atm/K·mol})\cdot(273.0 \text{ K})] / (1.00 \text{ atm})$$

$$V = 56 \text{ L}$$

Several useful expressions are derived from the ideal gas law (PV = nRT) that relates the molar mass and density of gases to the pressure and temperature, substituting a known expression for one of the ideal gas law variables.

Work (w) is *force multiplied by the distance traveled*:

$$w = Fd$$

By pushing the wall, a distance of L with a force of F, the gas has done work equal to FL.

When gas does work, it symbolizes a change in energy.

If a change in PV is equal to a change in energy, then PV is the total energy of the gas.

For the ideal gas law, this means that nRT is the expression for the total kinetic energy of the gas molecules.

The ideal gas law (PV = nRT) is related to the number of molecules (N) and the Boltzmann's constant (k), which has a value of 1.381×10^{-23} J/K:

$$PV = NkT$$

The number of moles (n) and the gas constant (R) are constant.

Other gas laws derive from the ideal gas law by holding specific variables constant:

Boyle's Law

Charles' Law

Combined gas law

Closed container law

Copyright © Sterling Test Prep.

Ideal gases follow KMT (isothermal process)

The *kinetic molecular theory* (KMT) described previously is based on a theoretical ideal gas.

An *ideal gas* follows the *five* assumptions of KMT, whereby gas molecules:

> move in random motion

> are point masses with no volume

> have no intermolecular forces

> exhibit elastic collisions

> kinetic energy of the molecules is proportional to temperature

Compared to an ideal gas, the behavior of a real gas is *complicated.*

By conceptualizing ideal gas behavior, real gas behavior is easier to comprehend.

In the 17th and 18th centuries, scientists realized several relationships between gases' properties.

Boyle's, Charles', and Avogadro's Laws relate the properties of pressure, volume, and temperature developed empirically as individual cases of the ideal gas equation: $PV = nRT$.

Boyle's Law (isothermal process)

In 1661, Robert Boyle (1627-1691) studied the compressibility of gases. He observed that *the volume of a fixed amount of gas at a given temperature is inversely proportional to the pressure exerted on the gas.*

For *Boyle's Law*, the temperature of a gas is held constant.

It states that an *increase in pressure* causes a *decrease in volume.*

Likewise, a *decrease in pressure* causes an *increase in volume.*

Boyle's Law is expressed as:

$$P_1V_1 = P_2V_2$$

where subscripts 1 and 2 refer to the same gas sample under two sets of pressure and volume conditions

A plot of volume *vs.* pressure for a gas illustrates that Boyle's Law is a particular case of the ideal gas law, where n (moles of gas) and T (temperature) are constant (figure below).

Any duplication (copies, uploads, PDFs) is illegal.

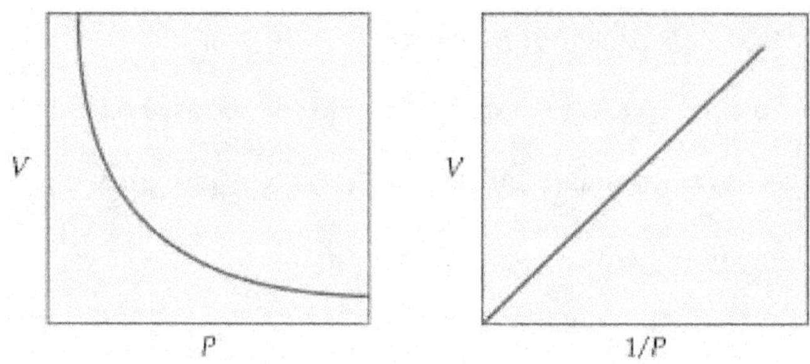

Boyle's Law: as pressure increases, volume decreases

For example, what is the new pressure in a container of gas in a 15.0 L container at a pressure of 5.00 atm when the volume decreases to 0.500 L?

Substitute the known quantities (P_1, V_1, and V_2) into Boyle's Law equation to solve P_2.

Note: the volume units need to be consistent. In this example, volumes are expressed in units of liters.

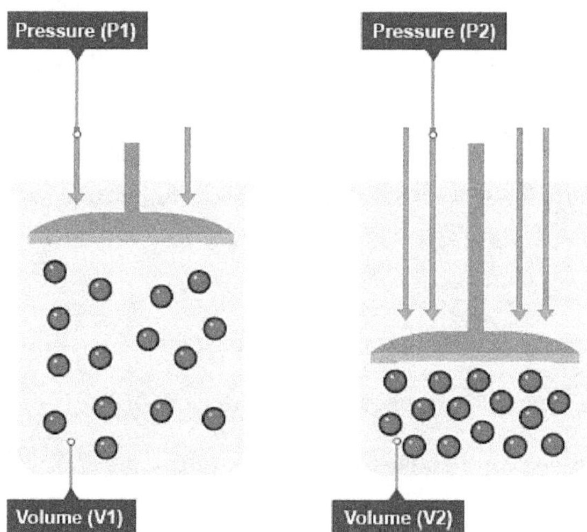

Boyle's Law:

$$P_1V_1 = P_2V_2$$

Solve for P_2:

$$P_1V_1 \,/\, V_2 = P_2$$

$$[(5.00 \text{ atm}){\cdot}(15.0 \text{ L})] \,/\, 0.500 \text{ L} = P_2$$

$$P_2 = 150 \text{ atm}$$

Copyright © Sterling Test Prep.

Charles' Law (isobaric process)

French scientist Jacques Charles (1746-1823) developed the relationship between temperature and gas volume.

Charles' Law states that *at constant pressure, the volume of a gas is directly proportional to its absolute temperature (Kelvin)*. It describes the behavior of a gas under *constant pressure*.

The volume and temperature are directly proportional; when the temperature increases, the volume increases; when the temperature decreases, the volume decreases.

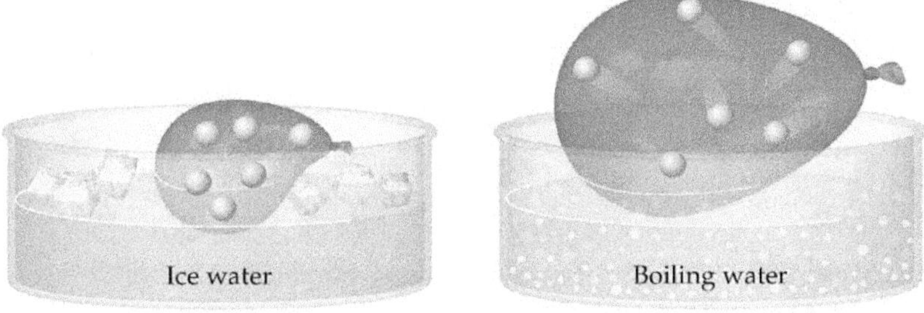

Charles' Law states that the volume of a gas is proportional to temperature

Charles' Law is expressed as:

$$\frac{V_1}{T_1} = \frac{V_2}{T_2}$$

where subscripts 1 and 2 refer to the same gas sample but under different temperature and volume conditions

A plot of *temperature vs. volume* for a gas illustrates that Charles' Law is another application of the ideal gas law, where n and P are constant, as shown.

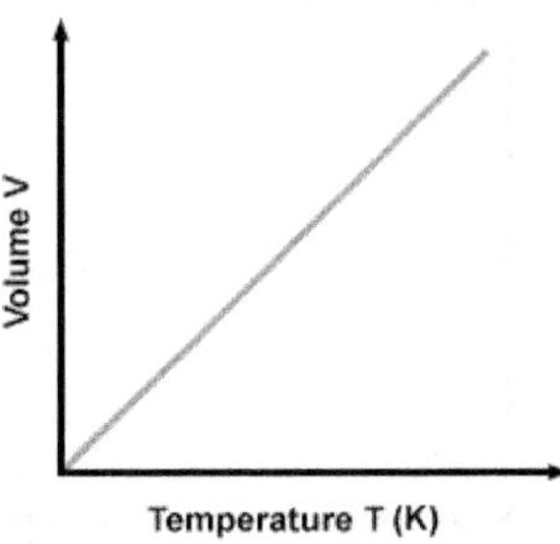

Charles' Law states that as temperature increases, volume increases

For example, if there is 25.0 L of gas at 0 °C, and the temperature is raised to 100 °C, what is the volume of the gas?

First, the temperature is converted to Kelvin.

$$T_1 = 0\ ^\circ\text{C} + 273 = 273\ \text{K}$$

$$T_2 = 100\ ^\circ\text{C} + 273 = 373\ \text{K}$$

Substitute the known quantities into the equation for Charles' Law.

$$V_1 / T_1 = V_2 / T_2$$

Rearrange and solve for V_2:

$$V_1\, T_2 / T_1 = V_2$$

$$(25.0\ \text{L}) \cdot (373\ \text{K}) / 273\ \text{K} = V_2$$

$$V_2 = 34.2\ \text{L}$$

Gay-Lussac's Law

In 1802, French chemist Louis Gay-Lussac (1778-1850) proposed *Gay-Lussac's Law* that states that *the pressure of a given sample of gas is directly proportional at a constant volume to its absolute temperature (Kelvin)*.

Gay-Lussac's Law is expressed by:

$$\frac{P_1}{T_1} = \frac{P_2}{T_2}$$

where subscripts 1 and 2 refer to the same gas sample but under different temperature and pressure conditions.

A plot of *temperature vs. pressure* for a gas illustrates that Gay-Lussac's Law is another particular case of the ideal gas law, where n and V are constant.

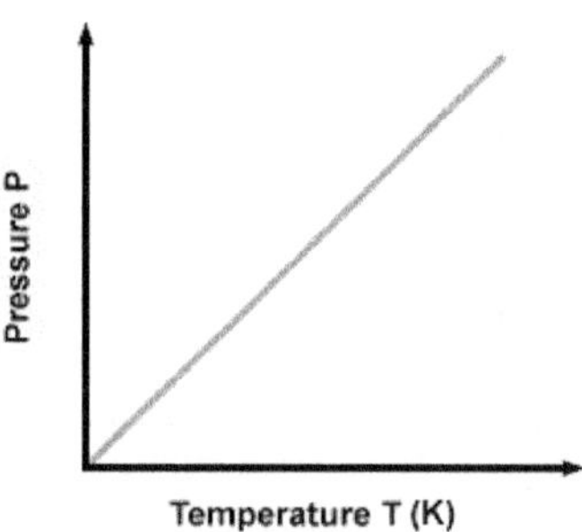

Gay-Lussac's Law states that as temperature increases, pressure increases

For example, suppose a gas is at 30.0 atm pressure and 100 °C, and the temperature changes to 400 °C. What is the new pressure of the gas?

Copyright © Sterling Test Prep.

Convert the temperature to units of Kelvin.

$$T_1 = 100\ °C + 273 = 373\ K$$

$$T_2 = 400\ °C + 273 = 673\ K$$

Gay-Lussac's Law solves P_2 by substituting the quantities into the equation.

$$\frac{P_1}{T_1} = \frac{P_2}{T_2}$$

$$P_2 = [(30.0\ atm)\cdot(673\ K)] / 373\ K$$

$$P_2 = 54.1\ atm$$

Closed container law (constant volume)

The *closed container law* describes the behavior of a gas under constant volume. In such cases, pressure and temperature are directly proportional:

$$\frac{P_i}{T_i} = \frac{P_f}{T_f}$$

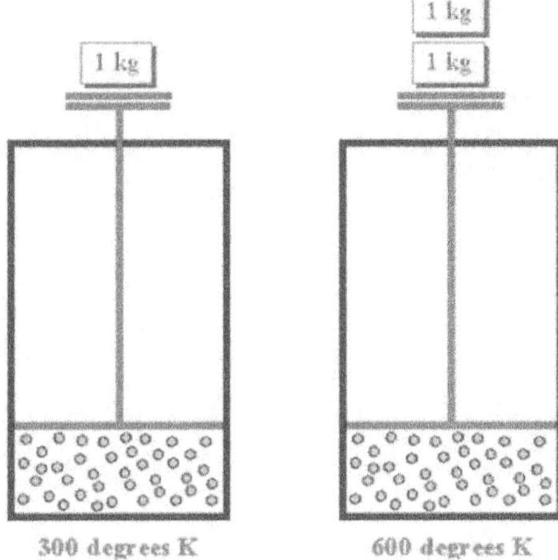

Avogadro's Law

The volume of a gas in a container is affected by pressure, temperature, and moles of gas.

The relationship between the quantity of gas and its volume was derived by Italian scientist Amadeo Avogadro (1776-1856).

Avogadro's Law states that *gases at a given temperature and pressure occupying a volume are directly proportional to the number of moles of gas present.*

Any duplication (copies, uploads, PDFs) is illegal.

Avogadro's Law is expressed as:

$$\frac{n_1}{V_1} = \frac{n_2}{V_2}$$

where n_1 and n_2 are moles of gas 1 and gas 2, and V_1 and V_2 are the volumes of gas 1 and gas 2

A plot of *volume vs. moles* of gas (shown below) illustrates that Avogadro's Law is a particular case of the ideal gas law where T and P are held constant.

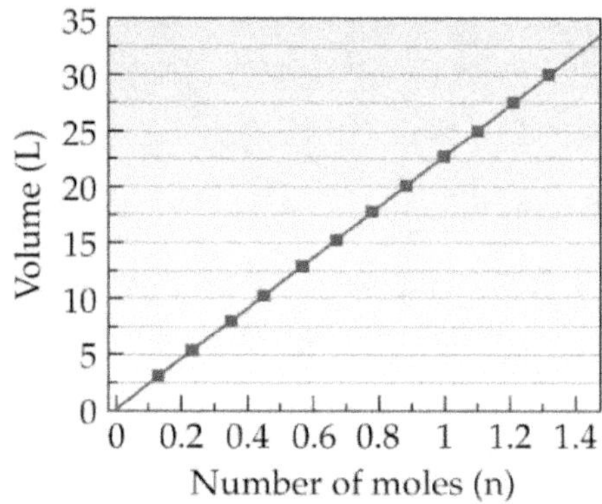

Avogadro's Law states that as moles of a gas increase, volume increases

For example, if 8.00 moles of a gas occupy a volume of 4.00 L at constant pressure and temperature, what volume of gas would 16.0 moles of this gas occupy at the same temperature and pressure?

Use *Avogadro's Law*

$$n_1 / V_1 = n_2 / V_2$$

where n_1 and n_2 are the moles of gas 1 and 2, and V_1 and V_2 are the volumes of gas 1 and gas 2

Solve for the V_2

$$V_2 = (V_1 \times n_2) / n_1$$

$$V_2 = [(4.00 \text{ L}) \cdot (16.0 \text{ mol})] / 8.00 \text{ mol}$$

$$V_2 = 8 \text{ L}$$

                                              Copyright © Sterling Test Prep.

Combined gas law

The *combined gas law* is an amalgam of the other gas laws.

The combined gas law is *not* the ideal gas law (PV = nRT); it is expressed as an equation to relate changes in temperature, volume, and gas pressure.

The *combined gas law* is:

$$\frac{P_1 V_1}{T_1} = \frac{P_2 V_2}{T_2}$$

where subscripts 1 and 2 refer to the same gas sample but under different temperatures, pressure, and volume

For example, suppose a gas is at 15.0 atm pressure, with a volume of 25.0 L and a temperature of 300.0 K. What would the volume of the gas be at standard temperature and pressure? (Use standard pressure = 1.00 atm and standard temperature = 273.0 K)

Use the *combined gas law* at standard temperature (T_2) and pressure (P_2):

$$(P_1 V_1) / T_1 = (P_2 V_2) / T_2$$

Solve for V_2:

$$(P_1 \times V_1 \times T_2) / (T_1 \times P_2) = V_2$$

$$V_2 = [(15.0 \text{ atm}) \cdot (25.0 \text{ L}) \cdot (273.0 \text{ k})] / [(300.0 \text{ K}) \cdot (1.00 \text{ atm})]$$

$$V_2 = 341 \text{ L}$$

Molar mass of a gas

The moles of a gas n can be expressed as the mass of gas in grams over the molar mass of gas:

$$n = m / MM$$

where MM = molar mass, m = mass of the gas in grams, and n = moles of gas

Substitute expression (m/MM) into the ideal gas law for moles (n):

$$PV = nRT$$

$$PV = mRT / MM$$

Multiplying each side by the molar mass (MM):

$$(MM)PV = mRT$$

This derived equation determines the molar mass of gas from experimental data, where the gas's mass, pressure, volume, and temperature are measured.

Any duplication (copies, uploads, PDFs) is illegal.

Divide each side of the above expression by the volume (V):

$$(MM)P = gRT / V$$

Density:

$$\rho = g/V$$

Substitute density for g/V in the derived equation:

$$(MM)P = \rho RT$$

This derived equation relates the gas's pressure, density, and temperature, like the other empirical gas laws.

For example, what is the molar mass of a gas with the density (ρ) of 1.855 g/L at 0.950 atm and 297.0 K.? (Use R = 0.0820 L·atm/K·mol)

Substitute the quantities into the derived equation above.

$$(MM)P = \rho RT$$

$$(MM) \cdot (0.950 \text{ atm}) = (1.855 \text{ g/L}) \cdot (0.0820 \text{ L·atm/K·mol}) \cdot (297.0 \text{ K})$$

Solve for molar mass (MM) by dividing each side of the equation by P (0.95 atm).

$$MM = [(1.855 \text{ g/L}) \cdot (0.0820 \text{ L·atm/K·mol}) \cdot (297.0 \text{ K})] / (0.950 \text{ atm})$$

$$MM = 47.6 \text{ g/mol}$$

Deviation of real-gas behavior

The *kinetic molecular theory* and the *ideal gas law* are approximations of gas behavior.

These theories are based on the theoretical ideal gas, while real gases have deviations from this behavior (five critical assumptions of an ideal gas).

When molecules are far apart (i.e., low pressure and high temperature), a real gas behaves more like an ideal gas.

When molecules are brought close (under high pressure and low temperature), gas molecules experience an intermolecular attraction. Therefore, the gas may deviate significantly from the behavior of ideal gas.

The most substantial deviations from the ideal occur at high pressure and low temperature. The gas molecules are "squished" at these conditions and thus experience intermolecular interactions. The molecular volume becomes significant when the total volume is *squished*.

The intermolecular attractions cause collisions to be sticky and inelastic.

Copyright © Sterling Test Prep.

Furthermore, gases *condense* into liquids at extremely high pressures and low temperatures.

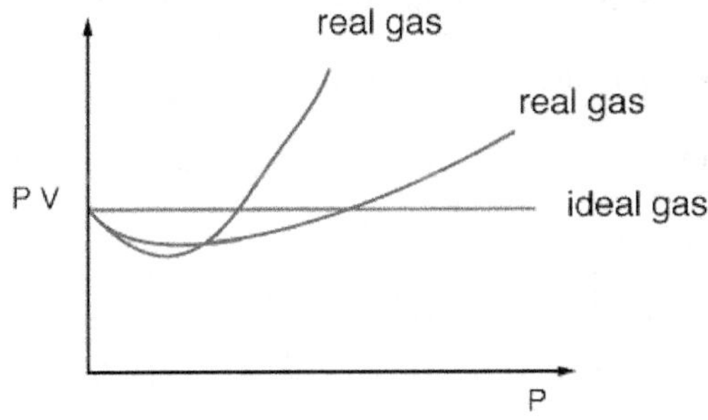

An ideal gas is a *point mass*, a particle so small that the particle's *volume is negligible*.

A real gas particle has a volume because real gases liquefy as they cool and cannot compress to zero volume; the molecules of a gas are not dimensionless points.

The *collisions* between gas particles are "elastic" with *no attractive or repulsive forces* for an ideal gas.

Thus, no energy is exchanged during the collisions.

For a real gas, collisions are *inelastic* (i.e., objects stick and lose kinetic energy).

Van der Waals equation describes deviations from ideal gas

In 1873, Dutch theoretical physicist Johannes Diderik van der Waals (1873-1923) derived a mathematical relationship describing real gases' *behavior* and *condensation to the liquid phase*.

The *ideal gas equation* for comparison:

$$PV = nRT$$

"Corrections" are made to the pressure (P) and volume (V) terms,

where P becomes:

$$P + (n^2a) / V^2$$

and V becomes:

$$V - nb$$

The *van der Waals equation* is expressed by:

$$[P + (n^2a) / V^2] \times (V - nb) = nRT$$

Since the collisions of real gases are inelastic, the term n^2a / V^2 corrects the interactions between gas particles.

Since real gas particles occupy space (volume), the nb term corrects the volume neglected by the ideal gas law.

Any duplication (copies, uploads, PDFs) is illegal.

The values of a and b are constant and are determined *experimentally* for each gas. The constants a and b would be given in a problem, and pressure or temperature can be easily solved.

Solving for the volume is nontrivial and involves solving a cubic polynomial equation.

For pressure, the van der Waals equation is rearranged by dividing by the volume term ($V - nb$):

$$P + (n^2 a) / (V^2) = (nRT) / (V - nb)$$

Subtract the intermolecular interaction term ($n^2 a / V^2$) from both sides of the equation:

$$P = [(nRT) / (V - nb)] - [(n^2 a) / (V^2)]$$

$$\text{Repulsion} \qquad \text{Attraction}$$

For example, use the real gas law to find the pressure of 2.00 moles of carbon dioxide (CO_2) gas at 298 K in a 5.00 L container. (Use van der Waals constants for CO_2 whereby $a = 3.592$ $L^2 \cdot atm/mol^2$ and $b = 0.04267$ L/mol).

Substitute the variables into the equation and solve for the pressure (P):

$$P = [(nRT) / (V - nb)] - [(n^2 a) / (V^2)]$$

$$P = [(2 \text{ mol}) \cdot (0.0821 \text{ L} \cdot atm) \cdot (298 \text{ K})] / [(5 \text{ L}) - (2.00 \text{ mol}) mol \cdot K \cdot (0.04267 \text{ L})]$$
$$- [(2 \text{ mol})^2 \cdot (3.592 \text{ L}^2 \cdot atm) / (5.00 \text{ L})^2 \cdot mol^2]$$

$$P = 9.38 \text{ atm}$$

Compare this value (9.38 atm) to the calculated pressure using the ideal gas law:

$$PV = nRT$$

$$P = nRT / V$$

$$P = [(2.00 \text{ mol}) \cdot (0.0821 \text{ L} \cdot atm) \cdot (298 \text{ K})] / (5.00 \text{ L}) mol \cdot K$$

$$P = 9.77 \text{ atm}$$

Although the results from the two equations are similar, they are not identical, illustrating the difference in the behavior of a real gas (not obeying the five stated assumptions) from an ideal gas.

Copyright © Sterling Test Prep.

Partial Pressure

Mole fraction

There is so much distance between the gas molecules that other gas molecules share the space.

Different gases combine to form homogeneous mixtures.

Each gas behaves *independently* and makes its unique contribution to the total pressure.

The *gas's partial pressure* is as if it were the *only gas* in the container.

The *mole fraction* (X_A) indicates the ratio between moles of a specific gas component and the total number of moles in the homogeneous mixture.

The symbol X_A expresses the component gas's mole fraction in the mixture.

$$X_A = \text{(moles of gas A)} / \text{(total moles of the gas mixture)}$$

Dalton's Law of partial pressures

When two or more nonreactive gases are present in the same container, they behave relatively independently.

Dalton's Law of partial pressures states that *the total pressure of a mixture of gases equals the sum of the individual gas pressures.*

It states that a mixture of gases' total pressure equals the sum of the individual gas components' partial pressures.

Dalton's Law of partial pressures is expressed by:

$$P_T = P_A + P_B + P_C \ldots$$

where P_T is the total pressure in the container, and P_A, P_B and P_C equal the partial pressures of gases A, B, and C

The partial pressure of a gas is related to its mole fraction by:

$$P_A = X_A P_T$$

where $X_A = \text{(moles of gas A)} / \text{(total moles of gas)}$

For example, suppose there is 1.0 L oxygen at 1.0 atm pressure in a container, 1.0 L nitrogen at 0.5 atm pressure in another container, and 1.0 L hydrogen at 3.0 atm pressure in a third container. What is the total pressure if the gases combine in a single 1.0 L container?

Any duplication (copies, uploads, PDFs) is illegal.

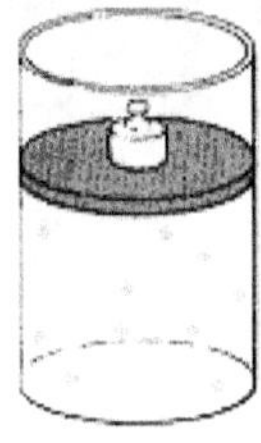 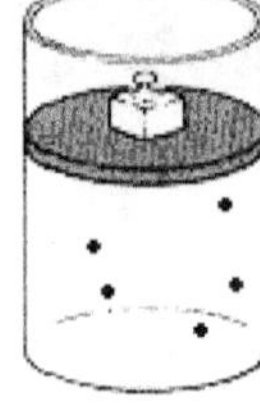 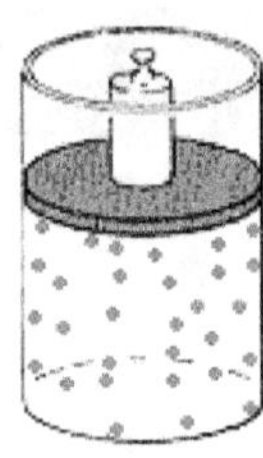

$$P_{O_2} = 1.0 \text{ atm} \qquad P_{N_2} = 0.5 \text{ atm} \qquad P_{H_2} = 3.0 \text{ atm}$$

For a single 1.0 L container, the total pressure is the sum of the partial pressures (X_X) of each gas component:

$$X_X = 1.0 \text{ atm} + 0.5 \text{ atm} + 3.0 \text{ atm} = 4.5 \text{ atm}$$

$$P_{Total} = P_{O_2} + P_{N_2} + P_{H_2}$$

$$P_{Total} = 4.5 \text{ atm}$$

Measuring vapor pressure

A critical application of Dalton's Law of partial pressure is to determine *vapor pressure* (i.e., the pressure exerted by a vapor in thermodynamic equilibrium with its liquid or solid phase).

The vapor pressure determines the amount of a water-insoluble gaseous reaction product or a slightly soluble gas such as hydrogen or oxygen.

A typical experimental method (illustrated below) to determine the amount of gas present is by collecting it over water and measuring the height of displaced water; this is accomplished by placing a tube into an inverted bottle, the opening of which is immersed in a larger container of water. The gas bubbles into the test tube and displace the water until the test tube is full.

The collected gas is not the only gas in the test tube since liquid water is in equilibrium with vapor, so the collected gas is a mixture of two gases: the gas collected and water vapor.

The partial pressure of water is the *vapor pressure of water* and is dependent on temperature.

The *volume* of gas collected consists of a mixture of the *gas collected* and *water vapor*.

Copyright © Sterling Test Prep.

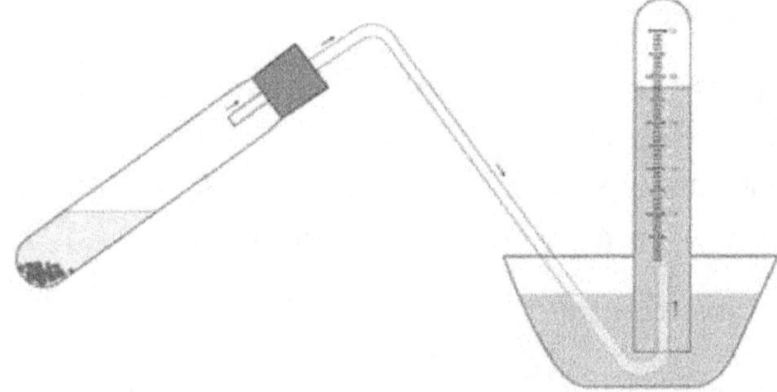

An experimental method to determine the amount of gas present

The total pressure is the sum of the two contributing partial pressures.

The *pressure of dry gas* collected is calculated from a table of water vapor pressure values at various temperatures. Partial pressure of the dry gas is calculated by subtracting the reference vapor pressure of H_2O.

The equalizing of the atmospheric pressure yields the partial pressure of the gaseous product collected. The amount of gas product can be determined with known volume and temperature.

Vapor pressure of water at specific temperatures

Temperature (°C)	Vapor Pressure (kPa)	Temperature (°C)	Vapor Pressure (kPa)
0	0.61	26	3.36
5	0.87	27	3.57
10	1.23	28	3.78
15	1.71	29	4.00
16	1.82	30	4.24
17	1.94	35	5.62
18	2.06	40	7.38
19	2.20	45	9.58
20	2.34	50	12.33
21	2.49	60	19.92
22	2.64	70	31.16
23	2.81	80	47.34
24	2.98	90	70.10
25	3.17	100	101.30

Any duplication (copies, uploads, PDFs) is illegal.

Determining partial pressure

For example, a small piece of zinc (Zn) reacts with dilute hydrochloric acid (HCl) to form hydrogen (H_2) gas, which is collected over water at 16.0 °C. The total pressure is adjusted to a barometric pressure of 100.24 kPa, and the volume of hydrogen gas is measured as 1,495 cm^3. What are the partial pressure and mass of the hydrogen gas? (Use $R = 8.314$ L·kPa·K^{-1}·mol^{-1})

The partial pressure of hydrogen gas:

$$P_{HCl} = 100.24 \text{ kPa} - 1.82 \text{ kPa}$$

$$P_{HCl} = 98.42 \text{ kPa}$$

To calculate the mass of the hydrogen gas, use the ideal gas equation ($PV = nRT$) and solve the number of moles.

For hydrogen (H_2) gas, multiply the mass of one hydrogen atom (1.008 g/mol) by two to obtain the molecular mass of 2.016 g/mol.

$$H_2 = 2 \times 1.008 \text{ g/mol}$$

$$H_2 = 2.016 \text{ g/mol}$$

Temperature and volume were given.

The partial pressure of hydrogen was calculated earlier; the ideal gas equation can now be used.

Determine the number of moles of H_2 gas:

$$PV = nRT$$

$$n = PV / RT$$

$$n = [(98.42 \text{ kPa})\cdot(1.495 \text{ L})] / [(8.314 \text{ L·kPa·K}^{-1}\text{·mol}^{-1})\cdot(289 \text{ K})]$$

$$n = 0.061 \text{ mol of } H_2 \text{ gas}$$

Calculate the mass of the H_2 gas formed:

$$mass = (\text{moles of } H_2 \text{ gas}) \times (\text{molecular mass of } H_2 \text{ gas})$$

$$mass = 0.061 \text{ mol } H_2 \text{ gas} \times 2.016 \text{ g/mol } H_2 \text{ gas}$$

$$mass = 0.123 \text{ g of } H_2 \text{ gas}$$

Copyright © Sterling Test Prep.

Notes for active learning

Notes for active learning

Copyright © Sterling Test Prep.

Practice Questions

1. Consider the phase diagram for H_2O. The termination of the gas-liquid transition at which distinct or liquid phases do NOT exist is the:

A. critical point

B. endpoint

C. triple point

D. condensation point

2. When liquids and gases are compared, liquids have [   ] compressibility compared to gases and a [   ] density.

A. lower... lower

B. higher ... higher

C. higher... lower

D. lower ... higher

3. How does a real gas deviate from an ideal gas?

    I. Molecules occupy a significant amount of space

    II. Intermolecular forces may exist

    III. Pressure is created from molecular collisions with the walls of the container

A. I only

B. II only

C. I and II only

D. II and III only

4. Under which conditions does a real gas behave most like an ideal gas?

A. High temperature and high pressure

B. High temperature and low pressure

C. Low temperature and low pressure

D. Low temperature and high pressure

5. Which of the following compounds has the highest boiling point?

A. CH_3OH

B. $CH_3CH_2CH_2CH_2CH_2OH$

C. $CH_3CH_2CH_2C(OH)HOH$

D. $CH_3CH_2OCH_2CH_2CH_3$

6. 15.0 liters of O_2 gas is at a temperature of 23 °C. If the temperature of the gas is raised to 45 °C at constant pressure, the new volume is:

A. 16.1 liters

B. 11.4 liters

C. 8.20 liters

D. 22.8 liters

7. The boiling point of a liquid is the temperature:

A. where sublimation occurs

B. where the vapor pressure of the liquid equals the atmospheric pressure over the liquid

C. equal to or greater than 100 °C

D. where the rate of sublimation equals evaporation

8. Which of the following statements about gases is correct?

A. Formation of homogeneous mixtures, regardless of the nature of non-reacting gas components

B. Relatively long distances between molecules

C. High compressibility

D. All of the above

9. What is the proportionality relationship between a gas's pressure and volume?

A. directly

B. inversely

C. pressure is raised to the 2nd power

D. pressure raised to the $\sqrt{2}$ power

10. How does the volume of a fixed sample of gas change if the pressure is doubled?

A. Decreases by a factor of 2

B. Increases by a factor of 4

C. Doubles

D. Remains the same

11. What is the ratio of the diffusion rate of O_2 molecules to the diffusion rate of H_2 molecules if six moles of O_2 gas and six moles of H_2 gas are placed in a large vessel, and the gases and vessels are at the same temperature?

A. 4:1

B. 1:4

C. 12:1

D. 1:1

E. 2:1

12. Under ideal conditions, which of the following gases is least likely to behave as an ideal gas?

A. CF_4

B. CH_3OH

C. N_2

D. O_3

Copyright © Sterling Test Prep.

Detailed Explanations

1. A is correct.

At a pressure and temperature corresponding to the *triple point* (point D on the graph) of a substance, the three states (gas, liquid and solid) exist in equilibrium.

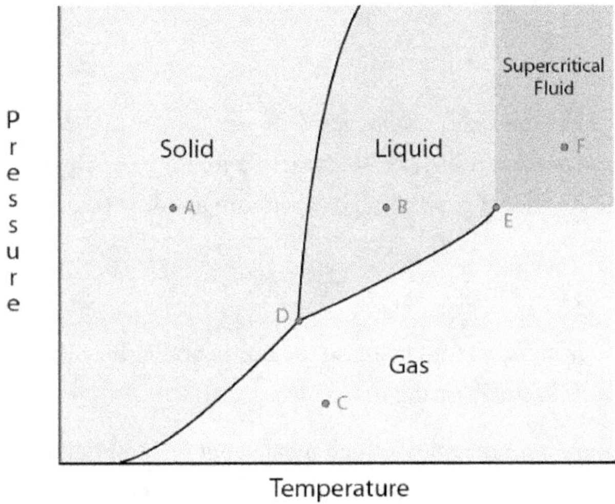

Phase diagram of pressure vs. temperature

The *critical point* (point E on the graph) is the phase equilibrium curve's endpoint, where the liquid and its vapor become indistinguishable.

2. D is correct.

Density = mass / volume

Gas molecules have a large amount of space between them; therefore, they can be pushed together, and thus, gases are very compressible.

There is a large amount of space between molecules in a gas, the extent to which gas molecules can be pushed together is much greater than the extent to which liquid molecules can be forced together.

Therefore, gases have greater *compressibility* than liquids.

Gas molecules are further apart than liquid molecules, so gases have a smaller density.

3. C is correct.

The molecules of an ideal gas *do not occupy a significant amount of space* and exert *no intermolecular forces*.

In contrast, real gas molecules occupy space and exert (weak attractive) intermolecular forces.

An ideal and real gas have pressure created by molecular collisions with the container walls.

4. B is correct.

The molecules of an ideal gas exert *no attractive forces*. Under these conditions, the molecules are far apart and exert little or no attractive forces on each other. Therefore, a real gas behaves nearly like an ideal gas at high temperature and low pressure.

5. C is correct.

Hydroxyl (~OH) groups significantly increase the boiling point because they form hydrogen bonds with ~OH groups of neighboring molecules.

Hydrocarbons are nonpolar molecules, which means that the dominant intermolecular force is London dispersion. This force gets stronger as the number of atoms in each molecule increases, and the stronger force increases the boiling point.

Branching of the hydrocarbon affects the boiling point, and straight molecules have slightly higher boiling points than branched molecules with the same number of atoms. The reason is that straight molecules can align parallel against each other.

All atoms in the molecules participate in London dispersion forces.

Another factor is *heteroatoms* (i.e., atoms other than carbon and hydrogen).

For example, the electronegative oxygen atom between carbon groups or in an ether (C–O–C) slightly increases the boiling point.

6. A is correct.

Charles' Law (or *law of volumes*) explains how, at constant pressure, gases behave when heated.

$$V \propto T$$

or

$$V / T = \text{constant}$$

Convert the initial temperature from Celsius to Kelvin:

$$23 \; ^{\circ}C = 273 + 23 = 296 \; K$$

Copyright © Sterling Test Prep.

Convert the initial temperature from Celsius to Kelvin:

$$45\,°C = 273 + 45 = 318\ K$$

Charles' Law:

$$(V_1\,/\,T_1) = (V_2\,/\,T_2)$$

Solve for the final volume:

$$V_2 = (V_1\,/\,T_1)\,/\,T_2$$

$$V_2 = [(15.0\ L)\,/\,(296\ K)] \times (318\ K)$$

$$V_2 = 16.1\ L\ of\ O_2$$

7. B is correct.

Boiling occurs when the vapor pressure of a liquid equals atmospheric pressure.

Vapor pressure is the pressure exerted by a vapor in equilibrium with its condensed phases (i.e., solid or liquid) in a closed system at a given temperature.

Atmospheric pressure is the pressure exerted by the weight of air in the atmosphere.

Vapor pressure is inversely correlated with the strength of the intermolecular force.

The molecules are more likely to stick together in liquid form from stronger intermolecular forces. Fewer molecules participate in the liquid-vapor equilibrium; therefore, the molecule boils at a higher temperature.

8. D is correct.

Gases form *homogeneous* mixtures, regardless of the identities or relative proportions of the component gases.

There is a relatively *large distance* between gas molecules (as opposed to solids or liquids where the molecules are much closer).

When pressure is applied to gas, its volume readily decreases, and thus, gases are highly compressible.

There are no attractive forces between gas molecules, which is why molecules of a gas can move about freely.

Any duplication (copies, uploads, PDFs) is illegal.

9. B is correct.

Boyle's Law (i.e., pressure-volume law) states that *pressure and volume are inversely proportional*:

$$(P_1 V_1) = (P_2 V_2)$$

or

$$P \times V = \text{constant}$$

If the volume of a gas increases, its pressure decreases proportionally.

10. A is correct.

Boyle's Law (i.e., pressure-volume law) states that *pressure and volume are inversely proportional*:

$$(P_1 V_1) = (P_2 V_2)$$

or

$$P \times V = \text{constant}$$

If the pressure of a gas increases, its volume decreases proportionally.

Doubling the pressure reduces the volume by half.

11. B is correct.

Graham's Law of Effusion states that the rate of effusion (i.e., escaping through a small hole) of a gas is inversely proportional to the square root of the molar mass of its particles.

$$\text{Rate 1 / Rate 2} = \sqrt{(\text{molar mass gas 1 / molar mass gas 2})}$$

The diffusion rate is the inverse root of the molecular weights of the gases.

Therefore, the rate of effusion is:

$$O_2 / H_2 = \sqrt{(2 / 32)}$$

rate of diffusion = 1: 4

12. B is correct.

Methanol (CH_3OH) is alcohol that participates in hydrogen bonding.

Therefore, this gas experiences the strongest intermolecular forces.

Copyright © Sterling Test Prep.

Notes for active learning

Notes for active learning

Copyright © Sterling Test Prep.

CHAPTER 2

Thermodynamics

- Thermochemistry
- Heat and Temperature
- Temperature Scales
- Heat Capacity and Specific Heat
- Calorimetry
- Heat Transfer
- Thermodynamic Systems
- Endothermic and Exothermic Reactions
- Thermodynamics
- Zeroth Law of Thermodynamics
- First Law of Thermodynamics
- Second Law of Thermodynamics
- Third Law of Thermodynamics
- Enthalpy H
- Bond Energy and Heat of Formation Relationship
- Gibbs Free Energy G
- Spontaneous Reactions and ΔG
- Thermal Expansion
- Kinetic Theory of Gases
- Phase Diagrams
- Colligative Properties
- Heat Transformation
- Pressure *vs.* Volume Phase Diagrams
- Practice Questions & Detailed Explanations

Copyright © Sterling Test Prep.

Thermochemistry

Applied thermochemistry

Thermochemistry:

- explains energy changes, particularly for energy exchange between the system and its surroundings;

- predicts reactant and product quantities during the reaction;

- predicts if a reaction is spontaneous or non-spontaneous, favorable, or unfavorable;

- combines concepts of thermodynamics and chemical bond energy.

Thermochemistry quantities

Thermochemistry commonly includes calculations of such *quantities* as:

heat capacity (c)

heat of combustion

heat of formation

enthalpy (H)

entropy (S)

Gibbs free energy (G)

calories

Energy changes in chemical reactions

Thermochemistry studies heat energy associated with chemical reactions and physical transformations.

Reactions may *release energy* (i.e., exothermic reactions) or *absorb energy* (i.e., endothermic reactions).

Phase change may *release or absorb energy* (e.g., boiling, melting).

Thermochemistry rests on *two* generalizations:

1. Energy change accompanying any transformation is *equal and opposite* to the reverse process's energy change. (Lavoisier and Laplace law proposed in 1780).

2. Hess' Law proposed in 1840 states that the *energy change of any transformation* is the same if the process occurs in one or many steps.

Any duplication (copies, uploads, PDFs) is illegal.

Systems conserve energy

German physicist Julius Robert Mayer (1814-1878) first proposed the *law of the conservation of energy* in 1842, now called the First Law of Thermodynamics.

Conservation of energy law states that *energy is neither created nor destroyed.*

These generalizations were the foundations for *specific heat, latent heat,* and *heat capacity*:

$$\Delta C_p = d\Delta H \,/\, dT$$

Integration of the expression $\Delta C_p = d\Delta H \,/\, dT$ evaluates the heat of reactions at different temperatures.

Copyright © Sterling Test Prep.

Heat and Temperature

Heat as energy

Heat is energy and exists independently from any medium; *heat is a way to transfer energy.*

Heat and temperature are often used interchangeably in everyday vernacular.

However, this usage of interchanging heat and temperature is incorrect.

Temperature as kinetic motion

Temperature measures *kinetic energy* (KE) in a specific medium and depends on the material.

Temperature and heat are connected principles, but heat is the *total energy* due to *molecular motion*. In contrast, temperature is the average measure of kinetic energy in a substance.

Some substances respond drastically to a change in thermal energy. This response depends on the material and heat capacity.

Thermal energy

The quantity of thermal energy transmitted from one body to another measures heat.

An object does *not* "contain" heat but can have a certain amount of thermal energy.

Heat can be measured in joules (J) and calories (cal).

On a microscopic level, thermal energy is related to atomic and molecular *vibrations*.

Matter is made of billions and trillions of small molecules; even in solids, these molecules and atoms are never entirely still.

The movement of molecules determines how much thermal energy they hold.

Any duplication (copies, uploads, PDFs) is illegal.

Notes for active learning

Copyright © Sterling Test Prep.

Temperature Scales

Fahrenheit and Celsius

Three commonly used temperature scales are *Fahrenheit*, *Celsius* and *Kelvin.*

Daniel Fahrenheit (1686-1736) proposed the classic English system of measuring temperature (i.e., °F). in 1724 by dividing the difference between the boiling and freezing point of water into 180 equal degrees (°F).

Anders Celsius (1701-1744) developed the modern metric measuring temperature by dividing the boiling and freezing point of water into 100 equal degrees (°C).

Celsius scale was originally the *centigrade scale.*

In the United States, the temperature unit is *Fahrenheit* (°F).

However, the SI unit for temperature is *Celsius* (°C) with water freezing at 0 °C and boiling at 100 °C (instead of the 32 °F and 212 °F).

Kelvin scale

William Thomas, 1st Baron Kelvin (1824-1907), developed the Kelvin (K) as the third temperature scale. The Kelvin scale is based on the concept of absolute zero (i.e., 0 K, –273.15 °C, or –459.67 °F) and is the most important science scale for temperature.

The size of units on the Kelvin scale is the *same* as for degrees on the Celsius scale.

The conversion between the two scales is:

$$°F = \frac{9}{5}°C + 32$$

Kelvin is another unit of temperature often used in scientific calculations. This scale is often used for gases, as usually, low temperatures are in question.

Absolute zero is the coldest theoretical temperature any substance can have, equal to 0 K.

0 K is equivalent to –273 °C!

Relative temperature scales

Three thermometers display equal temperatures and three standard temperature units are shown.

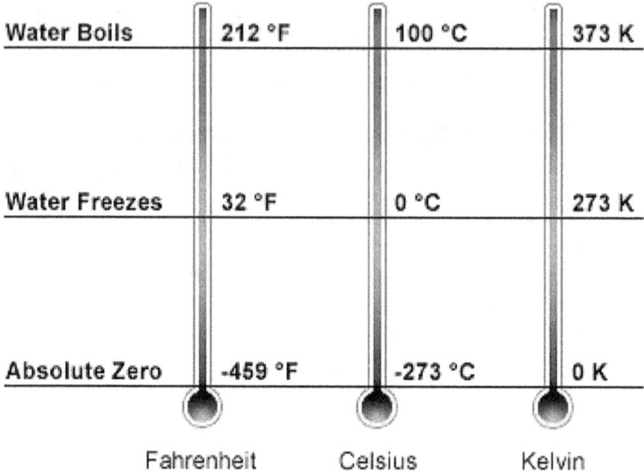

Comparison between the three temperature scales

Converting between scales

Temperature is the average amount of *kinetic energy* of the particles (i.e., molecules and atoms) of a substance.

Absolute zero when these particles theoretically have no KE and no movement.

Kelvin and Celsius use the same step in their degree measurements, making calculations easier.

For example, a temperature difference is the same in Kelvin as in Celsius, so when calculating a temperature difference in either of these units, the values do not need to be converted.

If required to convert between Kelvin and Celsius, the conversion can be performed by adding 273 to the Celsius temperature to get Kelvin:

$$K = {}^{\circ}C + 273$$

The temperature of absolute zero may never be attained. Scientists have achieved temperatures close to absolute zero but have never fully reached it.

In 2003, the lowest temperature recorded was by a group of scientists from MIT (Massachusetts Institute of Technology). Researchers successfully cooled sodium (Na) gas to a temperature of half a billionth of a degree above absolute zero but not at absolute zero.

Copyright © Sterling Test Prep.

Fahrenheit and Celsius temperature conversions

Convert between Fahrenheit and Celsius by:

$$°F = (1.8 \times °C) + 32$$

For example, what is the temperature in °F for 28 °C?

$$°F = (1.8 \times °C) + 32$$

$$°F = (1.8 \times 28 \text{ °C}) + 32$$

$$°F = (50.4) + 32$$

$$°F = 82.4$$

Therefore, 28 °C equals 82.4 °F

For example, what is the temperature in °C for 70 °F?

$$°F = (1.8 \times °C) + 32$$

Rearrange:

$$(1.8 \times °C) = °F - 32$$

$$(1.8 \times °C) = °F - 32$$

$$°C = (°F - 32) / 1.8$$

$$°C = (70 - 32) / 1.8$$

$$°C = 21$$

Therefore, 70 °F equals 21 °C

Kelvin and Celsius temperature conversions

Convert between Kelvin and Celsius by:

$$K = °C + 273$$

For example, what is the temperature in °C for 295 K?

$$K = °C + 273$$

Rearrange:

$$°C = K - 273$$

$$°C = 295 - 273$$

$$°C = 22$$

Therefore, 295 K = 22 °C.

Any duplication (copies, uploads, PDFs) is illegal.

Reference values among temperature scales

	K	°C	°F
Absolute zero	0	−273	−460
Freezing point of water / melting point of ice	273	0	32
Room temperature	298	25	77
Body temperature	310	37	99
Boiling point of water / condensation of steam	373	100	212

Copyright © Sterling Test Prep.

Heat Capacity and Specific Heat

Heat capacity, specific heat and molar heat capacity

	Heat capacity	Specific heat capacity	Molar heat capacity
Definition	Amount of heat required to raise temperature of a substance by 1 K (1 °C) Ability of a substance to absorb heat	Amount of heat required to raise the temperature of 1 g of a substance by 1 K (1 °C) Amount of heat added (or removed) to change the temperature	Amount of heat required to raise temperature of 1 mole of substance by 1 K (1 °C) At constant pressure unless noted
Equation	$C_H = q / \Delta T$	$C_S = q / m\Delta T$ C_S (specific) equals C_P (constant pressure)	$C_M = q / n\Delta T$
Intrinsicality	Intrinsic property	Non-intrinsic property	
Unit	J / °C	J / g °C or J / Kg K	J / mol °C or J / mol K
Property	Extensive property	Intensive property	Intensive property
Substance quantity	Depends on amount	Independent of amount	
Specificity	Non-specific	Highly specific	
Examples	1 L H_2O = 4182 J/kg °C 100 g Cu = 38.5 J/ °C	H_2O = 1 calorie 1 calorie = 4.18 J/g °C Cu = 0.385 J/g °C	

Any duplication (copies, uploads, PDFs) is illegal.

Heat capacity

Heat capacity (or *thermal capacity*) is a physical property of a substance's ability to hold heat.

Heat capacity (C_H) of a substance is the energy required to raise the temperature of a substance by *one degree Celsius*. It is the absorbed heat energy ratio to the resulting temperature change.

Heat capacity is a product of an *object's mass* (*m*) and a *substance's specific heat capacity (c)*.

When measuring the heat capacity of a substance, a variable must be held constant since the heat capacity of a system depends on the temperature, the pressure, and the volume of the system. These are noted with either a c_p or a c_v, respectively.

SI unit for heat capacity is the joule per Kelvin (J/K).

Isobaric and isochoric processes

Gases and liquids are typically measured at constant volume and subjected to pressure for a constant reading.

>*Isobaric process* (c_p) is when *pressure* is held constant, defining the *heat capacity*.

>*Isochoric process* (c_v) is when *volume* is held constant.

Heat capacity *vs.* specific heat capacity

Heat capacity and specific heat of substances are similar.

>*Specific heat* (C_S) is the ratio of energy absorbed to the temperature rise per unit of mass of the substance.

>*Heat capacity* is the ratio of energy absorbed to the rise in temperature.

Specific heat *c* is a characteristic of the substance and thus does not change.

Water has a relatively *high* specific heat capacity; 4.2 J of heat energy is needed to raise one gram of water 1 °C.

High specific heat of the water *helps modulate global temperatures*. Temperature fluctuations during day and night would be much larger if the water's heat capacity were low. This is evident in a desert where water is scarce. Daytime temperatures are high, yet nighttime temperatures can drop to near freezing.

Copyright © Sterling Test Prep.

Specific heat

Specific heat capacity (C_S) is the amount of heat required to raise the temperature of 1 gram of a substance by 1 °C.

Specific heat *c* is a characteristic of the substance and thus does not change. Metals tend to have *low* specific heat capacities.

Heat capacity and the specific heat of a substance are similar. *Specific heat capacity* differs from heat capacity because it relates to *the heat energy required to raise one gram of a substance by one degree Celsius.*

Specific heat capacity (or *specific heat*) is the heat capacity of 1 gram of a substance.

Molar heat capacity is the heat capacity of 1 mole of a substance.

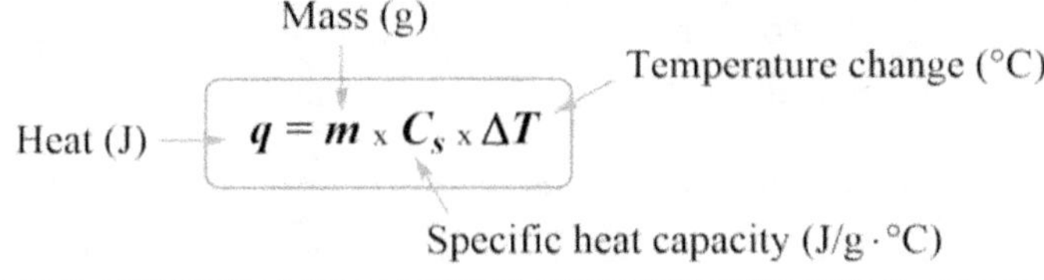

Specific heat capacity

Specific heat relates to heat and temperature

Units for specific heat are calories or joules per gram per °C. For example, the specific heat of water is 1 calorie (or 4.18 joules) per gram per Celsius degree.

The specific heat capacity of water is 4,180 joules per kilogram per degree Celsius (J/kg°C); 4,180 J are needed to raise the temperature of 1 kg of water by 1 °C.

Specific heat is related to heat and temperature by:

$$q = mC_S\Delta T$$

where q is the heat flow, m is mass measured in grams, C_S is the specific heat capacity of the substance, and ΔT is the temperature change.

Specific heat (C_S) of a substance, heat (Q), and temperature are related by:

$$Q = mC_S\Delta T$$

where Q is the heat transferred to the material (J), m is the mass of the object being heated (kg), and ΔT is the temperature change (K).

Any duplication (copies, uploads, PDFs) is illegal.

Calculating specific heat

For example, 3,200 J of heat is added to 1.0 kg of water at 10 °C. (Use specific heat of water C_S = 4,190 J/kg·°C)

Rearrange the specific heat equation to solve ΔT:

$$Q = mC_S\Delta T$$

$$\Delta T = \frac{Q}{mc}$$

$$\Delta T = \frac{3,200\,J}{1.0\,kg \cdot 4,190\,\frac{J}{kg} \cdot °C}$$

$$\Delta T = 0.76\ °C$$

where Q is the heat transferred to the material (J), m is the mass of the object being heated (kg), and ΔT is the change in temperature (K).

ΔT represents the temperature change. This value for the change (not the absolute value) must be either added to the initial temperature to determine the final temperature or subtracted from the final to find the initial.

A reaction must be contained within a system to observe all the products.

Specific heats of selected materials

Material	c(J/kg·K)
Aluminum	879
Concrete	850
Diamond	509
Glass	840
Helium	5,193
Water	4,181

Copyright © Sterling Test Prep.

Low *vs*. high specific heat

Substances such as metals (e.g., copper) have a *low specific heat* because it does not take much energy to transfer the heat and excite the molecules.

Raising the value of their average kinetic energy (raising their temperature).

Materials that are difficult to heat (e.g., rubber) have a much *higher specific heat* because more energy is required to raise the temperature by the same amount. The previous table shows that water has a much higher heat capacity than other common materials found on land.

Any duplication (copies, uploads, PDFs) is illegal.

Notes for active learning

Copyright © Sterling Test Prep.

Calorimetry

Calories

Calories relate heat to changes in temperature, which is often useful when solving problems in thermochemistry.

calorie (lowercase *c*) is the heat needed to raise the temperature of one gram of water by one degree Celsius (g/°C).

$$1\ cal = 1\ \frac{g}{°C} = 4.19\ J$$

Nutritional Calories

Food Calories (uppercase *C*) are not equivalent to heat calories. This resembles the relationship between kilograms and grams (1 Calorie = 1,000 calories).

Another difference is that while nutrition Calories measure stored energy just as calories do, they also measure the energy stored in the chemical bonds of food that are broken down and stored when digested. Calories measure energy in a substance and can be converted to equivalent conventional energy.

For example, the calories contained in 5 lbs. of spaghetti have enough energy to brew a pot of coffee; calories in one piece of cheesecake can power a 60 W incandescent light bulb for 1.5 hours; 222 Big Macs contain enough energy to drive a vehicle for almost 90 miles.

Some useful conversion relationships:

1 calorie = 4.2 J

1 Calorie (with capital C) = 1,000 calories = 1 kilocalorie

1 Calorie = 4,200 J

Nutrient Molecule in Food	Example	Cal/g	kJ/g
Carbohydrate	Table sugar, potatoes, flour	4	17
Protein	Meats, fish, beans	4	17
Fat	Oil, butter	9	38

Foods contain varying amounts of energy, as shown

Any duplication (copies, uploads, PDFs) is illegal.

Calorimeters

Calorimetry is an experimental method to calculate enthalpy changes involving heat flow.

Calorimeters measure the amount of energy in food samples experimentally. Food is placed inside a closed container surrounded by water and burned in the presence of oxygen. Water temperature change is related to the energy released, commonly reported in nutritional Calories (or kilojoules units). Note the term *Calories* in uppercase.

Even though cells within the body combust the molecules differently than a calorimeter, the calorimeter provides an accurate *caloric value* because of the same end products.

Calibration is an experiment in which hot water of known mass and temperature is poured into an insulated cup containing icy water of known mass and temperature. The composite sample equilibrates to a maximum temperature, and a calorimeter constant is determined below.

Calorimeters are ideally considered an insulated system. The heat liberated by the reaction process is transferred to the other substances within the calorimeter or the calorimeter itself. No heat is lost to the surroundings.

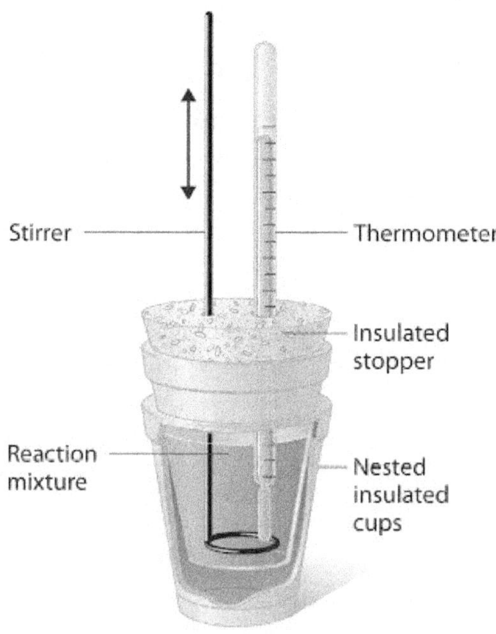

Components of a simple calorimeter

Since no thermal energy is lost in a calorimeter, the heat of a reaction can be measured by inserting a thermometer into a calorimeter and recording the temperature readings before and after the reaction. A stirrer ensures that all calorimeter contents are mixed and react uniformly.

Copyright © Sterling Test Prep.

Constant-pressure calorimeters

Constant-pressure and *constant-volume calorimeters* are two types of calorimeters.

Constant-pressure calorimeter is where heat flow is measured at constant atmospheric pressure.

A perfect calorimeter does not exist; no calorimeter completely insulates heat flow. The calorimeter itself, as part of the surroundings, absorbs some heat. Therefore, the calorimeter must be calibrated to obtain accurate results. Styrofoam cups are simple, constant-pressure calorimeters often used in introductory chemistry classes. Although a Styrofoam cup's temperature-insulating properties are reasonably good, the cup absorbs heat.

Calorimeter constant K_{cal}

Calorimeter constant K_{cal} is determined by a composite sample equilibrating at max. temp.

Basic principle of calorimetry:

heat lost by hot water = heat gained by cool water + heat gained by calorimeter

Determine the mathematical equations:

$$-m_{hot}C_{water}(T_{final} - T_{intial, hot}) = m_{cool}\, c_{water}(T_{final} - T_{initial, cool}) + K_{cal}\,(T_{final} - T_{initial, cool})$$

Sign is negative for heat lost by the hot water because heat loss is negative. Substitute known masses and temperatures into the expression. Since the cool water was in the cup, the calorimeter's initial and cool temperatures were the same.

For example, suppose 60.1 g of water at 97.6 °C is poured into a cup calorimeter containing 50.3 g of water at 24.7 °C. The final temperature of the combined water samples reaches 62.8 °C. What is the calorimeter constant? (Use c_{water} = 4.184 J/g°C)

$$-(60.1\text{ g})\cdot(4.184\text{ J/g°C})\cdot(62.8\text{ °C} - 97.6\text{ °C})$$
$$= [(50.3\text{ g})\cdot(4.184\text{ J/°C})\cdot(62.8\text{ °C} - 24.7\text{ °C})] + [K_{cal}\,(62.8\text{ °C} - 24.7\text{ °C})]$$

$$-(60.1\text{ g})\cdot(4.184\text{ J/g°C})\cdot(-34.8\text{ °C})$$
$$= [(50.3\text{ g})\cdot(4.184\text{ J/°C})\cdot(38.1\text{ °C})] + [K_{cal}\,(38.1\text{ °C})]$$

$$[8.75 \times 10^3\text{ J}] - [8.02 \times 10^3\text{ J}] = [K_{cal}\,(38.1\text{ °C})]$$

$$0.73 \times 10^3\text{ J} = [K_{cal}\,(38.1\text{ °C})]$$

$$[0.73 \times 10^3\text{ J}] / (38.1\text{ °C}) = K_{cal}$$

$$K_{cal} = 19.2\text{ J/°C}$$

Any duplication (copies, uploads, PDFs) is illegal.

Heat flow

After calibrating the calorimeter, the specific heat capacity will be determined.

A known quantity of metal is heated to a known temperature and placed in an insulated cup calorimeter containing a known quantity and temperature until the temperature equilibrates.

For example, suppose 28.2 g of an unknown metal is heated to 99.8 °C and placed into a cup calorimeter containing 150.0 g of water at 23.5 °C. The temperature of the water equilibrates at a final temperature of 25.0 °C.

What is the specific heat capacity of the metal? (Use the calorimeter constant of 19.2 J/°C).

Consider heat flow:

heat lost $_{hot\ metal}$ = heat gained $_{water}$ + heat gained $_{calorimeter}$

$-m_{hot}C_{water}(T_{final} - T_{intial,\ hot}) = m_{cool}\ c_{water}(T_{final} - T_{initial,\ cool}) + K_{cal}\ (T_{final} - T_{initial,\ cool})$

$-(28.2\ g){\cdot}(C_{metal}){\cdot}(25.0\ °C - 99.8\ °C) =$

$\quad = [(150.0\ g\ ){\cdot}(4.184\ J/g°C){\cdot}(25.0\ °C - 23.5\ °C)] + [19.2\ J/°C(25.0\ °C - 23.5\ °C)]$
$\quad\quad -(28.2\ g){\cdot}(C_{metal}){\cdot}(- 74.8\ °C)$

$\quad = [(150.0\ g\ ){\cdot}(4.184\ J/g°C){\cdot}(1.5\ °C)] + [19.2\ J/°C(1.5\ °C)]$

$2.11 \times 10^3\ g°C\ (C_{metal}) = (9.4 \times 10^2\ J) + (29\ J)$

$2.11 \times 10^3\ J\ (C_{metal}) = 969\ J$

Solving for specific heat capacity:

$C_{metal} = 969\ J\ /\ 2.11 \times 10^3\ g°C$

$C_{metal} = 0.459\ J/\ g°C$

                                    Copyright © Sterling Test Prep.

Constant-volume calorimeters

Constant-volume calorimeter is where heat flow is measured at constant volume.

Bomb calorimeter (*decomposition vessel*) is the most used constant-volume calorimeter.

Combustion is when oxygen reacts with a substance to produce carbon dioxide (CO_2), water (H_2O), and energy.

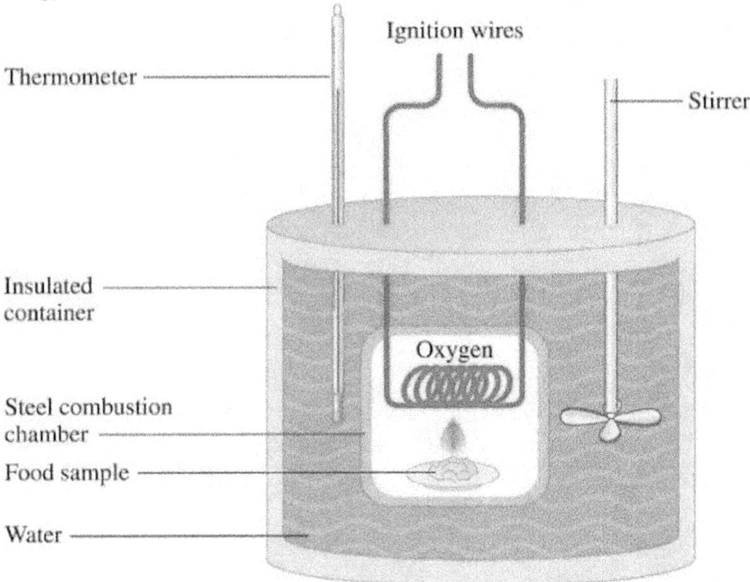

Bomb calorimeter: constant-volume calorimeter

Notes for active learning

Copyright © Sterling Test Prep.

Heat Transfer

Rate of heat transfer

When heat transfers by conduction, the rate of heat transfer (*H*) can be calculated in joules per second (J/s) or watts (W).

Amount of energy (heat) conducted per unit:

$$H = \frac{kA\Delta T}{t}$$

where ΔT is the temperature difference across the object (K), *A* is the cross-sectional area (m^2), *t* is the thickness of the material (m), and *k* is the thermal conductivity of the material (J/s·m·k).

Heat transfer (H) rate is a material characteristic usually provided in the problem.

Heat transfers between two objects by *conduction*, *convection*, or *radiation*.

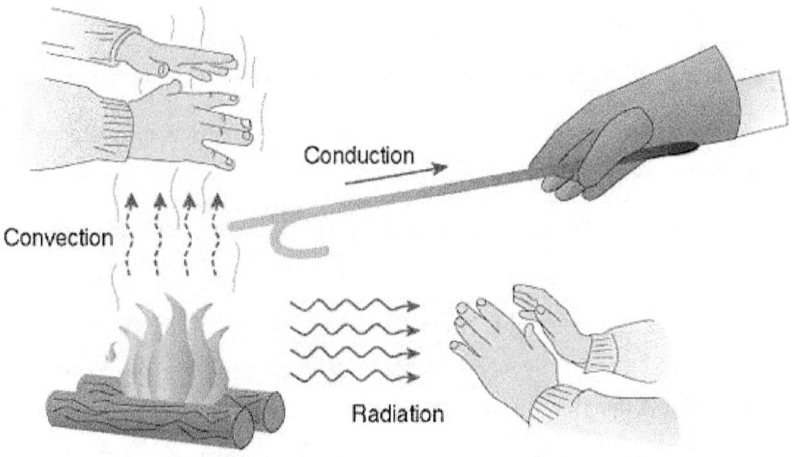

Heat transfer by convection, conduction, and radiation

Heat flows from hotter object

According to the First Law of thermodynamics, *heat always flows from a hotter to a colder system until the two systems are in thermal equilibrium*.

> *Conduction* is the flow of heat between two materials in *direct contact*.

> *Convection* is the flow of energized (e.g., air) molecules that settle and transfer their energy by *direct contact* using conduction.

> *Radiation* is the heat flow (i.e., energy) emitted by matter through *electromagnetic radiation* (e.g., photons or electromagnetic waves).

Any duplication (copies, uploads, PDFs) is illegal.

Thermal conductivities of selected materials

Material	k (J/s·m·K)
Aluminum	237
Concrete	1
Copper	386
Glass	0.9
Stainless Steel	16.5
Water	0.6

Thermal conductivities of common materials

Conduction transfers heat by contact

Conduction is heat transfer between substances in *direct contact* (i.e., must be touching).

When particles of a hotter substance vibrate, these molecules bump into nearby particles and transfer some energy.

Conduction is an essential means of heat transfer for solids, and metals are conductive. It occurs when heat is transferred through *direct contact*.

For example, on a frigid winter night, when a person wraps their hands around a cup of hot cocoa, their fingers become warm because they are in direct contact with the hot mug, which is in direct contact with the hot liquid. The same process is happening at a molecular level.

The liquid in the mug contains heat energy, and the molecules of the hot liquid are in constant motion (high kinetic energy). These molecules continuously collide with the molecules of the mug, increasing their motion. The collisions gradually transfer energy from the hot liquid molecules to the mug molecules.

The mug transfers heat to a person's hands through the same conduction mechanism. The person perceives energy transfer as increasing temperature; thus, their hands feel warm.

Copyright © Sterling Test Prep.

Calculating heat conduction

When calculating heat conduction through an object, it is essential to note that the cross-sectional area refers to where the heat is being conducted. For the mug example, the heat is being conducted through its outer curved surface.

When the mug is placed on a table, and the heat conduction to the table is calculated, the cross-sectional area is the area of the bottom of the mug instead of the area of the sidewalls. Therefore, the cross-sectional area is the circumference of the mug times its height. The thickness, in this case, is the thickness of the mug itself.

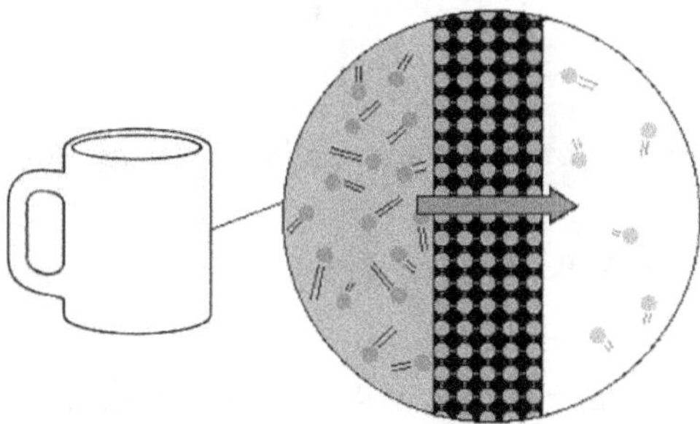

Conduction (heat transfer by direct contact) from the hot liquid to the walls of a mug

Convection transfers heat by motion of particles

Convection refers to the *physical flow of matter*. It is when heat flows energized molecules from one place to another through the movement of fluids.

Convective heat transfer occurs by diffusion (i.e., random Brownian motion of particles) and advection (i.e., larger-scale flowing currents transport matter or heat).

For example, convection occurs when warmer areas of a liquid or gas rise to colder areas, and the cooler liquid or gas replaces the warmer areas.

Gas is fluid, as it can flow and change shape depending on its container.

A pot of water is placed on a hot stove. The water at the bottom of the pot warms first. As the water warms, it rises and is replaced by cooler water. This circulation of the heated water transfers thermal energy from the bottom of the pot to the top.

This movement of molecules is a closed fluid flow pattern of *convection currents*.

Convection ovens work similarly with gas instead of liquid molecules. Air is blown past a heating element to the oven chamber. The hot element energizes the air molecules when passing through the heating element. Energized air molecules are circulated throughout the oven by airflow provided by the fan and continue until all molecules are energized to the same amount.

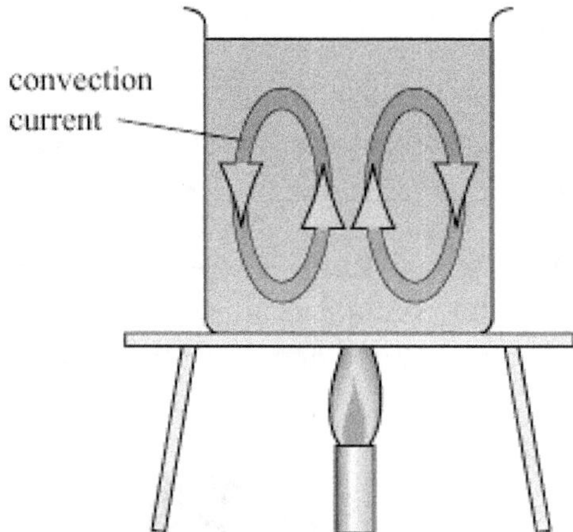

Convention current represented by the circulating water

Radiation transfers heat through electromagnetic waves

Radiation transfers heat *without* contact between the heat source and the heated object. It consists of electromagnetic waves traveling at the speed of light and can occur in a vacuum.

It has *no mass exchange*, and *no medium is required* to transfer heat.

Radiation transfers energy through electromagnetic waves emitted from all objects or surfaces with heat energy and does not require a medium; thus, radiation can occur in a vacuum.

Objects with more energy emit more radiation energy and do not require a medium; thus, radiation can occur in a vacuum.

For example, a hot piece of metal radiates heat as infrared electromagnetic waves. As the metal gets hot, it also radiates heat in the visual spectrum, thus making the metal appear "red hot."

Sun is a radiation source, releasing electromagnetic waves through space and the atmosphere to heat Earth. Similarly, holding hands next to a fire heats hands through radiation.

Radiation can be important even in situations with an intervening medium. An example is the heat transfer between a living entity and its surroundings.

For example, the Sun's thermal energy travels through the 93 million miles of space to warm the Earth, even passing through the mesosphere, the coldest part of Earth's atmosphere, which can reach temperatures of -173° C.

Power radiated by an object is related to its temperature:

$$P = e\sigma A(T^4 - T_C^4)$$

where P is the power radiated (W), e is the emissivity of the radiator, σ is Stefan's constant equal to 5.67×10^{-8} W/m^2·K, A is the area of the radiating surface (m^2), T is the temperature of the radiator (K), and T_C is the temperature of the surroundings (K).

Copyright © Sterling Test Prep.

Thermodynamic Systems

Thermodynamic variables

Thermodynamics defines variables related to energy of physical and chemical changes.

Thermodynamic variables:

ΔE or ΔU measures the *total internal energy* of a system

q measures the *heat* of a system

w measures the *work* of a system

ΔH measures the *enthalpy change* of a system – whether a process is exothermic or endothermic

ΔG measures the *Gibbs free energy change* of a system – whether a process is spontaneous or nonspontaneous

ΔS measures the *entropy change* of the system or surroundings – whether the randomness of the system or surroundings is increasing or decreasing

Steady-state processes occur without a change in the internal energy.

System and surroundings

Thermodynamic system is a body of matter or radiation confined by partitions, with defined permeabilities separating it from its surroundings.

System defines the parameters of the universe being studied, such as the substances involved in a chemical reaction or a phase change.

Surroundings are everything other than the system being studied.

Isolated, closed, and open systems

Thermodynamic system describes processes and matter within a boundary.

Isolated system has no heat transfer, work, or matter with its surroundings.

Closed system may transfer heat and work to its surroundings but *not* matter between the system and surroundings.

Open system can transfer heat, work, or matter between the system and surroundings.

Note: *heat* and *temperature* are different quantities.

Heat is a form of *thermal energy that can leave or enter a system*.

Temperature measures the *average kinetic energy* (KE) of particles in a system.

Characteristics of thermodynamic systems

System	Exchanges with surrounding	Total amount of energy	Example	Illustration
Open	Energy & Matter	Not remain constant	Solution in an open flask	Matter (water vapor) · Energy · Open system · Energy
Closed	Only Energy	Not remain constant	Solution in a sealed flask	Energy · Energy · Closed system
Isolated	Neither energy nor matter	Remains constant	Sealed flask in a thermos flask	Isolated system

Copyright © Sterling Test Prep.

State functions

State function describes the energy required to change the system into another equilibrium state.

State functions (i.e., state quantities) are a group of equations representative of the properties of a system, which depend *only* on the current state of the system, not the path used to achieve the current state.

State functions are quantities with values *not* dependent on the path used to measure the value.

State functions ($\Delta E°$, $\Delta H°$, $\Delta S°$, $\Delta G°$) depend only on the *initial and final states* of the system.

State function variables (e.g., pressure, volume, temp) are a function *of other state variables*.

Thermodynamic state functions include *enthalpy* (H), *entropy* (S), and *Gibbs free energy* (G).

Path functions

Process functions quantitatively describe the transition between equilibrium states of a thermodynamic system.

Path functions depend on the *path taken* to reach one state from another. Therefore, different routes give different quantities (e.g., work, heat).

Quantities, such as work (w) and heat (q), are not state functions, but are *path functions*, represented by *lowercase letters*.

Any duplication (copies, uploads, PDFs) is illegal.

Notes for active learning

Notes for active learning

Copyright © Sterling Test Prep.

Endothermic and Exothermic Reactions

Endothermic *vs.* exothermic processes

Endothermic processes experience heat gained by the system from the surroundings, which equals heat lost by the surroundings.

Exothermic processes cause the system's heat loss to equal the heat gain of the surroundings.

Some chemical reactions produce energy, while others require heat energy.

Energy is used to *form or break chemical bonds.*

Enthalpy change (ΔH) describes energy changes associated with forming or cleaving a chemical bond.

Unit for enthalpy (H) is the Joules (J).

Joule is the same unit used for energy; it can be expressed as energy per mol (J/mol).

Exothermic reactions release heat

Exothermic reactions, the energy needed to initiate a reaction is less than the released energy, *liberating heat.*

Exothermic reactions have a change in *enthalpy of less than zero.*

$$\Delta H < 0$$

Energy is *released* into the surroundings because stronger, more stable bonds form in the products than reactants.

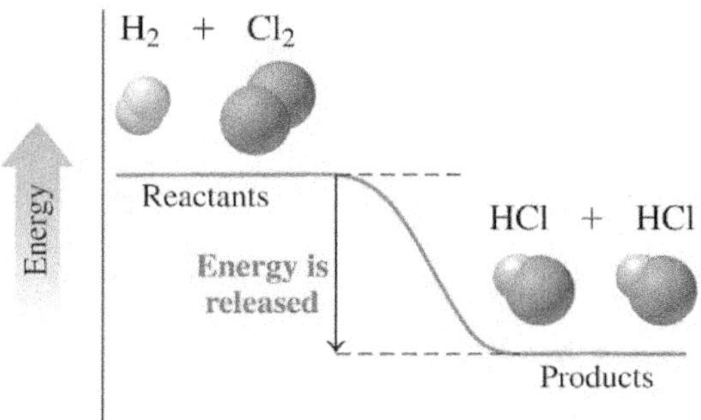

$$H_2\,(g) + Cl_2\,(g) \rightarrow 2\,HCl\,(g) + 185\text{ kJ}$$

In exothermic reactions, heat is released, and the sign of ΔH is negative ($-\Delta H$); the energy of the products is *less* than the energy of the reactants using bond-dissociation energies.

energy is released $\rightarrow \Delta H < 0$

$\Delta H < 0 \rightarrow$ enthalpy decreases

Exothermic reactions can be represented in a chemical reaction with *heat* on the product side:

$$A + B \rightarrow C + heat$$

For example, the chemical reaction for hydrogen and chlorine gas is:

$$H_2\,(g) + Cl_2\,(g) \rightarrow 2\ HCl\,(g) + 185\ kJ$$

$$\Delta H = -185\ kJ/mol\ (\text{heat released})$$

Heat is released (i.e., exothermic reaction), and ΔH is negative ($-\Delta H$).

Another example of an exothermic reaction is fuel burning (i.e., combustion).

Endothermic reactions absorb heat

Endothermic reactions have a change in *enthalpy greater than zero.*

$$\Delta H > 0$$

Endothermic reactions require (net) energy to break bonds.

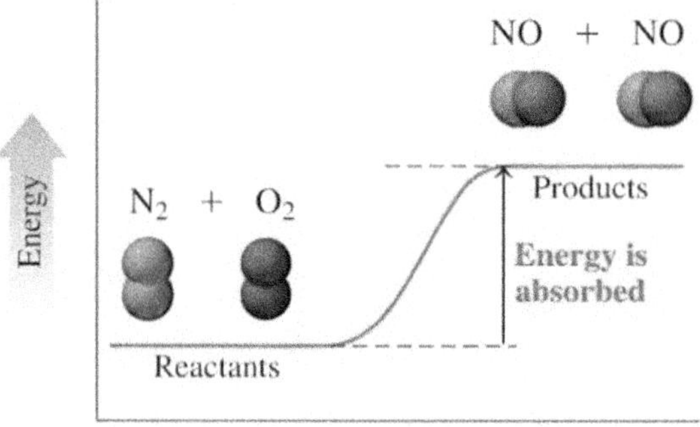

$$N_2\,(g) + O_2\,(g) + 181\ kJ \rightarrow 2\ NO\,(g)$$

In endothermic reactions, weaker bonds form in the products than in reactants.

For example, an endothermic chemical reaction represented with the heat as a reactant:

$$heat + NH_4NO_3\,(s) \rightarrow NH_4^+\,(aq) + NO_3^-\,(aq)$$

The above reaction is the ionic compound ammonium nitrate dissolving in water (H_2O), dissociating into constituent ions, NH_4^+ and NO_3^-.

Endothermic reactions include melting ice and cooking an egg (heat is required for processes).

In an endothermic reaction, heat is absorbed, and the sign of ΔH is ($+\Delta H$); the energy of the products is greater than the energy of the reactants.

Copyright © Sterling Test Prep.

Amount of energy absorbed is calculated using bond-dissociation energies.

energy is absorbed $\rightarrow \Delta H > 0$

$\Delta H > 0 \rightarrow$ enthalpy increases

Endothermic reactions can be represented in a chemical reaction with *heat* on the reactant side:

A + B + *heat* $\rightarrow$ C

For example, nitrogen gas (N_2) and oxygen gas (O_2) form nitrogen monoxide (NO):

N_2 (g) + O_2 (g) + 181 kJ $\rightarrow$ 2 NO (g)

$\Delta H = +181$ kJ (heat added)

Heat is absorbed (i.e., endothermic reaction), and ΔH is positive.

Exothermic and endothermic reactions comparison

Figure below depicts an endothermic (absorbing) and exothermic reaction (producing heat).

Reaction is exothermic on the left because heat is *produced and released* to the surroundings.

Reaction is *endothermic* on the right because heat is *absorbed* from the surroundings.

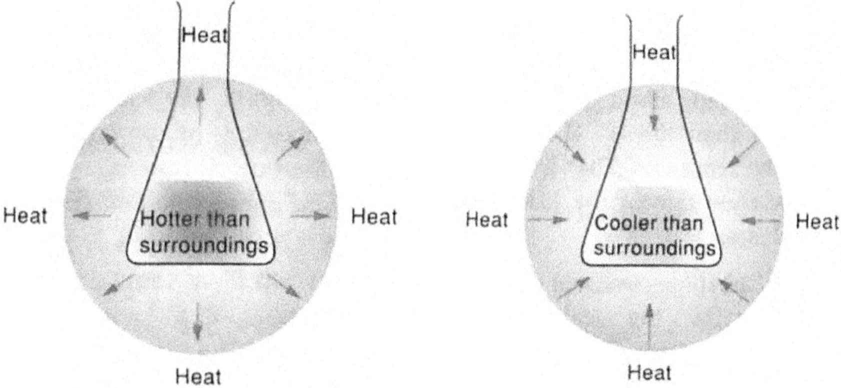

Exothermic process releases heat (left), and the endothermic process absorbs heat (right)

Reaction	Energy Change	Heat in the Equation	Sign of ΔH
Endothermic	Heat absorbed	Reactant side	Positive (+)
Exothermic	Heat released	Product side	Negative (−)

For example, is the following reaction endothermic or exothermic?

$$O_2N\text{–}NO_2 \rightarrow O_2N + NO_2$$

Reaction is endothermic (i.e., it absorbs heat) because it forms two products from a single reactant.

Breaking chemical bonds *requires energy*.

Copyright © Sterling Test Prep.

Thermodynamics

Heat, work and temperature

Thermodynamics relates heat, work, and temperature to energy, entropy, and properties of matter and radiation.

Thermodynamics answers several fundamental questions regarding *whether* a reaction occurs, to what *extent*, and whether it *releases or absorbs heat*.

An important concept in thermodynamics is the thermodynamic system, the region under study. Everything else is the surroundings.

Systems are separated from the surroundings by a boundary. Transfers of energy as work, heat, or matter between the system and surroundings occur through the boundary.

Four laws of thermodynamics

Thermodynamics states four laws characterizing heat and energy relationships by describing the behavior of physical quantities such as temperature, heat, work, energy, and entropy.

Zeroth Law of thermodynamics (or *concept of temperature*) states that **heat flows from the hotter body to the colder** *when placed in contact until thermal equilibrium.*

First Law of thermodynamics (or *conservation of energy*) states that *the total energy in a system remains constant as* **energy can neither be created nor destroyed**.

Second Law of thermodynamics (or *concept of entropy*) states that *the universe's natural order* **favors a net increase** *in the overall entropy (i.e., disorder)*.

Third Law of thermodynamics (or *entropy at absolute zero*) states that *the entropy of a perfectly ordered, pure crystalline solid is* **zero at the temperature of absolute zero** *(equal to 0 K, −273.15 °C or −459.67 °F)*.

Any duplication (copies, uploads, PDFs) is illegal.

Notes for active learning

Copyright © Sterling Test Prep.

Zeroth Law of Thermodynamics

Thermal equilibrium

Zeroth Law of Thermodynamics is the logical predecessor of the First and Second Laws of Thermodynamics.

Zeroth Law states that *if two systems are in thermal equilibrium with a third system, the two initial systems are also in thermal equilibrium.*

Three systems in thermal equilibrium contain the same amount of heat energy (same temperature) and do *not* exchange heat, fundamental principle underlying thermodynamics.

For example, if system A is in equilibrium with B, and system C is in equilibrium with B, then system A and C must be in equilibrium, and all three systems must be at the same temperature.

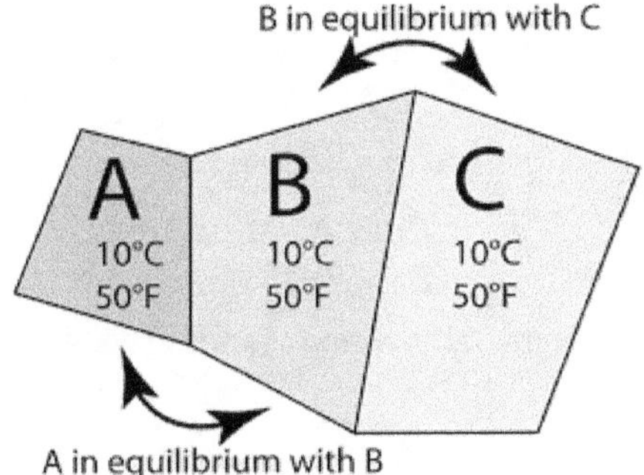

Zeroth Law of Thermodynamics with three systems in thermal equilibrium

However, it is vital to notice that although no net heat transfer occurs between systems in equilibrium, thermal energy is technically transferred between the related systems.

The exchange of thermal energy requires that every unit of energy passed from any system has the same value of energy passed back into the system. This holds even if the two systems contain materials with different specific heats.

This observation of thermal energy transfer means that there must be a property that can be considered the same upon which heat transfer depends and that property is temperature.

Concept of temperature

Zeroth Law of thermodynamics states that *heat flows from the hotter to the colder body when in contact until thermal equilibrium.*

It establishes that *temperature is a core, measurable property of matter.*

Separated by a permeable barrier, two systems are in thermal equilibrium, and *no heat* transfers.

After thermal equilibrium is achieved, no heat transfers, defining the *concept of temperature.*

Expressing thermal equilibrium

If two systems are in thermal equilibrium with a third system, each is in thermal equilibrium.

Thermal equilibrium is expressed by:

$$T_A = T_B$$

and

$$T_C = T_B$$

then

$$T_A = T_C$$

Zeroth Law is crucial for calculations of thermodynamics (i.e., heat and energy relationships).

   Copyright © Sterling Test Prep.

First Law of Thermodynamics

Conservation of energy

Energy, like matter, is always conserved; it can neither be created nor destroyed.

First Law of Thermodynamics states that *energy can neither be created nor destroyed, and the total energy in a system remains constant. Energy is conserved in a chemical reaction.*

First Law of thermodynamics relates to the *internal energy* of a substance.

Internal energy increases if heat is added to the system or if work is done on the system.

Disordered motion of particles

Change in total *internal energy* of a system equals contributions by heat (q) and work (w):

$$\Delta U = q - w$$

where ΔU is the change in internal energy, q is the contribution from heat and w from work.

Internal energy refers to the microscopic energy of the disordered motion of particles and is represented by the symbol (U).

Internal energy cannot be seen and has no apparent effect on the object's motion, but it does play a part in the thermodynamic properties of the substance. This principle is the foundation for the law of *conservation of energy*.

For example, a glass of water does not look like it contains kinetic or potential energy. Water molecules are whizzing around on the atomic scale, moving hundreds of m/s. If the water is dumped from the glass, its internal energy does not affect its motion.

Thermodynamic properties of the substance include potential and kinetic energy

Sign conventions

$\Delta U = q - w$ uses the following sign conventions:

q is *positive* if heat is *absorbed into* the system (i.e., heating)

q is *negative* if heat is *released from* the system (i.e., cooling)

w is *positive* if work is done *by* the system (i.e., compression)

w is *negative* if work is done *on* the system (i.e., expansion)

First Law of thermodynamics:

$$\Delta U = q + w$$

where q is *positive* if heat is *absorbed into* the system and *negative* if heat is *released from* the system; work (w) is positive if done *on the* system or negative if done *by* the system.

Heat (q) and work (w) absorbed or released

Use the sign conventions in the chart below:

q	+	Heat ***absorbed*** by the system	E_{system} increases
q	−	Heat ***released*** by the system	E_{system} decreases
w	+	Work done ***on*** the system	E_{system} increases
w	−	Work done ***by*** the system	E_{system} decreases

Heat or energy *flowing out* of the system is *negative*.

Heat or energy *flowing into* the system is *positive*.

Other natural sciences use the opposite sign convention; note the sign convention used.

Copyright © Sterling Test Prep.

Internal energy

The difference between the amount of heat added to the system (Q) and the work done by the system (W) is equal to the total change in internal energy (ΔU):

$$\Delta U = \Delta Q - \Delta W$$

As with other energy, the unit of internal energy is the joule (J), although it is also expressed using calories (cal).

By convention, work done *on a system* is positive, and work done *by a system* is negative.

Occasionally, the above equation is written as $\Delta U = \Delta Q + \Delta W$, and ΔW is defined as the work done on the system. Be sure to note which form is being used.

Calculating internal energy

The quantities in $\Delta U = \Delta Q + \Delta W$ can determine what a thermodynamic system is experiencing.

For example, in a chemical reaction, if ΔQ is a quantity less than zero ($\Delta Q < 0$), the reaction is losing heat. This reaction is *exothermic,* and heat is a product.

If ΔQ is a quantity greater than zero ($\Delta Q > 0$), the reaction absorbs heat from its surroundings. This is an *endothermic* reaction, and heat is a reactant.

Internal energy is expressed:

$$\Delta U = q + w$$

where ΔU is the change in internal energy, q is the contribution from heat and w from work.

For example, consider internal energy, heat, and work. For a process in which 3.4 kJ of heat flows out of the system while the system in the surroundings does 4.8 kJ of work, what is the internal energy?

Calculate internal energy:

$$\Delta U = q + w$$

$$\Delta U = -3.4 \text{ kJ} + (-4.8 \text{ kJ})$$

$$\Delta U = -8.2 \text{ kJ}$$

where ΔU is the change in internal energy, q is *negative* as heat is *released from* the system, and work (w) is negative when done *by* the system.

Use the appropriate heat (q) and work (w) signs.

Energy has flowed from the system to the surroundings by -8.2 kJ.

Notes for active learning

Notes for active learning

Copyright © Sterling Test Prep.

Second Law of Thermodynamics

Concept of entropy (S)

Second Law of thermodynamics is that *the universe favors a net increase in the overall entropy.*

> *Isolated system*s increase in entropy and *never decrease.*

> *Open systems* decrease entropy as entropy increases for the surroundings.

Second Law of Thermodynamics states *that in an isolated system (no transfer of heat, work, or matter), a process increases entropy or stays constant, but entropy never decreases.*

Entropy is a measure of the randomness or disorder of a system. Entropy measures the system's thermal energy per unit temperature, which is unavailable for useful work.

Work requires ordered molecular motion.

Open systems transfer energy

An exception to the Second Law of Thermodynamics (i.e., entropy cannot decrease) is an *open system* where heat, work, or matter can be transferred to surroundings; the system's entropy decrease is related to an increase in the entropy of the surroundings.

Second Law states that *heat flows spontaneously* from a hot object to a cold object but never spontaneously flows in the opposite direction. A bowl of ice never gets colder in a warm room; the ice always gets warmer.

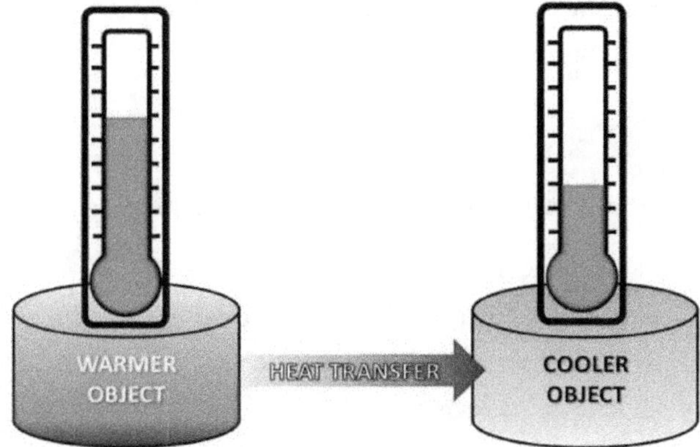

Heat flows spontaneously from a hot object to a cold object

Entropy measures disorder

Entropy measures disorder within a system and likelihood—how likely a system resembles the same state twice or how likely a reaction is to happen.

Entropy of a natural system never decreases, as it is the natural tendency of all things to move towards maximum disorder—maximum entropy.

For example, gas expands to fill space with more degrees of freedom to arrange particles—more *multiplicities*.

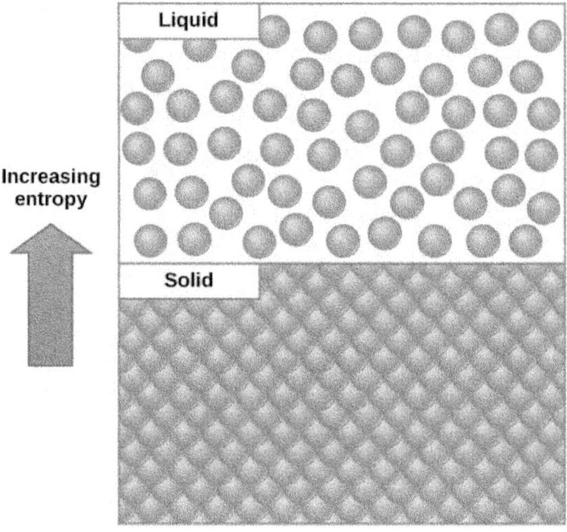

A solid has less entropy than a liquid, which has much less entropy than a gas

According to the Second Law of Thermodynamics (i.e., entropy can stay constant or increase but never decrease), entropy is given by:

$$\Delta S \geq \frac{Q}{T}$$

where ΔS is the change in entropy of a process (J/K), Q is heat transferred (J), and T is the temperature (K).

If a system strives to increase entropy (i.e., disorder), it does not mean that the system cannot become more orderly. This means that to do so, outside energy must be transferred into the system, and it cannot be done on its own.

Copyright © Sterling Test Prep.

Nonspontaneous and spontaneous processes

*Nonspontaneous process*es require energy (or work) for the system to decrease entropy.

Spontaneous processes move towards randomness without energy input to increase entropy.

For example, a valve connects a container with air and another container in a vacuum. When the valve is opened, the air flows from the container to the container in a vacuum. No work or energy is needed for this to happen. The process is *spontaneous*, and entropy *increases*.

For the reverse to occur, energy must be *added to the system* to re-pressurize the original container with air. This process is an example of a *nonspontaneous* reaction.

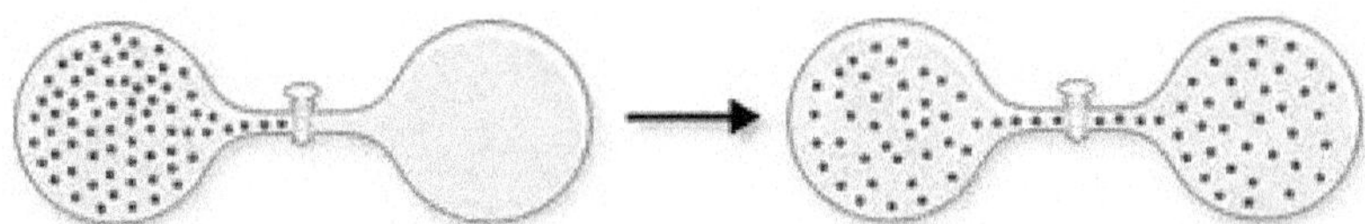

Air flows from the container to the container in a vacuum

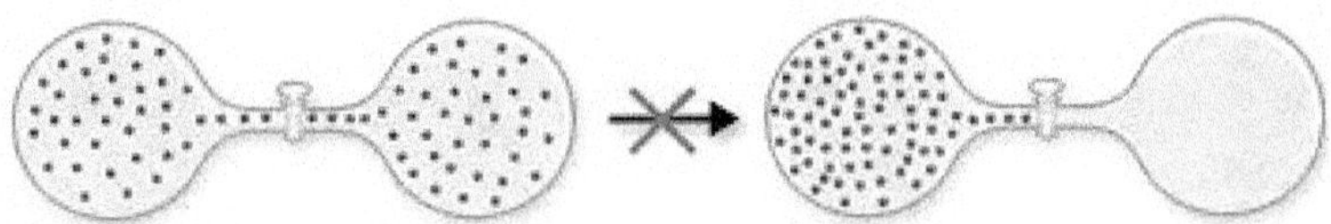

Nonspontaneous reactions do not occur without the input of energy

The universe's entropy always *increases* when a *spontaneous process* occurs.

Entropy of the universe is expressed as:

$$\Delta S_{universe} > 0$$

Reversible and irreversible processes

Thermodynamic processes are *reversible* or *irreversible*.

Reversible processes are idealized, while the irreversible process frequently occurs in nature.

Reversible processes

$$\Delta S = q\,/\,T$$

Irreversible processes

$$\Delta S > q\,/\,T$$

where ΔS is the change in entropy, q is the contribution from heat, and T is the temperature (K).

Natural processes are *irreversible*, so entropy change is greater than the heat transfer over temperature. Due to the irreversible nature of processes, the universe's entropy is increasing.

Reactions at equilibrium are *reversible* with no net change to the system or surroundings and proceed in both directions.

Reversible processes reverse direction if some change is made to some system property:

$$\Delta S_{total} = \Delta S_{system} + \Delta S_{surroundings}$$

ΔS of the system equals in magnitude, but *opposite in sign*, to the ΔS of the surroundings.

Entropy related to heat and enthalpy

Entropy change of surroundings is related to heat (q) flow, enthalpy (ΔH), and temperature:

$$\Delta S_{surroundings} = -\,q\,/\,T$$

or (at constant pressure)

$$\Delta S_{surroundings} = -\,\Delta H\,/\,T$$

Combining these two equations:

$$\Delta S_{surroundings} = \Delta S_{system} - (\Delta H_{system}\,/\,T)$$

where ΔS is the change in entropy, ΔH for enthalpy, q is the contribution from heat, and T is the temperature (K).

ΔS from tabulated values

ΔS is calculated from standard molar entropies as for ΔH and ΔG by heats of formation.

Equation to calculate ΔS from tabulated values:

$$\Delta S^{\circ}_{rxn} = \Sigma\, n(\Delta S^{\circ}_{f}\text{ of products}) - \Sigma\, n(\Delta S^{\circ}_{f}\text{ of reactants})$$

where n represents moles of each reactant (or product) as a balanced chemical equation, and ΔS is the change in entropy.

For example, determine the entropy change (ΔS) for the above chemical equation as (ΔH):

$$16\ H_2S\,(g) + 8\ SO_2\,(g) \rightarrow 16\ H_2O\,(l) + 3\ S_8\,(s)$$

ΔS°_{rxn} as a measure of the entropy change of a system:

$$\Delta S^{\circ}_{rxn} = [(16\text{ mol}){\cdot}(S^{\circ}_{f}\text{ of }H_2O\,(l)) + (3\text{ mol}){\cdot}(S^{\circ}_{f}\text{ of }S_8\,(s))]$$
$$- [(16\text{ mol}){\cdot}(S^{\circ}_{f}\text{ of }H_2S\,(g)) + (8\text{ mol}){\cdot}(S^{\circ}_{f}\text{ of }SO_2\,(g))]$$

Copyright © Sterling Test Prep.

From the table of standard *entropies of formation*, substitute:

$$\Delta S^\circ_{rxn} = \Sigma\, n(\Delta S^\circ_f \text{ of products}) - \Sigma\, n(\Delta S^\circ_f \text{ of reactants})$$

$$\Delta S^\circ_{rxn} = [(16\text{ mol})\cdot(69.91\text{ J/mol}\cdot\text{K}) + (3\text{ mol})\cdot(31.80\text{ J/mol}\cdot\text{K})]$$
$$- [(16\text{ mol})\cdot(205.79\text{ J/mol}\cdot\text{K}) + (8\text{ mol})\cdot(161.92\text{ J/mol}\cdot\text{K})]$$

$$\Delta S^\circ_{rxn} = [1{,}118\text{ J/K} + 95.40\text{ J/K}] - [3{,}292.6\text{ J/K} - 1{,}295.4\text{ J/K}]$$

$$\Delta S^\circ_{rxn} = -3{,}375\text{ J/K}$$

where ΔS°_{rxn} is a measure of the entropy change of a system, and the negative sign accounts for the heat of the surroundings opposite in sign to the heat of the system.

Entropy calculations

For example, the reaction enthalpy (ΔS_{surr}) was −1,876 kJ at a standard temperature of 298 K. The entropy change was calculated as −3,375 J/K (or −3.375 kJ/K). What is the entropy change of the surroundings?

Using the equation:

$$\Delta S_{surroundings} = -\,\Delta H\,/\,T$$

$$\Delta S_{surroundings} = -(-1{,}876\text{ kJ})\,/\,298\text{ K}$$

$$\Delta S_{surroundings} = 6.30\text{ kJ/K or } \Delta S_{surroundings} = 6.30 \times 10^3\text{ J/K}$$

where $\Delta S_{surroundings}$ is change in entropy, ΔH is change in enthalpy, and T is temperature (K).

As a *positive* value, the entropy of the surroundings *increased*.

Total entropy change is the sum of the entropy change of the system and the surroundings:

$$\Delta S_{total} = \Delta S_{system} + \Delta S_{surroundings}$$

$$\Delta S_{total} = -3.375 \times 10^3\text{ J/K} + 6.30 \times 10^3\text{ J/K}$$

$$\Delta S_{total} = 2.92 \times 10^3\text{ J/K}$$

where ΔS_{total} is the total entropy change, ΔS_{system} is the entropy change of the system, and $\Delta S_{surroundings}$ is the entropy change of the surroundings.

Entropy measures degrees of freedom

Entropy (S) measures *randomness* (or *disorder*) from the number of ways (i.e., *degrees of freedom*) a thermodynamic system can be arranged.

Entropy (or disorder) of the universe is *increasing*.

Entropy uses energy divided by temperature; the unit is Joules per Kelvin (J/K).

Entropy is expressed as *entropy per unit mass* (J/K·kg) or *unit amount of substance* (J/K·mol).

For example, imagine taking a piece of paper, tearing it into small pieces, and throwing them in the air. How much energy is needed to restore it to its original form?

Consider tearing paper as *spontaneous* and putting it together as a *nonspontaneous* process.

Temperature and entropy relationship

The higher the entropy and lower the energy (i.e., temperature), the more stable the system.

Higher entropy + lower energy (lower temperature) → *more stability*

Lower entropy + higher energy (higher temperature) → *less stability*

As *temperature increases, entropy increases*:

$T = 0$ K

particles are in equilibrium lattice positions, and S is relatively low

$T > 0$ K

molecules vibrate, S increases

As T increases, more violent vibrations occur, and the entropy (S) increases.

Entropy for gasses, liquids, and crystals

Entropy differences between phases of matter (i.e., solid, liquid, gas) are related to the *degrees of freedom* of kinetic energy (KE).

Solids have low entropy because particles forming a crystal solid are packed tightly.

Particles of a solid are confined to vibrating in fixed positions and limited in *degrees of freedom*.

Liquids are intermediate in entropy as intermolecular forces hold particles as a liquid but have increased *degrees of freedom* to move.

Gases are high in entropy because the gas particles have complete *freedom* to move with a few particles arranged in an orderly fashion.

Low-pressure gas experiences little to no intermolecular forces of attraction or repulsion.

Copyright © Sterling Test Prep.

Entropy of phase changes

Entropy changes accompanying phase changes can be estimated, at least qualitatively.

Freezing decreases entropy as the relatively disordered liquid becomes a well-ordered solid.

Boiling results in a substantial increase in entropy as the liquid becomes a more highly disordered gas.

Sublimation (i.e., from a solid to a gas) is the phase transition with the greatest entropy change for a substance.

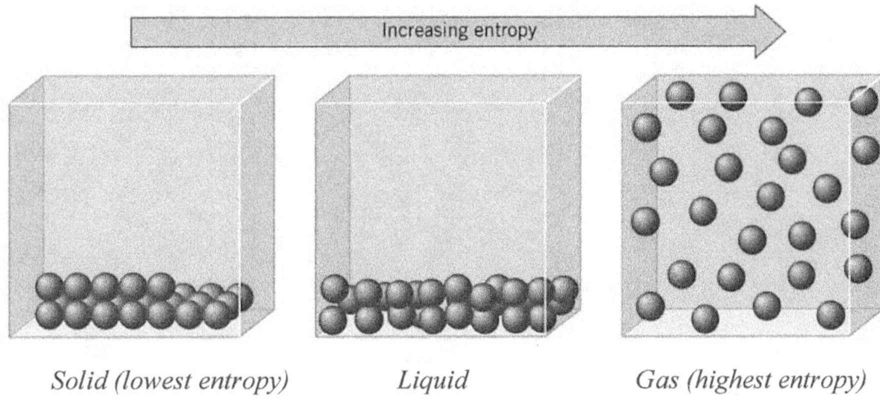

The figure displays a crystalline solid, liquid, and gas with entropy (i.e., randomness) increasing to the right.

Entropy for gases

Entropy *increases* as gas volume *increases* (i.e., decreasing pressure).

Consider a gas separated from a vacuum by an impermeable partition (figure on the left). Removing the partition (center image) allows more energy distribution.

Gas expands to achieve higher particle distribution (figure on the right). Gas molecules' configuration is *random* and *more positive S* (more disorder).

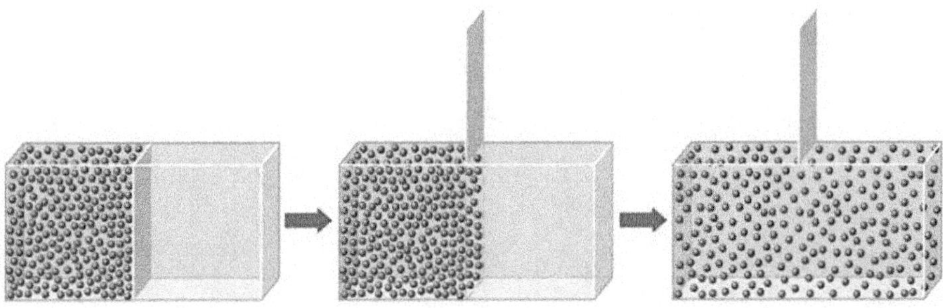

Gas particles are highly ordered on the left (low S) and random on the right (high S)

Any duplication (copies, uploads, PDFs) is illegal.

Number of particles affects entropy reactions, producing more particles as a *positive* ΔS.

Adding particles to a system (below on the right) increases the number of ways energy can be distributed.

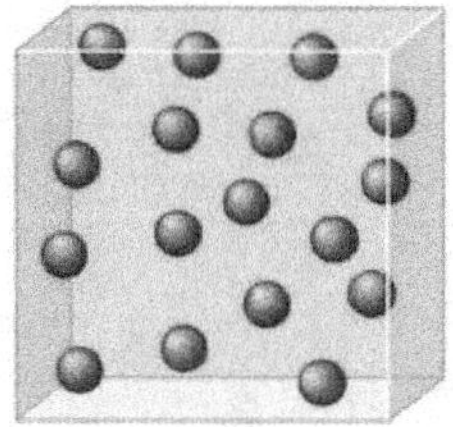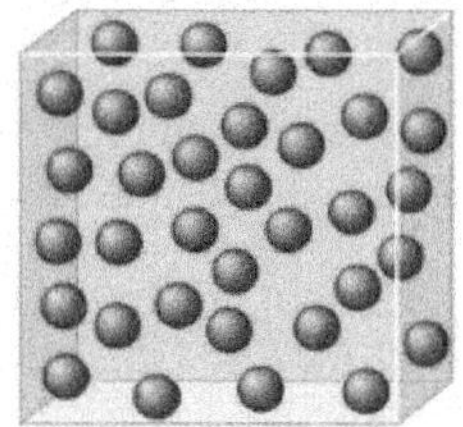

Low entropy   *High entropy: more disorder*

Copyright © Sterling Test Prep.

Third Law of Thermodynamics

Absolute zero

Third Law of thermodynamics states that *as the temperature of a system approaches absolute zero, the entropy approaches a constant value.*

Typically, this constant value is zero. In this definition, the entropy of a system is related to the number of microstates possible.

Microstates refer to the atomic configurations possible with the equation:

$$S = k \cdot ln\,(\Omega)$$

where S is the entropy (J/K), k is Boltzmann's constant (1.381×10^{-23} J/K), and Ω is the number of microstates.

Only one microstate is possible at absolute zero, and so Ω equals 1:

$$S = k \cdot ln(1) = k \cdot 0 = 0$$

The entropy of a system at absolute zero is zero.

Minimum energy state

However, some systems have more than one minimum energy state as the temperature approaches absolute zero: the entropy levels off at a value other than zero.

Cooling any process to absolute zero in a finite number of steps is impossible.

In left image, $T = 0$ can be reached following infinite steps (step lines between X_1 and X_2).

In the image on the right, $T = 0$ is reached by a finite number of steps, each getting closer to zero but never to zero.

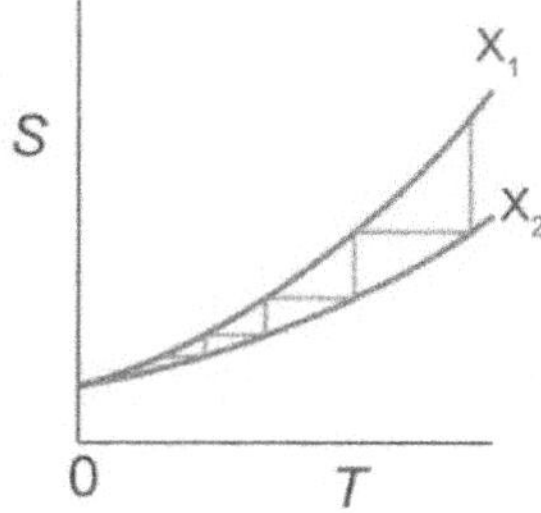

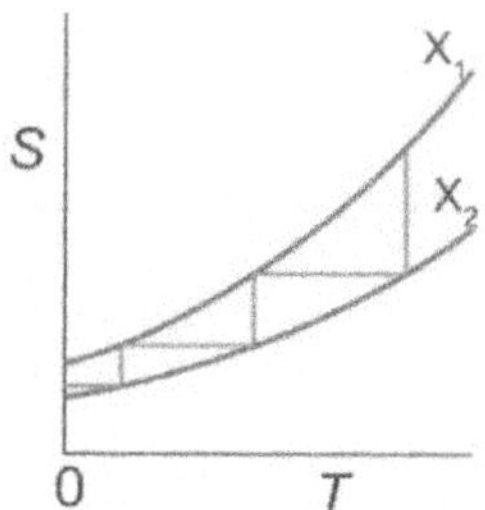

Entropy vs. temperature illustrating the step to approach absolute zero

Entropy at absolute zero

Third Law of Thermodynamics states that *the entropy of a perfectly ordered, pure crystalline solid is zero at absolute zero.*

At *absolute zero* (the coldest temperature possible on the Kelvin scale equals 0 K, –273.15 °C, or –459.67 °F), the particles' average kinetic energy (KE) equals zero.

Particles' average kinetic energy (KE) increases as entropy increases.

With increasing temperature (i.e., KE), substances transition to liquid and then gaseous state.

For a substance, the gas state has the highest entropy.

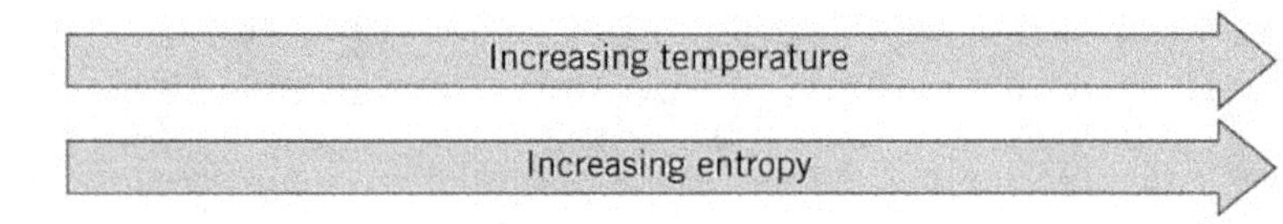

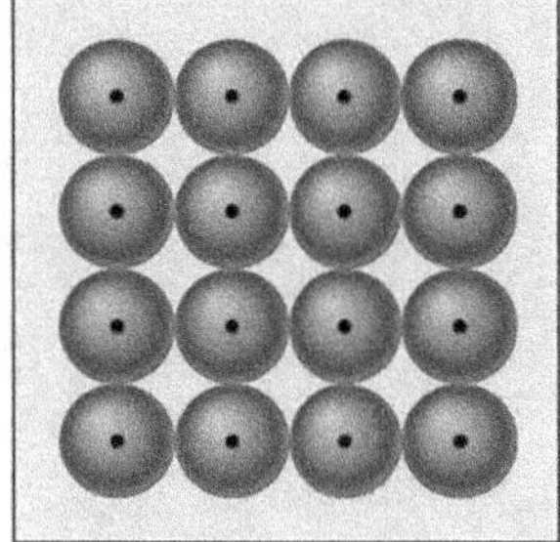

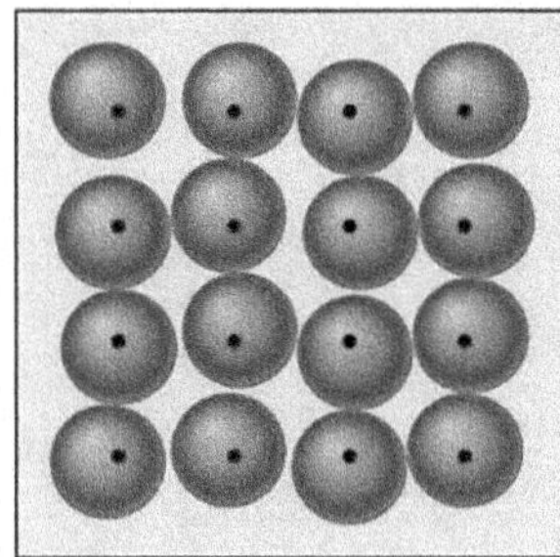

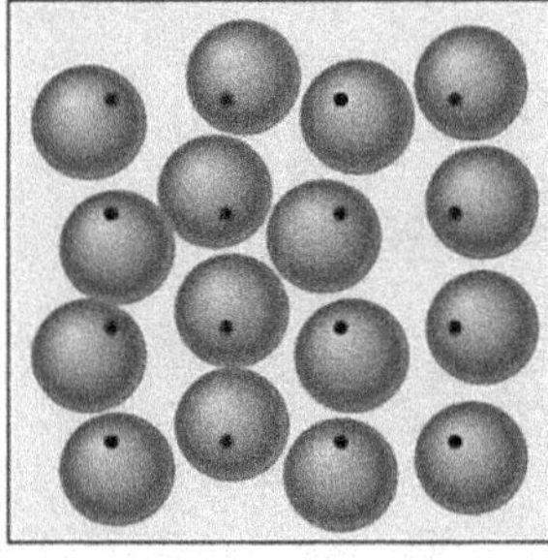

Crystalline solids have low KE with particles arranged orderly.

Liquid particles vibrate from higher KE, and entropy (S) is moderate.

Gas particles have large KE, and entropy (S) is large.

Copyright © Sterling Test Prep.

Determining ΔS for reactions

For example, determine the sign of ΔS for the reaction:

$$CaCO_3\,(s) + 2\,H^+\,(aq) \rightarrow Ca^{2+}\,(aq) + H_2O + CO_2\,(g)$$

$$\Delta n_{gas} = 1\text{ mol} - 0\text{ mol}$$

$$\Delta n_{gas} = 1\text{ mol}$$

Since Δn_{gas} is positive, ΔS is positive.

For example, determine the sign of ΔS for the reaction:

$$2\,N_2O_5\,(g) \longrightarrow 4\,NO_2\,(g) + O_2\,(g)$$

$$\Delta n_{gas} = 4\text{ mol} + 1\text{ mol} - 2\text{ mol}$$

$$\Delta n_{gas} = 3\text{ mol}$$

Since Δn_{gas} is positive, ΔS is positive.

For example, determine the sign of ΔS for the reaction:

$$OH^-\,(aq) + H^+\,(aq) \longrightarrow H_2O$$

$$\Delta n_{gas} = 0\text{ mol}$$

$$\Delta n_{gas} = 1\text{ mol} - 2\text{ mol}$$

$$\Delta n_{gas} = -1\text{ mol}$$

Since Δn is negative, ΔS is negative.

Any duplication (copies, uploads, PDFs) is illegal.

Notes for active learning

Copyright © Sterling Test Prep.

Enthalpy

Energy changes

Enthalpy (ΔH) of a system is the amount of heat used at constant pressure in a system or reaction and is given in SI units of Joules.

Energy changes in the system can be written as:

$$\Delta U = \Delta H - P\Delta V$$

where ΔH is change in system enthalpy (J), P is pressure (Pa), and ΔV is volume change (m^3).

During an isobaric process (pressure is held constant), such as that conducted in an open container, the change in enthalpy equals the amount of heat transferred during the process. Most reactions do not generate many gaseous products, so little work is associated with the reaction.

Energy change of the system equals the change in enthalpy:

$$\Delta E \approx \Delta H$$

Enthalpy is equal to the transfer of thermal energy (heat); it is only valid at *constant pressure*.

Enthalpy changes can occur at standard temperatures and pressures.

Standard state refers to substances in their most stable form (i.e., the lowest-energy state).

Standard state in thermodynamics refers to one-atmosphere pressure (1 atm) and a temperature of 25 °C (298 K).

For example, oxygen is O_2 (diatomic gas) at the standard state, and carbon is C (solid graphite).

Standard heats of reaction (ΔH)

Standard heat of reaction (*standard enthalpy of reaction* ΔH°_{rxn}) is the change in heat content when one mole of the matter is transformed by a chemical reaction under standard conditions (1 atm and 298 K).

Theoretical and experimental methods calculate the *enthalpy change* of a chemical reaction.

General equation to calculate the heat of a reaction (ΔH) from the difference (delta or Δ) in the energy of the reactants and products is:

$$\Delta H = H_{products} - H_{reactants}$$

Standard heats of formation (ΔH°_f)

Standard heat of formation (*standard enthalpy of formation* ΔH°_f) is the change in heat content when one mole of a compound or molecule forms from constituent elements in standard states.

Standard enthalpy of formation has been determined for many molecules, and the ΔH°_f for the most common molecules are reported and tabulated data.

When choosing ΔH°_f from a table, choose the value corresponding to the appropriate state of matter (solid, liquid, aqueous, gas).

Standard heat of formation for an element in its standard state equals 0 kJ/mol.

For example, calculate the heat of reaction from the heat of formation and determine if the process is exothermic or endothermic for the balanced chemical equation:

$$16 \; H_2S \; (g) + 8 \; SO_2 \; (g) \rightarrow 16 \; H_2O \; (l) + 3 \; S_8 \; (s)$$

From thermodynamic quantities, use values for the heat of formation of each component in the reaction; set the equation to calculate the heat of the reaction:

$$\Delta H = H_{products} - H_{reactants}$$

$$\Delta H^\circ_{rxn} = [16 \; mol(\Delta H^\circ_f \text{ of } H_2O \; (l)) + 3 \; mol(\Delta H^\circ_f \text{ of } S_8 \; (s))]$$
$$- [16 \; mol(\Delta H^\circ_f \text{ of } H_2S \; (g)) + 8 \; mol(\Delta H^\circ_f \text{ of } SO_2 \; (g))]$$

$$\Delta H^\circ_{rxn} = [16 \; mol(-285.8 \text{ kJ/mol}) + 3 \; mol(0 \text{ kJ/mol})]$$
$$- [16 \; mol(-20.2 \text{ kJ/mol}) + 8 \; mol(-296.8 \text{ kJ/mol})]$$

$$\Delta H^\circ_{rxn} = [-4{,}573 \text{ kJ} + 0 \text{ kJ}] - [-323 \text{ kJ} + (-2{,}374 \text{ kJ})]$$

$$\Delta H^\circ_{rxn} = -4{,}573 \text{ kJ} + 323 \text{ kJ} + 2374 \text{ kJ}$$

$$\Delta H^\circ_{rxn} = -1{,}876 \text{ kJ}$$

ΔH°_{rxn} is negative, and the reaction is exothermic; the system releases heat to the surroundings.

Enthalpy for exothermic and endothermic reactions

Exothermic reactions have a change in enthalpy of less than zero.

$$\Delta H < 0$$

Endothermic reactions have a change in enthalpy greater than zero.

$$\Delta H > 0$$

Since enthalpy (H) is a state function, it represents a system property; it can be used to describe a reaction or an equation for a reaction.

Copyright © Sterling Test Prep.

Calculating the heat of reaction (ΔH)

Calculate the heat of reaction (ΔH) using these steps:

1) Identify the given and needed quantities.

2) Write an expression using the heat of reaction and molar mass needed.

3) Write the conversion factors, including the heat of the reaction.

4) Set the expression to calculate heat.

Hess' Law for heat of summation

In 1840, Swiss-born Russian chemist Germain Hess (1802-1850) published Hess' Law of constant heat summation.

Hess' Law describes an essential relationship in physical chemistry that *total enthalpy change (ΔH) during the complete course of a chemical reaction is independent of the number of steps.*

Hess' Law states that the change of enthalpy in a reaction is equal to the sum of the products' enthalpy minus the sum of the reactants' enthalpy:

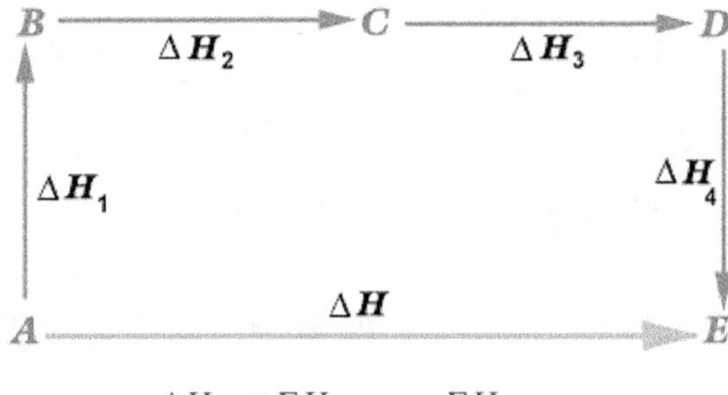

$$\Delta H_{rxn} = \Sigma H_{products} - \Sigma H_{reactants}$$

Enthalpy is a *state function* (i.e., depends only on initial and final states, *not* the reaction's path).

Hess' Law of Heat Summation is expressed

$$\Delta H^\circ_{rxn} = \Sigma\, n(\Delta H^\circ_f \text{ of products}) - \Sigma\, n(\Delta H^\circ_f \text{ of reactants})$$

where n represents the moles of each reactant or product in the balanced chemical equation, and the Greek letter Σ means the sum of the variables.

Hess' Law states that the total energy change throughout a reaction is the same, regardless of whether the reaction occurs in one step or several steps.

Calculating heat of summation

For example, how much heat (kJ) is absorbed when 1.65 grams of NO gas is produced in the following reaction?

$$N_2 \, (g) + O_2 \, (g) \rightarrow 2 \, NO \, (g)$$

$$\Delta H = +181 \text{ kJ/mol}$$

Step 1: State the given and needed quantities.

Given:

> 1.65 grams of NO

> $\Delta H = +181$ kJ/mol

Solve:

> Heat absorbed in kJ

> $N_2 \, (g) + O_2 \, (g) \rightarrow 2 \, NO \, (g)$ $\Delta H = +181$ kJ/mol

> 1.65 g   ? kJ

Step 2: Write an expression for the heat of the reaction and the molar mass needed.

> grams of NO $\rightarrow$ moles of NO $\rightarrow$ kilojoules of energy

> *molar mass*   *heat of reaction*

Step 3: Write the conversion factors, including the heat of the reaction.

> 1 mole of NO = 30.01 g of NO:

> 30.01 g NO / 1 mol NO   or   1 mol NO / 30.01 g NO

> ΔH for 2 moles of NO = +181 kJ/mol

therefore:

> 2 mol NO / +181 kJ/mol   or   +181 kJ/mol / 2 mol NO

Step 4: Set the expression to calculate the heat.

> 1.65 g NO × (1 mol NO/30.01 g NO) × (+181 kJ/mol/2 mol NO)

> 1.65 g ~~NO~~ × (1 ~~mol NO~~/30.01 ~~g NO~~) × (+181 kJ/mol/2 ~~mol NO~~)

> 1.65 × (1/30.01) × (+181 kJ/mol/2) = 4.98 kJ/mol

4.98 kJ/mol of nitrogen gas is absorbed when 1.65 g of nitrogen monoxide is produced from nitrogen and oxygen.

Copyright © Sterling Test Prep.

Quantity relationships in chemical reactions

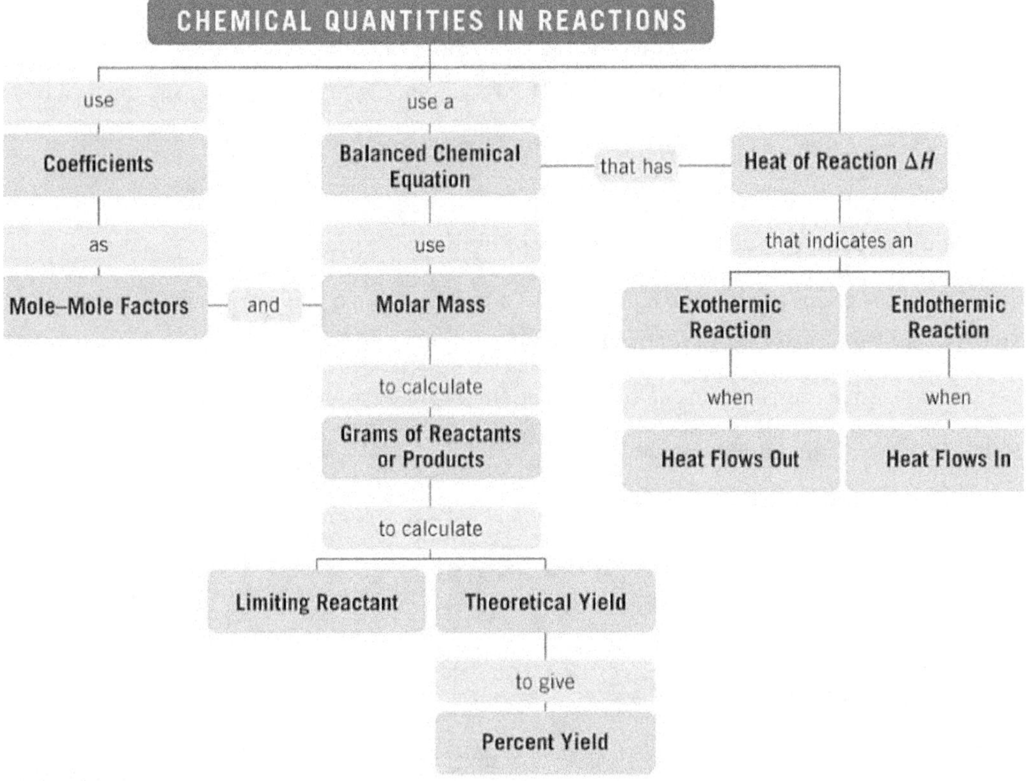

Interrelationship of chemical processes and quantitative outcomes

Thermochemical systems to calculate enthalpy

In addition to Hess' Law of Heat Summation, alternative methods can calculate enthalpy.

For example, if a series of reactions whose enthalpies are known to create an overall reaction that gives the reaction sought, the sum of those enthalpies is the enthalpy of the entire reaction.

Calculating enthalpy change by *thermochemical systems of equations* is below.

For example, from Hess' Law, determine the enthalpy for the reaction:

$$2\,N_2\,(g) + 5\,O_2\,(g) \;\longrightarrow\; 2\,N_2O_5\,(g) \qquad\qquad \Delta H = ?$$

Use the following heats of reaction:

$$\text{Equation (1)}\; 2\,H_2\,(g) + O_2\,(g) \;\longrightarrow\; 2\,H_2O\,(l) \qquad \Delta H = -571.7\ \text{kJ}$$

$$\text{Equation (2)}\; N_2O_5\,(g) + H_2O\,(l) \;\longrightarrow\; 2\,HNO_3\,(l) \qquad \Delta H = -92\ \text{kJ}$$

$$\text{Equation (3)}\; N_2\,(g) + 3\,O_2\,(g) \;\longrightarrow\; 2\,HNO_3\,(l) \qquad \Delta H = -348.2\ \text{kJ}$$

Manipulate the above equations, so sum the desired equation. Inspect the compounds in each equation and choose an equation with a compound appearing once in the equations.

Oxygen (O_2) is in equations (1) and (3), so it is *not* suitable to begin solving the problem.

Dinitrogen pentoxide (N_2O_5) is in equation (2), so equation (2) could begin solving the problem.

Nitrogen is in equation (3), so equation (3) could begin solving the problem.

For example, solve the problem using nitrogen (N_2) and equation (3). Equation (3) contains one mole of nitrogen as a reactant, and two moles of nitrogen are needed.

Since N_2 is a reactant in both, the equation is not reversed, but equation (3) must be multiplied by two for the required two moles of nitrogen (N_2).

What is done to manipulate the chemical equation is applied to the heat of the reaction.

Heat of reaction is multiplied by two:

$2 \times$ Equation (3):

$$2\,N_2\,(g) + 6\,O_2\,(g) + 2\,H_2O\,(g) \rightarrow 4\,HNO_3\,(l)$$

$$\Delta H = 2(-348.7\ \text{kJ})$$

$$\Delta H = -696.4\ \text{kJ}$$

Dinitrogen pentoxide (N_2O_5) is introduced since only equation (2) contains (N_2O_5).

In equation (2), N_2O_5 appears as one mole of reactant.

Reverse the equation and multiply by two.

The sign of the heat of reaction is reversed and doubled:

$-2 \times$ Equation (2):

$$4\,HNO_3\,(l) \longrightarrow 2\,N_2O_5\,(g) + 2\,H_2O\,(l)$$

$$\Delta H = -2(-92\ \text{kJ})$$

$$\Delta H = +184\ \text{kJ}$$

Four moles of nitric acid cancel, which is convenient since it is not in the equation sought.

$2 \times$ Equation (3):

$$2\,N_2\,(g) + 6\,O_2\,(g) + 2\,H_2O\,(g) \longrightarrow 4\,\cancel{HNO_3\,(l)}$$

$$\Delta H = 2(-348.7\ \text{kJ})$$

$$\Delta H = -696.4\ \text{kJ}$$

Copyright © Sterling Test Prep.

$-2 \times$ Equation (2):

$$\text{4 } \cancel{HNO_3\,(l)} \longrightarrow 2\,N_2O_5\,(g) + 2\,H_2O\,(l)$$

$$\Delta H = -\mathbf{2}(-92 \text{ kJ})$$

$$\Delta H = +184 \text{ kJ}$$

Cancel two moles of H_2 and H_2O and one mole of O_2 by reversing equation (1).

Reverse the sign for the heat of reaction:

$2 \times$ Equation (3):

$$2\,N_2\,(g) + \cancel{6}\,O_2\,(g) + 2\,H_2O\,(g) \longrightarrow \text{4 } \cancel{HNO_3\,(l)}$$

$$\Delta H = \mathbf{2}(-348.7 \text{ kJ})$$

$$\Delta H = -696.4 \text{ kJ}$$

$-2 \times$ Equation (2):

$$\text{4 } \cancel{HNO_3\,(l)} \longrightarrow 2\,N_2O_5\,(g) + 2\,H_2O\,(l)$$

$$\Delta H = -\mathbf{2}(-92 \text{ kJ})$$

$$\Delta H = +184 \text{ kJ}$$

$-1 \times$ Equation (1):

$$2\,H_2O\,(l) \longrightarrow 2\,H_2\,(g) + O_2\,(g)$$

$$\Delta H = -\mathbf{1}(571.7 \text{ kJ})$$

$$\Delta H = +571.1 \text{ kJ}$$

Summing the chemical equations gives the desired reaction.

Summing the heat of reactions gives the heat of the reaction for the *overall process*:

$$2\,N_2\,(g) + 5\,O_2\,(g) \longrightarrow 2\,N_2O_5\,(g)$$

$$\Delta H = +59 \text{ kJ}$$

ΔH values used in calculations need not be for actual, measurable processes.

Values from the table of heats of formation have been calculated *via* Hess' Law and can be used for further calculations.

Any duplication (copies, uploads, PDFs) is illegal.

Notes for active learning

Copyright © Sterling Test Prep.

Bond Energy and Heat of Formation Relationship

Bond-dissociation energy

Bond-dissociation energy is the standard enthalpy change when a bond is cleaved, and it measures a bond's strength.

Bond energy refers to the heat required to break the chemical bonds of *one mole of gaseous molecules* to give *separate gaseous atoms*. Bond energy depends on the types of atoms bonding.

Since energy *input* is required to pull two atoms apart, bond-dissociation energy and bond energy are positive quantities associated with an *endothermic* process. (i.e., absorb energy)

Bond dissociation is the reverse process of *bond formation*; the energy is negative and associated with an exothermic process.

For example, the H–H single bond has a bond energy of 436 kJ per mole, but the H–O single bond has a bond energy of 464 kJ per mole.

Common bond energies

Bond	Bond Energy (kJ/mole)	Bond	Bond Energy (kJ/mole)
H–H	436	N–N	159
H–C	414	O–O	138
H–N	389	Cl–Cl	243
H–O	464	C=O	803
H–F	569	N=O	631
H–Cl	431	O=O	498
C–O	351	C≡C	837
C–C	347	N≡N	946

The bond energies in the above chart can be used to estimate $\Delta H°_f$

Bond energy for calculating the heat of formation

For example, calculate $\Delta H°_f$ for CH_3OH (*g*) in the following reaction.

$$C(g) \quad + \quad 4H(g) \quad + \quad O(g) \xrightarrow{\quad\quad} \boxed{4}$$

$$\boxed{1} \uparrow \qquad \boxed{2} \uparrow \qquad \boxed{3} \uparrow \qquad\qquad \downarrow$$

$$C(s) \quad + \quad 2H_2(g) \quad + \quad \tfrac{1}{2}O_2(g) \longrightarrow CH_3OH(g)$$

$$C\,(g) \quad + \quad 4\,H\,(g) \quad + \quad O\,(g) \qquad\qquad \text{Step 4}$$

$$\text{Step 1} \uparrow \qquad \text{Step 2} \uparrow \qquad \text{Step 3} \uparrow \qquad\qquad \downarrow$$

$$C\,(s) \quad + \quad 2\,H_2\,(g) \quad + \quad \tfrac{1}{2}\,O_2\,(g) \quad \rightarrow \quad CH_3OH\,(g)$$

Step 1: **break C−C bonds** (positive energy)

Step 2: **break H−H bonds** (positive energy)

Step 3: **break O−O bond** (positive energy)

Step 4: **form four C−H bonds** (negative energy)

$$\Delta H°_f\,CH_3OH\,(g) = \Delta H°_f\,C\,(g) + 4 \times \Delta H°_f\,H\,(g) + \Delta H°_f\,O\,(g) - \Delta H_{atom}\,CH_3OH\,(g)$$

$$\Delta H°_f\,CH_3OH\,(g) = [716.7 + (4 \times 217.9) + 249.2]\,kJ - \Delta H_{atom}\,CH_3OH\,(g)$$

$$\Delta H°_f\,CH_3OH\,(g) = 1{,}837.5\,kJ - \Delta H_{atom}\,CH_3OH\,(g)$$

$$\Delta H°_f\,CH_3OH\,(g) = 1{,}837.5\,kJ - 3D_{C-H} + D_{C-O} + D_{O-H}$$

$$\Delta H°_f\,CH_3OH\,(g) = 1{,}837.5\,kJ - (3 \times 412) + 360 + 463$$

$$\Delta H°_f\,CH_3OH\,(g) = 1{,}837.5\,kJ - 2{,}059\,kJ$$

$$\Delta H°_f\,CH_3OH\,(g) = -222\,kJ$$

$\Delta H°_f$ of CH_3OH (*g*) is calculated as –222 kJ using bond energies.

According to the chart, $\Delta H°_f$ of CH_3OH (*g*) was experimentally determined to be –201 kJ/mol; estimated bond energies are within 10% (–222 kJ/mol *vs.* –201 kJ/mol).

Copyright © Sterling Test Prep.

Gibbs Free Energy (*G*)

Reaction spontaneity

In 1870, Josiah Willard Gibbs (1839-1903) expressed *Gibbs free energy* (*G*) as the maximum amount of energy present in molecules available *to do non-expansion work* at constant temperature and pressure.

Free energy change (ΔG) is the difference of free energy in the product and reactant molecules.

Gibbs free energy ΔG equation:

$$\Delta G = \Delta H - T\Delta S$$

where *G* is Gibbs free energy, measured in Joules (J).

Calculating Gibbs free energy

Because ΔG depends on signs, it can be calculated from estimated values.

If ΔG is equal to *zero*, the reaction is at *equilibrium*.

If ΔG is *negative*, the reaction is *spontaneous*.

If ΔG is positive, the reaction is *nonspontaneous*.

For example, if ΔG is *negative* at a given temperature and pressure, the reaction is *spontaneous*.

Chemical bonds may break in the reactants when molecules react, and others form in products. Breaking bonds requires energy, while making bonds releases energy, free energy exchanges during a reaction. Not all chemical bonds have the same strength or require the same energy.

Gibbs free energy change (ΔG) equals the maximum energy produced during a reaction harnessed as work.

Spontaneous *vs.* nonspontaneous reactions

Table approximates ΔG and whether the reaction is spontaneous or nonspontaneous.

ΔH	ΔS	Result
-	+	Spontaneous at *all* temperatures
+	+	Spontaneous at *high* temperatures
-	-	Spontaneous at *low* temperatures
+	-	Not spontaneous at *any* temperatures

Any duplication (copies, uploads, PDFs) is illegal.

Reaction energy diagrams

A chemical reaction's thermodynamics can be represented on a *reaction energy diagram* showing the energy of the reactants, products, activation energy, and free energy change ΔG.

Energy appears on the *y*-axis, and the reaction progress (time) from the starting reactants to the final products appears on the *x*-axis.

Each reaction has an *energy of activation* (E_{ac}) or *energy barrier* before the reactants transform into products.

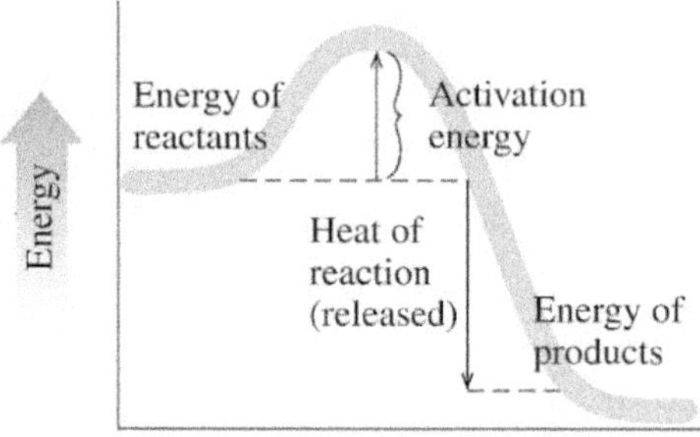

Reaction diagram of energy vs. time (reaction progress)

Gibbs free energy equation

There are several methods to calculate Gibbs free energy ΔG.

Gibbs free energy is a state function, so one method is analogous to the methods for enthalpy and entropy.

Gibbs free energy expression of ΔG°_f products *minus* ΔG°_f reactants:

$$\Delta G^{\circ}_{rxn} = \Sigma\, n(\Delta G^{\circ}_f \text{ of products}) - \Sigma\, n(\Delta G^{\circ}_f \text{ of reactants})$$

where *n* represents moles of each product or reactant given by the balanced chemical equation coefficients.

Like enthalpy, the standard Gibbs free energy of formation (ΔG°_f) of elements in their standard states is 0 kJ/mol.

Notice the similarity of Gibbs free energy to *Hess' Law of constant heat summation*.

Copyright © Sterling Test Prep.

Spontaneous Reactions and ΔG

ΔG for forward and reverse reactions

Gibbs free energy (ΔG) determines a reaction as spontaneous or nonspontaneous.

$\Delta G < 0$, the reaction is spontaneous in the *forward* reaction

$\Delta G = 0$, the reaction is at *equilibrium* (i.e., rate of forward = rate of reverse reaction)

$\Delta G > 0$, the reaction is spontaneous in the *reverse* direction

ΔG for spontaneous and nonspontaneous reactions

Spontaneous describes a process occurring independently without the input of energy from outside the system, related to the Second Law of Thermodynamics.

ΔG is negative, the reaction is *spontaneous* (or exergonic).

ΔG is positive, the reaction is *nonspontaneous* (or endergonic).

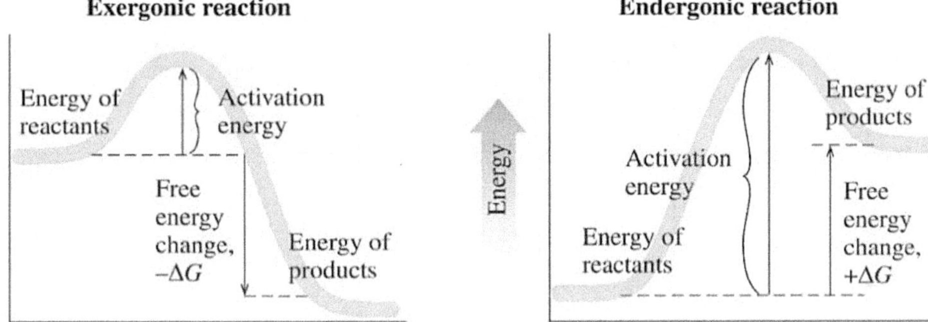

Entropy measures disorder

Entropy (S) is a thermodynamic quantity measuring the spread of a system's energy or disorder.

An increase in entropy of a system corresponds to an increase in the system's randomness or disorder.

$\Delta S° < 0$ represents a decrease in entropy

$\Delta S° > 0$ represents an increase in entropy

For example, forming sugar crystals in a supersaturated sugar water solution *increases entropy*.

Entropy increases because the crystals form spontaneously, resulting in an entropy increase.

Gibbs free energy relates enthalpy, temperature, and disorder

Enthalpy, temperature, and entropy relate free energy by:

$$\Delta G^\circ = \Delta H^\circ - T\Delta S^\circ$$

where ΔG° is the change in free energy, ΔH° is the change in enthalpy, T is temperature, and ΔS° is the change in entropy

An exothermic reaction may *not* be spontaneous because a large, negative ΔS makes it nonspontaneous.

Do *not* assume an endothermic reaction is not spontaneous because a large, *positive* ΔS can make it *spontaneous*.

ΔH	ΔS	Sign of ΔG	Spontaneous
–	+	$\Delta G = (-) - [T\,(+)] = -$	Always, regardless of T
+	–	$\Delta G = (+) - [T\,(-)] = +$	Never, regardless of T
+	+	$\Delta G = (+) - [T\,(+)] = ?$	Depends; spontaneous at high T, $-\Delta G$
–	–	$\Delta G = (-) - [T\,(-)] = ?$	Depends; spontaneous at low T, $-\Delta G$

Table displays how ΔH, ΔS, and T affect spontaneity.

Copyright © Sterling Test Prep.

Temperature-controlled reactions

Temperature-controlled reactions are spontaneous at some temperatures but not at others in the pie chart's bottom left and top right quadrants.

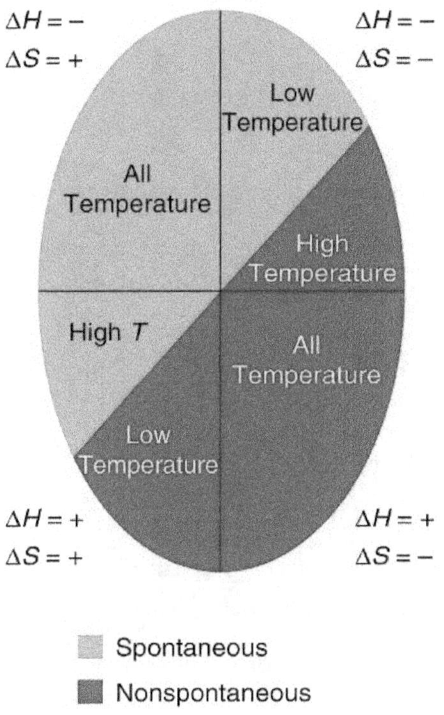

Temperature-controlled reactions

For example, if the enthalpy change is −1,876 kJ, the entropy change is −3.375 kJ/K, and the temperature is 298 K (standard temperature), is reaction spontaneous? (Units must be consistent).

Entropy is often expressed in units of J/K and converted to kJ/K.

Substituting these quantities into the equation for *Gibbs free energy*:

$$\Delta G° = \Delta H - T\Delta S$$

$$\Delta G° = -1876 \text{ kJ} - [(298 \text{ K}) \cdot (-3.375 \text{ kJ/K})]$$

$$\Delta G° = -870.2 \text{ kJ}$$

Since $\Delta G°$ is *negative*, the reaction is *spontaneous*.

Any duplication (copies, uploads, PDFs) is illegal.

Calculating $\Delta G°$ by chemical equilibrium

Use chemical equilibrium to calculate $\Delta G°$:

$$\Delta G° = -RT \ln K$$

$$\Delta G° = -2.303RT \log K$$

where R is the ideal gas law constant (R = 8.314 J/mol·K), T is the temperature in Kelvin, and K is the equilibrium constant

ΔG implies the *direction* and *extent* of a chemical reaction; it does NOT imply a *rate* for the reaction.

Spontaneous reactions may not occur quickly as they may take years, depending on kinetics (i.e., rate).

Calculating $\Delta G°$ by electrochemistry

Using electrochemistry to calculate $\Delta G°$:

$$\Delta G° = -nF\text{E}°$$

where n represents the moles of electrons transferred, F is Faraday's constant (96,485 C/mol), and E° is the standard potential for an electrochemical cell

Calculations of Gibbs free energy involve substituting appropriate quantities into the expression.

Copyright © Sterling Test Prep.

Thermal Expansion

Thermal expansion of materials

Thermal expansion is the tendency of matter to *change volume* with temperature changes. Most materials expand when the temperature rises and contract when the temperature falls.

Thermal expansion occurs through *heat transfer*.

When material changes temperature, it shrinks or expands, depending upon the temperature difference.

As the temperature increases, the molecules within the material gain energy and vibrate at higher rates. This vibration increases the distance between the molecules and causes the material to expand.

If the material does not experience a phase change (e.g., liquid to gas), this expansion of material can be related to the change in temperature.

Coefficients of thermal expansion (α)

Thermal expansion coefficient is an empirically derived constant that describes the expansion of a given material during a temperature rise and depends on the geometry of the material.

Coefficient of thermal expansion (α) is the degree of expansion divided by the change in temperature.

In solid materials, the magnitude of thermal expansion can be calculated to account for:

linear expansion:	$\Delta L / L_0 = \alpha \Delta T$	
area expansion:	$\Delta A / A_0 = 2\alpha \Delta T$	
volumetric expansion:	$\Delta V / V_0 = 3\alpha \Delta T$	

All three calculations require the thermal coefficient of the material. The value of the coefficient α depends on the type of expansion.

Linear expansion

The *coefficient of linear expansion* is a measure of the expansion (or contraction) per unit of length of a material that occurs when the temperature is increased or decreased by 1 °C. This term is also known as *expansivity*.

The *coefficient of thermal expansion* (α) is given as:

$$\alpha_{linear} = \frac{\Delta L}{l_i \cdot \Delta T}$$

Any duplication (copies, uploads, PDFs) is illegal.

where α_{linear} is the coefficient of thermal expansion (K^{-1}), ΔL is the change in length (m), l_i is the initial length (m), and ΔT is the change in temperature (K)

Thermal expansion equation for linear expansion is:

$$\Delta L = \alpha L_0 \Delta T$$

where ΔL is the change in length (m), L_0 is the initial length (m), ΔT is the change in temperature (K), and α is the coefficient of linear expansion(K^{-1})

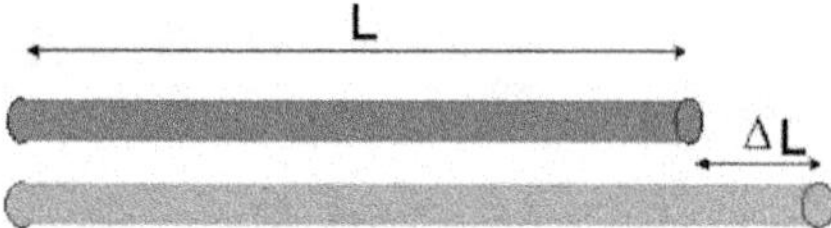

A bar of material is at an initial temperature and has a length denoted as L.
If the bar is heated such that its temperature rises, it expands linearly by some length ΔL.

For example, a metal rod of 7.00 m is heated to 35 °C. If the rod's length expands to 7.12 m after some time, calculate the linear expansion coefficient (Use room temperature = 27 °C).

Calculate the change in length ΔL from the initial $L_0 = 7.00$ m and expanded length $L = 7.12$ m.

$$\Delta L = 7.12 \text{ m} - 7.00 \text{ m}$$

$$\Delta L = 0.12 \text{ m}$$

calculate ΔT from the initial room temperature $T_0 = 27$ °C and the heated $T = 35$ °C.

$$\Delta T = 35 \text{ °C} - 27 \text{ °C}$$

$$\Delta T = 8 \text{ °C}$$

convert ΔT to Kelvin

$$\Delta T = 8 + 273$$

$$\Delta T = 281 \text{ K}$$

linear expansion expression:

$$\Delta L \,/\, L_0 = \alpha \Delta T$$

isolate α

$$\alpha = \Delta L \,/\, (L_0 \times \Delta T)$$

$$\alpha = 0.12 \text{ m} \,/\, (7 \text{ m} \times 281 \text{ K})$$

$$\alpha = 6.1 \times 10^{-5}$$

The coefficient of linear expansion α for the material is 6.1×10^{-5}.

Copyright © Sterling Test Prep.

Area expansion

Objects can undergo area expansion due to the same principles.

Thermal expansion equation for area expansion is:

$$\Delta A = \gamma A_0 \Delta T$$

where ΔA is the change in the area (m^2), and γ is the coefficient of area expansion (K^{-1}).

The *coefficient of area expansion* is expressed as:

$$\alpha_{area} = \frac{\Delta A}{A_i \cdot \Delta T}$$

where α_{area} is the coefficient of thermal expansion (K^{-1}), ΔA is the change in the area (m^2), A_i is the initial area (m^2), and ΔT is the change in temperature (K)

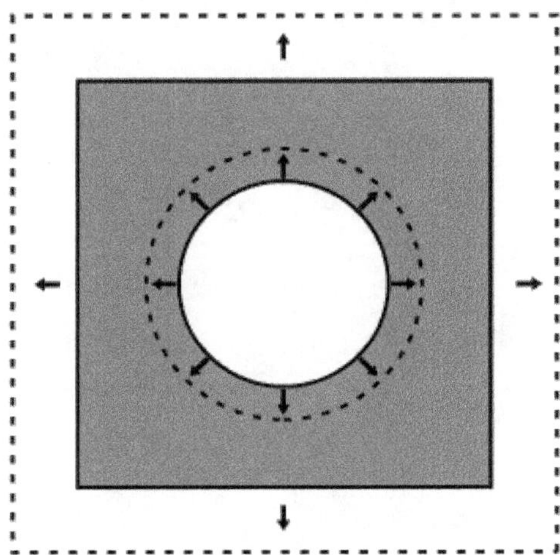

Volume expansion

In general, when a substance is heated, the molecules increase kinetic energy and move more; thus, greater average separation is maintained, increasing volume (expansion).

Volume expansion:

$$\Delta V = \beta V_0 \Delta T$$

where ΔV is the volume change (m^3), and β is the coefficient of volume expansion (K^{-1}).

The *coefficient of volume expansion* is expressed as:

$$\alpha_{volume} = \frac{\Delta V}{V_i \cdot \Delta T}$$

where α_{volume} is the coefficient of thermal expansion (K^{-1}), ΔV is the change in volume (m^3), V_i is the initial volume (m^3), and ΔT is the change in temperature (K).

Any duplication (copies, uploads, PDFs) is illegal.

Notes for active learning

Copyright © Sterling Test Prep.

Kinetic Theory of Gases

Macroscopic properties of gas

Thermal energy (or *internal energy*) is the sum of the *kinetic energies* of the particles. Thermal energy of a gas is directly proportional to the molecules' kinetic energy (KE).

The higher the kinetic energy of the molecules, the more thermal energy the system possesses.

The lower the kinetic energy of the molecules, the less thermal energy the system possesses.

The figure below illustrates a closed container of gas where the gas molecules are in random motion throughout the container.

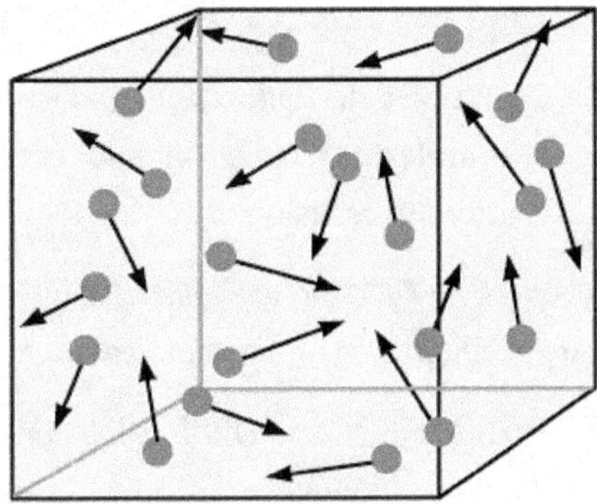

Gas molecules exhibit kinetic energy from random motion in the container

Kinetic Molecular Theory

Kinetic Molecular Theory of Gases relates the microscopic values of atomic kinetic energy to quantities like temperature and pressure.

Kinetic molecular theory (KMT) uses atomic theory to describe the behavior of gases. This theory developed over 100 years, culminating in an 1857 paper by German physicist Rudolf Clausius (1822-1888).

Kinetic molecular theory explains the macroscopic properties of a gas (e.g., pressure and temperature) in terms of its microscopic components (e.g., molecules or atoms). It originated in the ancient idea that matter consists of tiny invisible atoms in rapid motion. The idea was revived and developed between the 17th and 19th centuries to explain gases' properties.

Five assumptions for gases

Five premises for the kinetic theory of gases:

1. Gases are comprised of molecules. Molecules may be treated as perfect spheres of mass (*point masses*), and the space between each of these masses is many times greater than the diameter of the gas molecules.

2. Motion of each gas molecule is *random*, with no order or pattern to either the direction or magnitude of the velocity.

3. Molecules follow *Newton's Laws of Motion*. Each gas molecule moves in a straight line with a constant velocity. If gas molecules collide, the molecules exert an equal but opposite force on the other.

4. Collisions of molecules are *perfectly elastic*; they *lose no kinetic energy*.

   Therefore, since intermolecular attractive and repulsive forces are nonexistent, the gas molecules' total kinetic energy (KE) remains *constant* before and after the collision.

5. Average kinetic energy (*KE*) for gas particles is proportional to the absolute temperature, regardless of the chemical identity or atomic mass.

   At 0 K (absolute zero or –273 °C), the molecules (and orbiting electrons) are not moving and have no volume.

These rules display gases as ideal substances and only approximate their actual behavior. However, their description is remarkably accurate. From these assumptions, laws are derived to describe the behavior of gases.

Boltzmann's constant

Final assumption (#5) for KE is expressed by:

$$KE = \tfrac{1}{2}\, mv^2$$

$$\tfrac{1}{2}\, mv^2 = (3/2)\, k_{\mathrm{B}}T$$

So,

$$KE = (3/2)\, k_{\mathrm{B}}T$$

where *KE* is kinetic energy, *m* is mass, *v* is velocity, *T* is the temperature (Kelvin), and k_B is *Boltzmann's constant* ($k_{\mathrm{B}} = 1.38 \times 10^{-23}$ J/K).

Boltzmann's constant relates macroscopic and microscopic behavior of gases to their particles.

Copyright © Sterling Test Prep.

Root-mean-square speed

KE = (3/2) $k_\text{B}T$ equation is significant because it states that a gas particle's average kinetic energy is proportional to its absolute temperature. Increasing the temperature of the gas particles increases the total speed and overall energy.

Gas atoms are infinitesimally small; nearly impossible to measure a particle's speed accurately.

Speed of gas particles is defined by the *root-mean-square speed* (u_rms) by the equation:

$$u_\text{rms} = \sqrt{\frac{3RT}{MM}}$$

where R is ideal gas constant, T is absolute temperature (K), and MM is molar mass of gas.

Root-mean-square speed written with the *Boltzmann constant, k_B*:

$$u_\text{rms} = \sqrt{\frac{3kT}{m}}$$

where k is the Boltzmann constant, T is the absolute temperature (K), and m is the mass of one gas molecule.

Boltzmann distribution plot

Plotting each gas particle's speed distribution at a given temperature shows a slightly asymmetric curve. This speed-distribution curve is the *Maxwell-Boltzmann distribution curve.*

Boltzmann distribution plot shows that the curve's peak corresponds to the most probable speed.

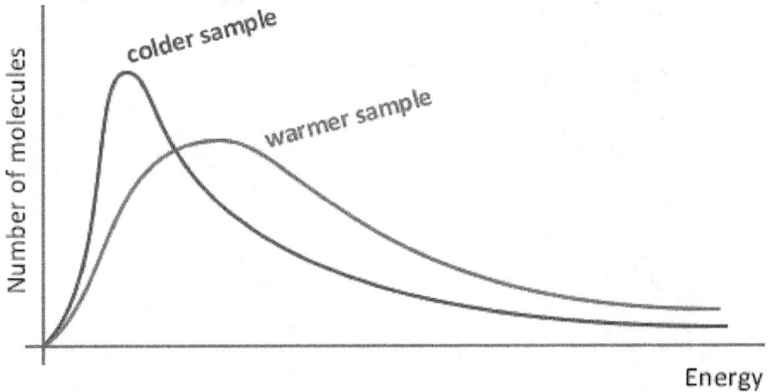

Maxwell-Boltzmann Distribution curves of molecular speeds at two temperatures

If the temperature (K) remains constant, the total kinetic energy remains unchanged.

However, the energy is distributed in several ways, and the gas particles travel at many speeds. This distribution changes continually as the gas atoms collide with each other and with the container walls.

Any duplication (copies, uploads, PDFs) is illegal.

The curve flattens (i.e., the net area under the curve is greater) and shifts to the right at higher temperatures, indicating that more gas molecules move at higher speeds and possess more kinetic energy.

Diffusion and effusion

Diffusion is when a substance (solute or particle) spreads from a high-concentration region to a lower concentration. It is explained using the kinetic theory of motion.

According to the kinetic theory, heavier gases diffuse more slowly than lighter gases because of the difference in their travel speed.

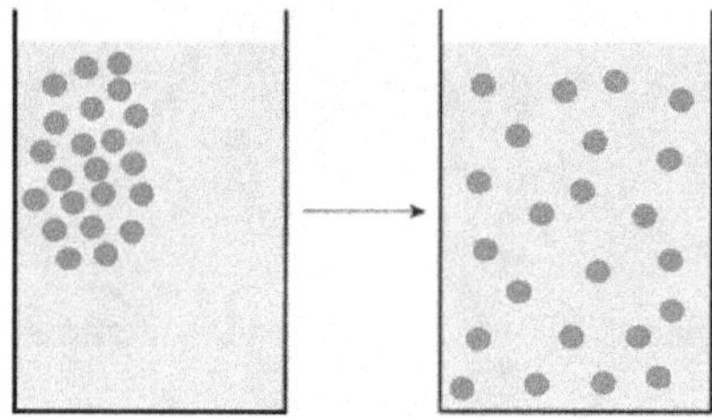

Diffusion of solutes in a solvent

Effusion is the flow of gas particles under pressure from one compartment to another through a small opening, as shown below.

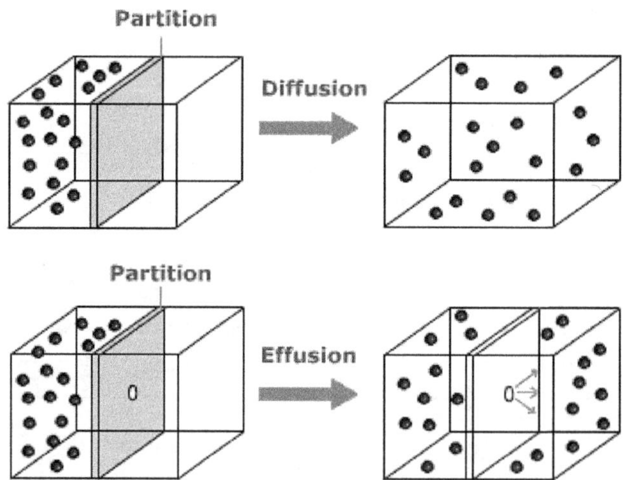

Diffusion and effusion of gas particles

Copyright © Sterling Test Prep.

Graham's Law

Scottish chemist Thomas Graham (1805-1869) developed *Graham's Law*, which states that *at constant temperature and pressure conditions, the rates at which two gases diffuse are inversely proportional to the square root of their molar masses.*

Graham's Law is expressed by:

$$\frac{r_1}{r_2} = \sqrt{\frac{MM_2}{MM_1}}$$

where r_1 and r_2 are the diffusion rates of gas 1 and gas 2, respectively, and MM_1 and MM_2 are the molar masses of the gases.

Graham's Law applies to the effusion of gas particles, and the equation for effusion is the same as the equation for diffusion.

Ideal Gas Law

Ideal Gas Law explains the relationship between pressure (P), volume (V), and temperature (T).

Gas law is expressed as:

$$PV = nRT$$

where n is number of moles, and R is universal gas constant with a value of 8.314 J/ mol·K.

One mole is equal to 6.023×10^{23} molecules. Technically, a mole is the number of hydrogen atoms in one gram of hydrogen. Atoms are tiny, so counting them in moles is more practical than counting them individually. Ideal Gas Law applies to the pressure a gas exerts on a cylinder with a moving wall.

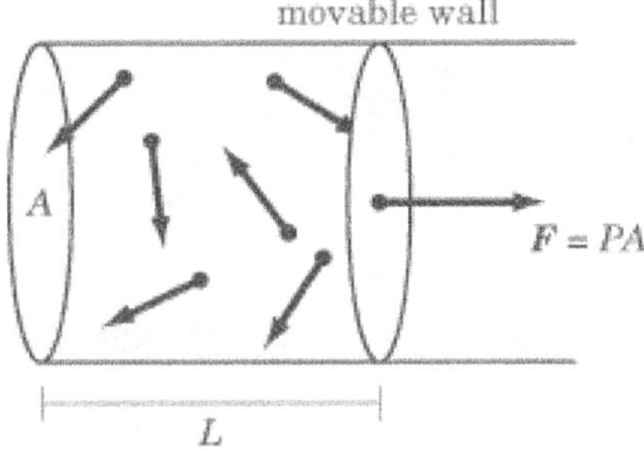

Force that the gas exerts on the wall equals F = PA

Since pressure equals $P = F/A$, the force the gas exerts on the wall equals $F = PA$.

If this force moves the wall back a length of L, the cylinder volume increases by $\Delta V = LA$.

By solving for A and substituting this into the equation for force, the result is:

$$F = P\Delta V L \text{ or } P\Delta V = FL$$

Any duplication (copies, uploads, PDFs) is illegal.

Work is *force multiplied by distance traveled*. By pushing the wall at a distance of L with a force of F, the gas has done work equal to FL.

When gas does work, it symbolizes a change in energy.

If a change in PV is equal to a change in energy, then PV is the total energy of the gas.

For the ideal gas law, nRT is the expression for the total kinetic energy of the gas molecules.

Ideal gas law (PV = nRT) can be related to the number of molecules (N) and Boltzmann's constant (k), which has a value of 1.381×10^{-23} J/K:

$$PV = NkT$$

Other gas laws derive from the Ideal Gas Law by holding specific variables constant.

The number of moles (n) and the gas constant (R) are constant, so four other gas laws result from the Ideal Gas Law:

>   Boyle's Law

>   Charles' Law

>   Combined Gas Law

>   Closed Container Law

Ideal gases follow KMT (isothermal process)

As described previously, *kinetic molecular theory (KMT)* is based on a theoretical ideal gas.

Ideal gases follow *five* assumptions of KMT in which gas molecules:

>   move in random motion

>   are point masses with no volume,

>   there are no intermolecular forces,

>   collisions are elastic and

>   kinetic energy of molecules is proportional to temperature.

Compared to an ideal gas, the behavior of a real gas is complicated. By conceptualizing ideal gas behavior, real gas behavior is easier to comprehend.

In the 17th and 18th centuries, scientists realized several relationships between gases' properties. Boyle's, Charles', and Avogadro's Law relate pressure, volume, and temperature properties.

These laws were developed empirically and are individual cases of the ideal gas equation:

>   PV = nRT

Copyright © Sterling Test Prep.

Boyle's Law (isothermal process)

In 1661, Robert Boyle (1627-1691) studied the compressibility of gases. He observed that *the volume of a fixed amount of gas at a given temperature is inversely proportional to the pressure exerted on the gas.*

Boyle's Law applies when the temperature of a gas is held constant. It states that an increase in pressure causes a decrease in volume or that a decrease in pressure causes an increase in volume.

Boyle's Law is expressed as:

$$P_1 V_1 = P_2 V_2$$

where subscripts 1 and 2 are same gas sample under two sets of pressure and volume conditions.

A plot of volume *vs.* pressure for a gas illustrates that Boyle's Law is a particular case of the ideal gas law, where n (moles of gas) and T (temperature) are constant (figure below).

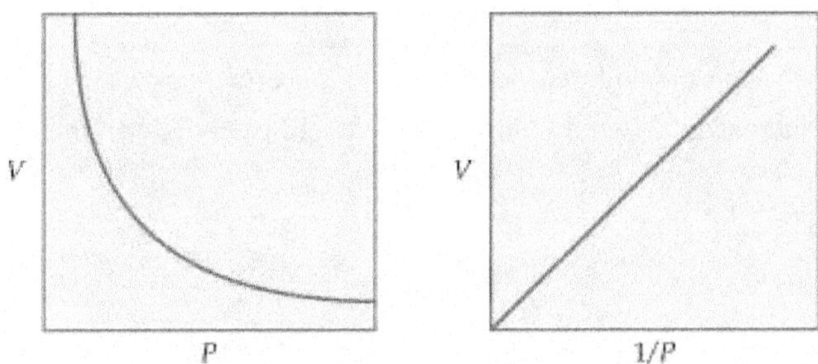

Boyle's Law: as pressure increases, volume decreases

For example, what is the new pressure in a container of gas in a 15.0 L container at a pressure of 5.00 atm when the volume decreases to 0.500 L?

Substitute the known quantities (P_1, V_1, and V_2) into Boyle's Law equation to solve P_2. Note: the volume units need to be consistent. In this example, volumes are expressed in units of liters.

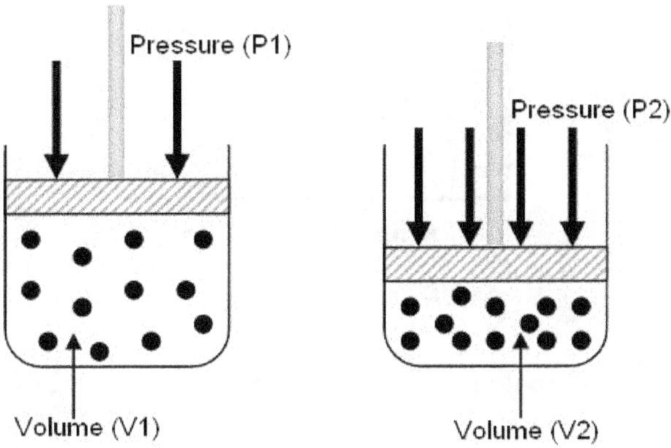

Boyle's Law is where n (moles of gas) and T (temperature) are constant

Any duplication (copies, uploads, PDFs) is illegal.

Boyle's Law:

$$P_1V_1 = P_2V_2$$

Solve for P_2:

$$P_1V_1 / V_2 = P_2$$

$$[(5.00 \text{ atm}) \cdot (15.0 \text{ L})] / 0.500 \text{ L} = P_2$$

$$P_2 = 150 \text{ atm}$$

Charles' Law (isobaric process)

French scientist Jacques Charles (1746-1823) developed the relationship between temperature and gas volume.

Charles' Law states that *at constant pressure, the volume of a gas is directly proportional to its absolute temperature (Kelvin)*. It describes the behavior of a gas under constant pressure.

In this relationship, volume and temperature are *directly proportional*; when temperature increases, volume increases; when the temperature decreases, the volume decreases.

Charles' Law is expressed as:

$$\frac{V_1}{T_2} = \frac{V_2}{T_2}$$

where subscripts 1 and 2 refer to the same gas sample at different temperature and volume.

Temperature vs. volume plot for a gas illustrates that Charles' Law is another application of the ideal gas law, where n and P are constant, as shown below.

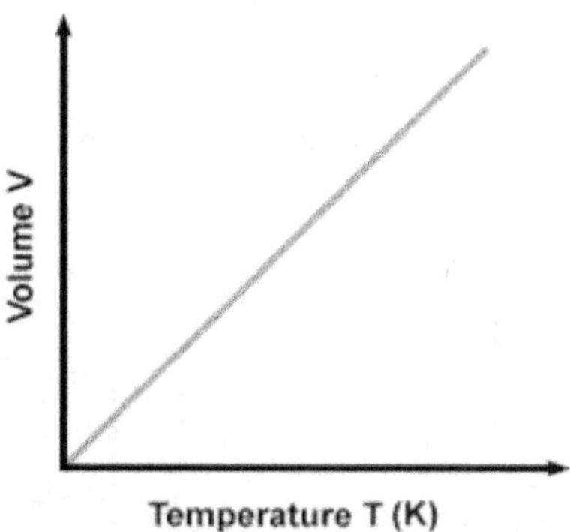

Charles' Law states that as temperature increases, volume increases

For example, if there is 25.0 L of gas at 0 °C, and the temperature is raised to 100 °C, what is the volume of the gas?

Copyright © Sterling Test Prep.

First, the temperature is converted to Kelvin.

$$T_1 = 0 \ °C + 273 = 273 \ K$$

$$T_2 = 100 \ °C + 273 = 373 \ K$$

Substitute the known quantities into the equation for Charles' Law.

$$V_1 / T_1 = V_2 / T_2$$

Rearrange and solve for V_2:

$$V_1 \ T_2 / T_1 = V_2$$

$$(25.0 \ L)·(373 \ K) / 273 \ K = V_2$$

$$V_2 = 34.2 \ L$$

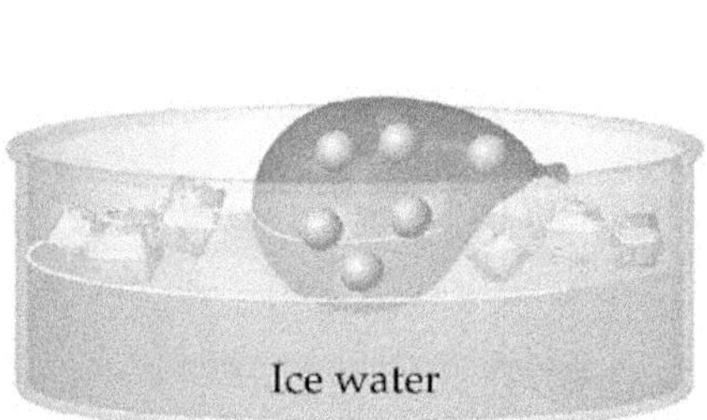

Volume and temperature are directly proportional, when temperature increases, volume increases

Gay-Lussac's Law

In 1802, French chemist Louis Gay-Lussac (1778-1850) proposed *Gay-Lussac's Law,* which states that *the pressure of a given sample of gas is directly proportional at a constant volume to its absolute temperature (Kelvin).*

Gay-Lussac's Law is expressed by:

$$\frac{P_1}{T_1} = \frac{P_2}{T_2}$$

where subscripts 1 and 2 refer to the same gas sample with different temperatures and pressure.

Temperature vs. pressure plot for a gas illustrates that Gay-Lussac's Law is another particular case of the ideal gas law, where n and V are constant.

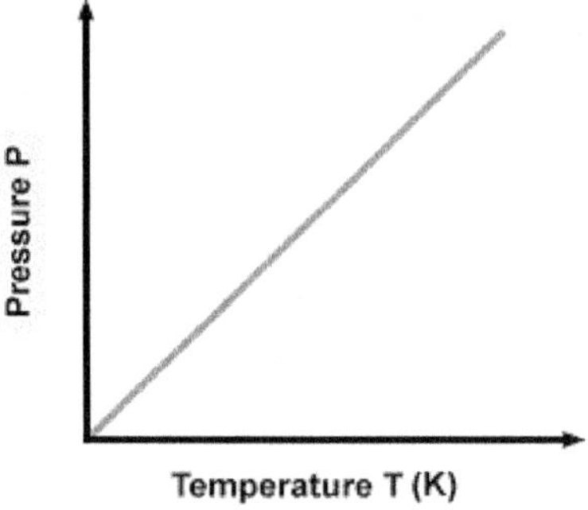

Gay-Lussac's Law states that as temperature increases, pressure increases

For example, suppose a gas is at 30.0 atm pressure and 100 °C, and the temperature changes to 400 °C. What is the new pressure of the gas?

Convert the temperature to units of Kelvin.

$$T_1 = 100\ °C + 273 = 373\ K$$

$$T_2 = 400\ °C + 273 = 673\ K$$

Gay-Lussac's Law solves P_2 by substituting the quantities into the equation.

$$\frac{P_1}{T_1} = \frac{P_2}{T_2}$$

$$P_2 = [(30.0\ atm)\cdot(673\ K)] / 373\ K$$

$$P_2 = 54.1\ atm$$

Closed Container Law (constant volume)

Closed Container Law describes the behavior of a gas under constant volume. In such cases, pressure and temperature are directly proportional:

$$\frac{P_i}{T_i} = \frac{P_f}{T_f}$$

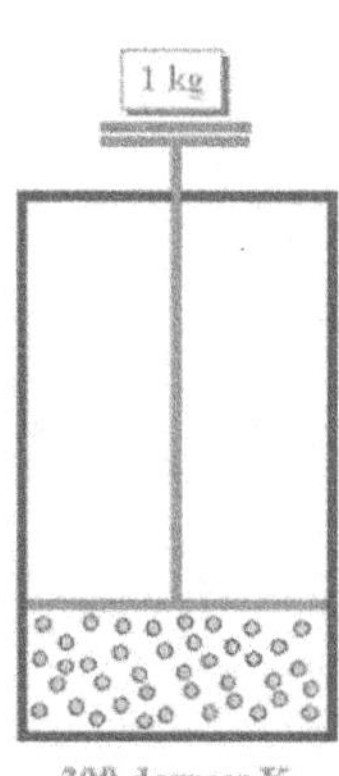

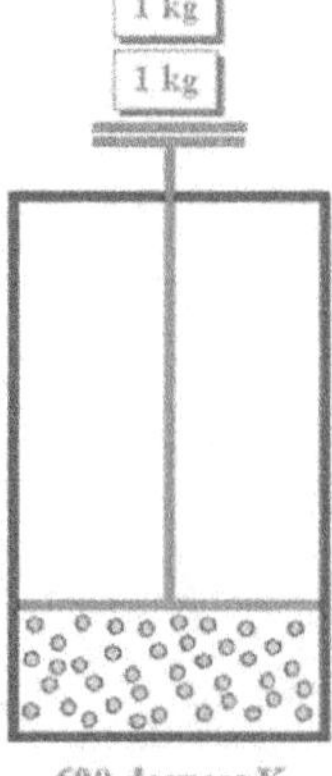

Copyright © Sterling Test Prep.

Combined Gas Law

Combined Gas Law amalgamates Boyle's and Charles' Laws. The combined gas law is *not* the ideal gas law (PV = nRT); it is expressed as an equation to relate changes in temperature, volume, and gas pressure.

Combined Gas Law relates the pressure, volume, and temperature of a gas and is expressed as:

$$\frac{P_i V_i}{T_i} = \frac{P_f V_f}{T_f}$$

where subscripts 1 and 2 refer to the same gas sample but under different temperatures, pressure, and volume.

For example, suppose a gas is at 15.0 atm pressure, with a volume of 25.0 L and a temperature of 300.0 K. What would the volume of the gas be at standard temperature and pressure? (Use standard pressure = 1.00 atm and standard temperature = 273.0 K)

Use the *combined gas law* at standard temperature (T_2) and pressure (P_2):

$(P_1 V_1) / T_1 = (P_2 V_2) / T_2$ solve for V_2:

$(P_1 \times V_1 \times T_2) / (T_1 \times P_2) = V_2$

$V_2 = [(15.0 \text{ atm}) \cdot (25.0 \text{ L}) \cdot (273.0 \text{ k})] / [(300.0 \text{ K}) \cdot (1.00 \text{ atm})]$

$V_2 = 341 \text{ L}$

Avogadro's Law

The volume of a gas in a container is affected by pressure, temperature, and moles of gas.

The relationship between the quantity of gas and its volume was derived by Italian scientist Amadeo Avogadro (1776-1856).

Avogadro's Law states that *gases at a given temperature and pressure occupying a volume are directly proportional to the number of moles of gas present.*

Avogadro's Law is expressed as:

$$\frac{n_1}{V_1} = \frac{n_2}{V_2}$$

where n_1 and n_2 are moles of gas 1 and gas 2, and V_1 and V_2 are the volumes of gas 1 and gas 2.

A plot of *volume vs. moles* of gas (shown below) illustrates that Avogadro's Law is a particular case of the ideal gas law where T and P are held constant.

Any duplication (copies, uploads, PDFs) is illegal.

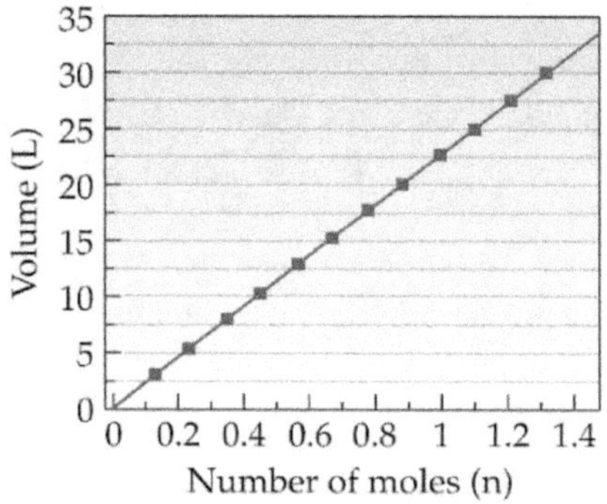

Avogadro's Law states that as moles of a gas increase, volume increases

For example, if 8.00 moles of a gas occupy a volume of 4.00 L at constant pressure and temperature, what volume of gas would 16.0 moles of this gas occupy at the same temperature and pressure?

Use *Avogadro's Law*

$$n_1 / V_1 = n_2 / V_2$$

where n_1 and n_2 are moles of gas 1 and gas 2, and V_1 and V_2 are the volumes of gas 1 and gas 2.

Solve for the V_2

$$V_2 = (V_1 \times n_2) / n_1$$

$$V_2 = [(4.00 \text{ L}) \cdot (16.0 \text{ mol})] / 8.00 \text{ mol}$$

$$V_2 = 8 \text{ L}$$

Copyright © Sterling Test Prep.

Phase Diagrams

Phases of substances

Most substances exist in solid, liquid, or gas form.

Solids have a definite shape and volume (e.g., ice). The molecules in a solid vibrate at a fixed position and a solid cannot be compressed.

Liquids have a definite volume but take their container shape (e.g., water). The molecules move about but are close and bound by intermolecular forces.

Gases have neither a definite volume nor a definite shape; they take the container's volume and shape (e.g., steam or water vapor). The molecules move independently and are not held together by intermolecular forces, making gas easily compressible.

Phases at equilibrium

Phase diagrams indicate the relationship between temperature, pressure, and associated phases, plotting pressure, temperature, or volume for thermodynamically distinct phases (solid, liquid, or gas) at equilibrium.

Phase of a substance (i.e., solid, liquid, or gas) mostly depends on temperature and pressure.

If the pressure is exceptionally high, a substance is less likely to be in a gas phase because the pressure brings the molecules so close that the substance either liquefies or sublimes.

Phase change diagram

When an ice block is placed in the Sun, it slowly absorbs thermal energy through radiation and undergoes phase changes. The ice cube changes from solid to liquid to gas.

> *Melting point* of a substance is the temperature at which it changes from a solid to a liquid, and it has the same temperature range as the *freezing point*; the reverse reaction occurs at the same temperature.

> *Boiling point* (or vaporization) of a substance is the temperature at which it changes phase from a liquid to a gas, the same temperature range as the *condensation point*.

From the graph below, once the ice cube absorbs enough heat to melt (i.e., reaches 0 °C), its temperature remains constant.

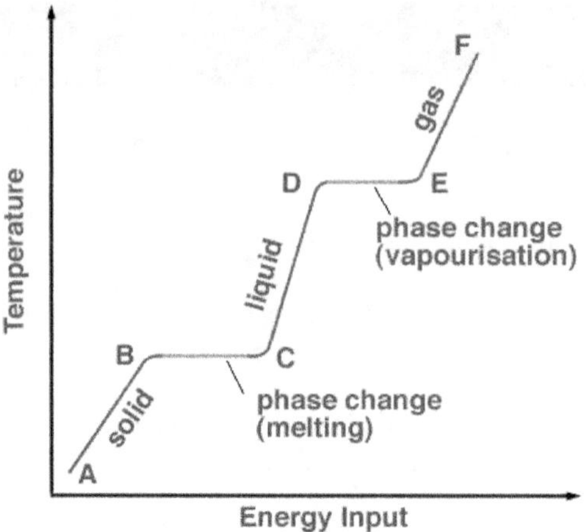

Phase diagram relates temperature changes and states

Melting is points B to C on the graph where all molecules have reached a temperature of 0 °C.

Ice melts only when this occurs, as illustrated by the graph's horizontal plateau from B to C.

When liquid water changes to a gas, the liquid absorbs enough heat to *vaporize* (i.e., at 100 °C).

Liquid temperature remains constant from point D to E until all molecules reach 100 °C.

As illustrated by the *horizontal plateau* from D to E, the *liquid vaporizes* only when this occurs.

Heat during phase changes

Although the energy (heat) input increases with temperature, it does not remain constant as the temperature does during a phase change. Technically, heat converts the potential energy stored within each atom to kinetic energy before changing phase.

As energy is continuously added to the process, the substance uses the added heat to transition each molecule *before increasing* the temperature.

Pressure *vs.* temperature relationship

Phase diagram graphically displays the phases of a substance of pressure *vs.* temperature.

The graph is split into three disproportionate parts, representing the substance's three phases.

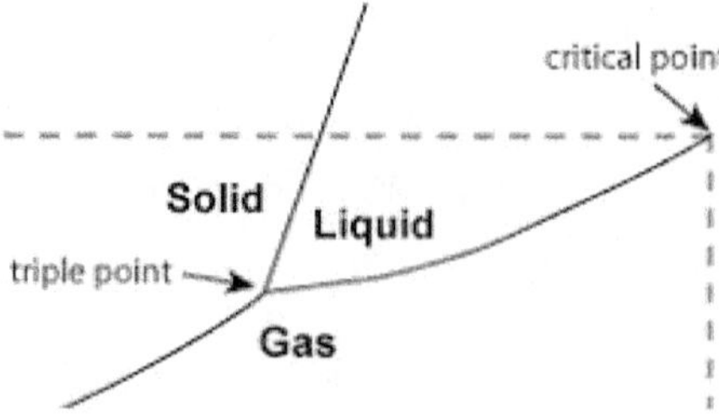

PV phase diagram with triple point and critical point labeled. Dashed horizontal line shows the critical pressure, and vertical dashed line shows the critical temperature.

     Copyright © Sterling Test Prep.

Triple point, critical point, and critical temperature

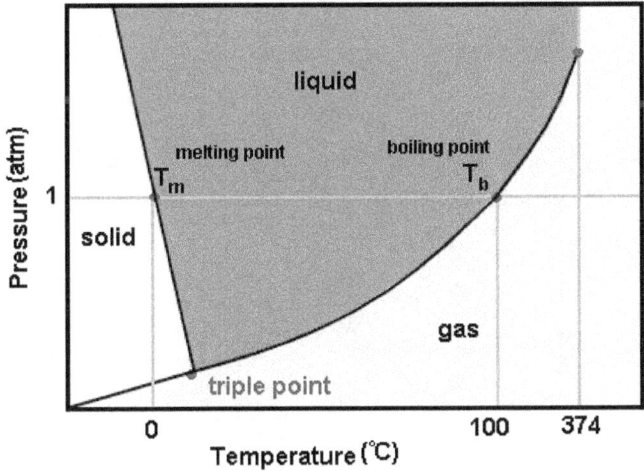

PV phase diagram with triple point and critical point labeled.

Melting point and *boiling point* are found by drawing a horizontal line at 1 atm of atmospheric pressure. The horizontal line intersects with a line on the phase diagram; the substance changes phases at that temperature.

Triple point is temperature and pressure at which three phases of matter coexist in equilibrium.

Critical point is temperature and pressure at which liquids and gases become indistinguishable.

Critical temperature is when a liquid cannot form, no matter the pressure.

Phase line boundaries

Solid-liquid boundary is where the *solid and liquid* exist in equilibrium.

Solid-gas boundary is where *solid and gas* exist in equilibrium.

Liquid-gas boundary is where *liquid and gas* exist in equilibrium.

Positive slope graphs

Solid-liquid phase boundary (or fusion curve) for most substances, has a positive slope (i.e., rise over run), so the *melting point increases with pressure*. This is true when the solid phase is denser than the liquid phase.

The greater the pressure, the closer the molecules of the substance are brought to each other, increasing the effect of intermolecular forces. Thus, the substance requires a higher temperature for its molecules to have enough energy to break the fixed pattern of the solid phase and enter the liquid phase.

A similar concept applies to liquid–gas phase changes.

Negative slope graphs

Water is an exception, which has a solid-liquid boundary with a negative slope so that the melting point decreases with pressure.

Ice (solid water) is less dense than liquid water, as ice floats on water.

At a molecular level, ice is less dense because it has a more extensive network of hydrogen bonding, which requires a greater separation of water molecules.

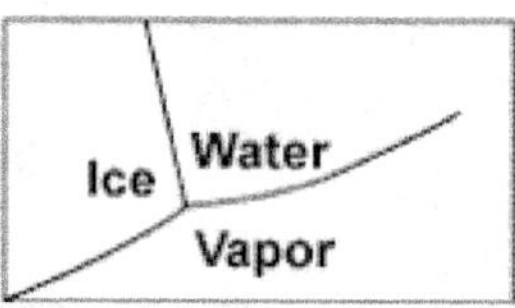

Phase diagram of pressure vs. pressure for water differs because the solid-liquid boundary is slanted to the left. Water (liquid) is denser than ice (solid), if the pressure increases, ice melts into water.

Copyright © Sterling Test Prep.

Colligative Properties

Solute *vs.* solvent ratio

Colligative properties depend on the ratio of solute particles to solvent but *not* on the type of chemical species.

Colligative properties include:

1) the relative lowering of vapor pressure,

2) the elevation of boiling point,

3) the depression of freezing point, and

4) osmotic pressure.

Colligative properties calculations incorporate the van 't Hoff factor (i), which measures a solute's effect on the solution's colligative properties.

van 't Hoff factor is:

$$\frac{[\text{concentration of particles produced when a substance dissolves}]}{[\text{concentration of a substance calculated from its mass}]}$$

Concentration is converted to reflect the total number of particles in the solution. (e.g., 5 g of NaCl dissociates into Na^+ and Cl^-).

For example, glucose ($C_6H_{12}O_6$) has an i value of 1 because it does not dissociate in solution. Nonelectrolytes do not dissociate in solution (e.g., CH_4 and O_2). The salt NaCl has an i value of 2 because it dissociates into 1 Na^+ and 1 Cl^- (2 resulting ions per one reactant molecule).

Any duplication (copies, uploads, PDFs) is illegal.

Boiling point elevation ($\Delta T_b = K_b m$)

Adding a solute to a solvent stabilizes the solvent in the liquid phase, thus lowering the solvent molecules' tendency to move to the gas or solid phases. Therefore, the boiling point increases.

The *boiling point elevation* is proportional to the decrease of vapor pressure in a dilute solution:

$$\Delta T_b = k_b \cdot m \cdot i$$

where ΔT_b is the increase in boiling point, k_b is the molal boiling point constant (a given value determined experimentally), m is molality (mol solute/kg solvent), and i is the van 't Hoff factor.

For example, what is the boiling point elevation if 6.4 g of ammonia is dissolved in 0.3 kg of water? (Use the k_b for water = 0.52 °C/m and the molar mass of ammonia = 17.031 g/mol)

Convert the mass of ammonia (NH_3) into moles:

molecular mass = mass / moles

moles = mass / molecular mass

moles = (6.4 g) / (17.031 g/mol)

moles = 0.38 mol NH_3

Calculate the molality of the solution:

molality = (moles of solute) / (mass (in kg) of solvent)

m = (0.38 mol NH_3) / (0.3 kg of water)

m = 1.27 m

Calculate the boiling point elevation:

$$\Delta T_b = k_b \cdot m \cdot i$$

$$\Delta T_b = (0.52 \text{ °C/m}) \cdot (1.27 \text{ m}) \cdot (1)$$

$$\Delta T_b = 0.66 \text{ °C}$$

The boiling point increased by 0.66 °C due to adding 6.4 g of NH_3.

Copyright © Sterling Test Prep.

Freezing point depression ($\Delta T_f = K_f m$)

Solute particles in mixtures increase the strength of intermolecular (e.g., dipole-dipole) bonds. It is more difficult to boil the solution (i.e., boiling point elevation) or freeze it (i.e., freezing point depression).

Freezing point depression (i.e., lowering the freezing point) is calculated by the following equation: (note the negative sign indicates that the change is a decrease).

$$\Delta T_f = -k_f \cdot m \cdot i$$

where ΔT_f is the decrease in freezing point, k_f is the molal freezing point constant (a given value determined experimentally), m is the molality (mol solute/kg solvent), and i is the van 't Hoff factor (given since it is determined experimentally).

For example, a 48.0 g sample of a nonelectrolyte (does not dissociate in solution) is dissolved in 500.0 g of water to produce a solution with a freezing point of -3.5 °C. What is the molar mass of the compound? (Use k_f of water = 1.86 °C/m, the freezing point of pure water = 0 °C)

The freezing point of water = 0 °C, and the freezing point depression, $\Delta T_f = 3.5$ °C.

Calculate the moles using the freezing point depression expression:

$$\Delta T_f = k_f \cdot m \cdot i$$

$$3.5 \text{ °C} = (1.86 \text{ °C/m}) \cdot (x / 0.5 \text{ kg}) \cdot (1)$$

$$3.5 \text{ °C} = (3.72 \text{ °C/m}) \cdot (x)$$

$$3.5 \text{ °C} / (3.72 \text{ °C/m}) = x$$

$$x = 0.94 \text{ mol nonelectrolyte}$$

Calculate molar mass by dividing the sample's mass by moles in the sample.

$$\text{Molar mass} = \text{mass} / \text{moles}$$

$$\text{Molar mass} = 48.0 \text{ g} / 0.94 \text{ mol}$$

$$\text{Molar mass} = 51.1 \text{ g/mol}$$

The molar mass of the unknown nonelectrolyte is 51.1 g/mol.

Notes for active learning

Copyright © Sterling Test Prep.

Heat Transformation

Fusion and vaporization

Latent heat (*latent energy* or *heat of transformation*) is the energy released or absorbed by a thermodynamic system during a constant-temperature process — usually a first-order phase transition (i.e., constant temperature during "mixed-phase" states).

Latent heat is the energy supplied or extracted to change the state of a substance without changing its temperature (i.e., first-order phase transition).

Examples include *latent heat of fusion* (solid to liquid) and *latent heat of vaporization* (liquid to gas) for phase changes when a substance condenses or vaporizes at a specified temperature and pressure.

Heat and phase changes

Energy is released for:

> *Condensation*: gas changes phase into liquid
>
> *Freezing*: liquid transforms into a solid
>
> *Deposition*: gas changes phase directly into solid

Reverse processes use the same amount of energy.

> *Vaporization* (reverse of condensation): liquid changes phase into a gas
>
> *Melting* (reverse of freezing): solid changes phase into liquid
>
> *Sublimation* (reverse of deposition): solid changes phase directly into a gas

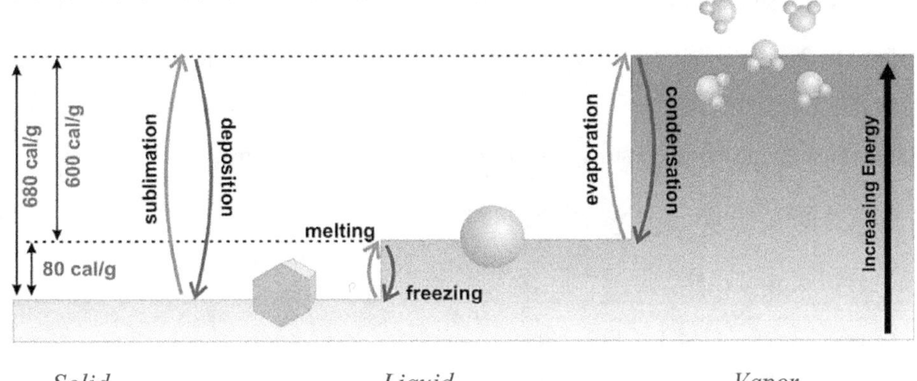

Latent heat

Latent heat is the amount of heat required for a phase change. Latent heat has been reported for most substances. In general, only computing the total heat is required for the reaction.

Heat required to convert a substance from one phase to another is:

$$Q = m \cdot l$$

where l is the latent heat (J/kg), and m is the mass of the substance (kg).

Latent heats are sometimes given in molar heat or joules per mol (J/mol). Divide by the molecular weight of the substance if given the mass in grams.

If any of these processes are reversed, the latent heat is negative (reverse process).

Heat of fusion (ΔH_{fusion})

Heat of fusion (ΔH_{fusion}) is the energy to melt a given quantity of solid to liquid at a constant temperature; the energy required to melt the solid times the number of moles.

Heat of fusion (ΔH_{fusion}) is the energy to melt a given quantity of solid to liquid at a constant temperature; the energy required to melt the solid times the number of moles.

Heat of fusion ΔH_{fusion} is expressed as:

$$q = n \times \Delta H_{fusion}$$

where q is heat energy, n is the number of moles of the solid, and ΔH_{fusion} is the heat of fusion.

Most substances have reported reference values for the heat of fusion (ΔH_{fusion}).

If the particles become excited (solid to a liquid, liquid to a gas), the change in entropy is always positive because heat is added to the system.

For this process, use latent heat of fusion. The total heat required is the heat of fusion.

If the particles become less excited (gas to a liquid, liquid to solid), the change in entropy is always negative because heat is removed from the system.

A negative change in entropy results in a more ordered molecular structure.

The total heat required for this process is the *heat of vaporization*.

Heat of vaporization (ΔH_{vap})

Heat of vaporization (ΔH_{vap}) is the energy to vaporize a given quantity of liquid to gas at a constant temperature.

It is the energy required to vaporize the liquid times the number of moles of that liquid.

Most substances have reported reference values for the heat of vaporization (ΔH_{vap}).

 Copyright © Sterling Test Prep.

Heat of vaporization ΔH_{vap} is expressed as:

$$q = n \times \Delta H_{vap}$$

where q is heat energy, and n is number of moles of the liquid, and ΔH_{vap} is heat of vaporization.

Sublimation is when a solid bypasses the liquid phase and transitions directly into the gas phase.

The heat required for *sublimation* sums up the *heat of fusion* and the *heat of vaporization*.

Deposition is the reverse reaction of sublimation.

$$Q_{sub} = Q_{fus} + Q_{vap}$$

Total heat of reaction

For a process with multiple phase changes, the total heat of the reaction is the sum of the separate heat needed for each phase change and the addition of heat.

For these problems, consider the First Law of Thermodynamics, which states that heat must be conserved (heat gained = heat loss), and the equation for heat ($Q = mc\Delta T$), where c is the specific heat capacity, and m is the mass.

For example, when calculating the change in heat (change in entropy) when 2 kg of ice at 0 °C is warmed to 120 °C, a few equations are needed using the known values.

- melting point = 0 °C

- boiling point = 100 °C

- latent heat of fusion = 3.33×10^5 J/kg

- latent heat of vaporization = 2.26×10^6 J/kg

- specific heat capacity = 4,186 J/kg °C

Heat required to phase change the *ice into liquid water*:

$$Q = m \, l_{fusion}$$

$$Q = (2 \text{kg}) \cdot (3.33 \times 10^5 \text{ J/kg}) = 6.66 \times 10^5 \text{ J}$$

Heat required to heat the *water to its boiling point*:

$$Q = mc\Delta T$$

$$Q = (2 \text{ kg}) \cdot (4186 \text{ J/kg °C}) \cdot (100 \text{ °C} - 0 \text{ °C}) = 837{,}200 \text{ J}$$

Any duplication (copies, uploads, PDFs) is illegal.

Heat required to phase change the *liquid into gas:*

$$Q = m \, l_{vap} = (2\text{kg}) \cdot (2.26 \times 10^6 \text{ J/kg})$$

$$Q = 4.52 \times 10^6 \text{ J}$$

Heat required to heat the water to 120 °C:

$$Q = mc\Delta T$$

$$Q = (2 \text{ kg}) \cdot (4{,}186 \text{ J/kg °C}) \cdot (120 \text{ °C} - 100 \text{ °C}) = 167{,}440 \text{ J}$$

Total heat required for the process to occur is the *sum of the separate heats*:

$$Q_{total} = Q_{fusion} + Q_{heat\_1} + Q_{vap} + Q_{heat\_2}$$

$$Q_{total} = 6.66 \times 10^5 + 837{,}200 + 4.52 \times 10^6 + 167{,}440$$

$$Q_{total} = 6{,}190{,}640 \text{ J}$$

Calculating heat of fusion and heat of vaporization

Heat of fusion and heat of vaporization use units of kJ/mol.

However, they may be expressed as J/g, where energy is obtained by multiplying the latent heat by the substance's mass rather than moles.

For example, how much kJ is required when 31.5 g of H_2O melts at its melting point of 0 °C? (Use molar mass for H_2O = 18 g/mol and molar heat of fusion for water ΔH_{fusion} = 6.02 kJ/mol).

Use molar mass relationship:

$$\text{mass (g)} = \text{moles (mol)} \times \text{molar mass (g/mol)}$$

Rearrange to find the number of moles:

$$\text{moles (g)} = \text{mass (g)} / \text{molar mass (g/mol)}$$

$$\text{moles} = (31.5 \text{ g}) / (18 \text{ g/mol})$$

$$\text{moles} = 1.75 \text{ mol}$$

Use heat of fusion (ΔH_{fusion}):

$$q = n \times \Delta H_{fusion}$$

$$q = 1.75 \text{ mol} \times 6.02 \text{ kJ/mol}$$

$$q = 10.54 \text{ kJ}$$

10.54 kJ heat (i.e., positive value) is required to melt 31.5 g of water.

Copyright © Sterling Test Prep.

Pressure *vs.* Volume Phase Diagrams

PV graphs

Pressure vs. volume diagram (*PV* graph) describes system pressure and volume changes.

PV diagrams show the thermodynamic process by graphing pressure against volume (P *vs.* V).

Work done by the system is equal to the area under the curve.

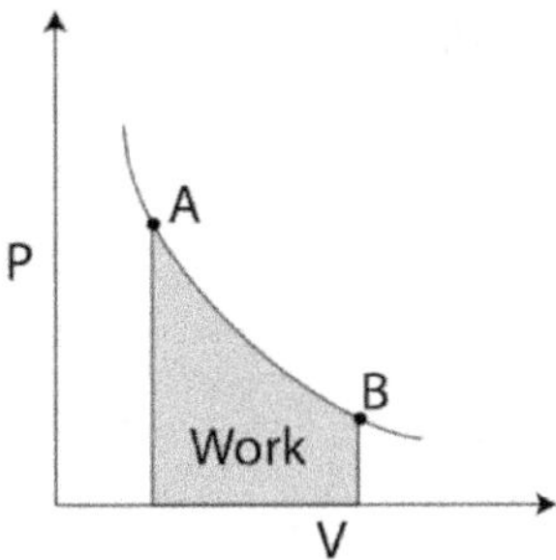

Area under the curve on the above PV diagram equals the work done

Cyclic pressure *vs.* volume reactions

In cyclic reactions, the work done is the *area enclosed* in the graph.

These reactions are often comprised of different processes, as shown below.

work is *done **by*** the gas when the volume increases and pressure decreases.

work is *done **on*** the gas when the volume decreases and pressure increases.

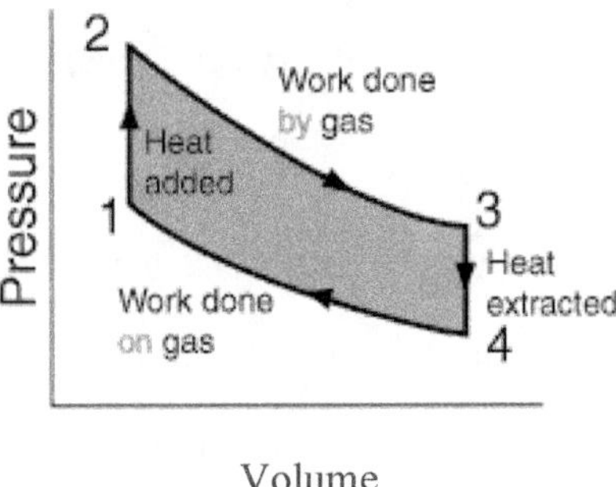

*Points 4 to 1 show work done **on** the gas; points 2 to 3 work done **by** the gas*

Any duplication (copies, uploads, PDFs) is illegal.

Calculating work done by gas

Work can be calculated for the gas phase of a gas and its container.

Gases may *expand* or *compress*; the *pressure* of the gas and *volume changes* determine work:

Figure illustrates:

> *expansion* is work done ***by*** the system ($\Delta E = q - w$)

> *compression* is work done ***on*** the system ($\Delta E = q + w$)

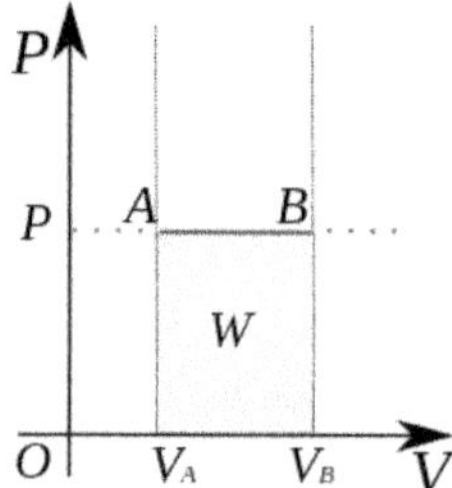

Shaded area (W) represents the work done by the gas

Positive and negative volume changes

Volume change (ΔV) is the final minus initial volume of gas:

$$\Delta V = V_{final} - V_{initial}$$

Volume is increased to *expand* a gas (ΔV is *positive*).

> Gas (of the system) must do work (w) on surroundings; thus, work must be *negative*.

Volume is decreased to compress a gas (ΔV is *negative*).

> Surroundings must do work (w) on the system; thus, work must be *positive*.

Thermodynamic work cycles

Processes may labeled *adiabatic, isothermal, isobaric,* and *isochoric.*

Adiabatic process has no transfer of heat ($Q = 0$); the change in energy equals the work.

Adiabatic processes are often confused with the isothermal process because it is sometimes assumed that no heat transfer means no change in temperature. However, this is not true—in isothermal instances, heat energy must be allowed into (or out of) the reaction to maintain a constant temperature.

Adiabatic processes, unlike an isothermal reaction, occur quickly and have a PV graph with the *same shape* but are *much steeper* than isothermal processes.

Copyright © Sterling Test Prep.

Adiabatic processes

Adiabatic process is when energy transfer within the system results in *work only*.

For example, a rapid expansion or contraction of gas is nearly adiabatic.

Adiabatic processes have *no heat* transfer.

Adiabatic processes *cannot decrease entropy*.

temperature is constant

$q = 0$

$\Delta E = w$

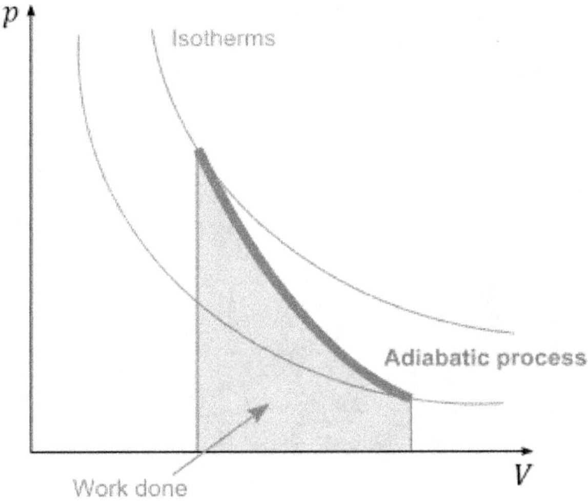

Adiabatic process PV diagrams of work with isothermal lines called isotherms

Adiabatic processes include a tire pump or a sharp expulsion of breath.

Isochoric processes

Isochoric process (or *constant-volume process*, *isovolumetric process*, and *isometric process*) is when the volume of the closed system remains *constant*; these include those inside a closed and rigid container or a constant volume thermometer.

volume is constant

$w = 0$

$\Delta E = q$

Thermodynamic process is the addition or removal of heat; the isolation of the contents of the container establishes the closed system, and the inability of the container to deform imposes the *constant-volume* condition.

Any duplication (copies, uploads, PDFs) is illegal.

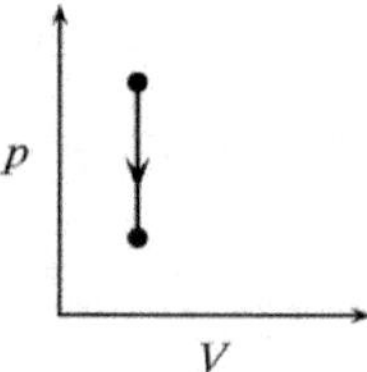

Isochoric process PV diagrams with constant volume show a vertical line

Isochoric processes include heating (or cooling) contents of a sealed, inelastic container.

Isothermal processes

Isothermal reaction keeps temperature constant ($\Delta T = 0$). The change in internal energy must stay the same, as an increase in thermal energy (heat) increases temperature ($\Delta U = 0$).

temperature is constant

$\Delta T = 0$

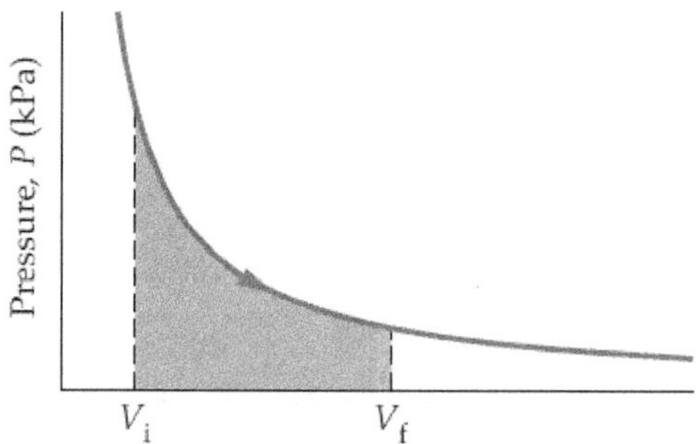

Isothermal process PV diagram at constant temperature

Isothermal process is when the temperature of the system remains constant; typically, when a system is in contact with an outside thermal reservoir. A change in the system allows the system to adjust temperature via the reservoir through heat exchange.

Copyright © Sterling Test Prep.

Isobaric processes

Isobaric process is when the pressure of the system remains constant.

Isobaric reactions keep the pressure constant; work equals the pressure times the volume change ($W = P\Delta V$).

pressure is constant

$$\Delta P = 0$$

$$w = P\Delta V$$

Examples of isobaric reactions include a piston in an engine or a flexible container open to Earth's atmosphere.

When pressure is constant, the PV graph of an isobaric process is a horizontal line.

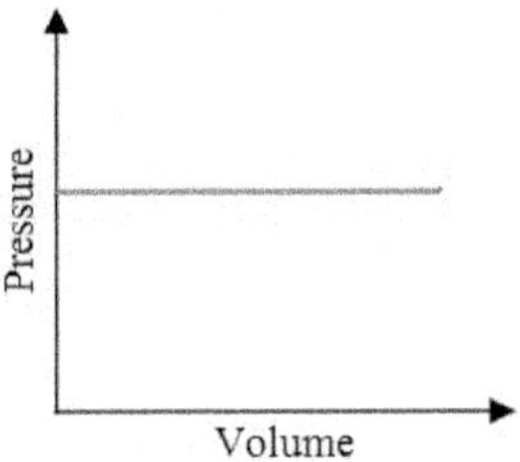

Isochoric process is an isovolumetric process because the volume is constant.

Since there is no movement, the work is zero, and the change in energy is due to the amount of heat added to the system.

$$W = 0$$

$$\Delta E = Q$$

Heat transferred to the system does work and *changes* the internal energy (U) of the system.

Thermodynamic work processes summary

Process	Constant	PV Diagram	Ideal Gas Law	First Law of Thermodynamics
Isobaric	Pressure	Horizontal line	$V \, \alpha \, T$	$\Delta U = Q + W$
Isochoric	Volume	Vertical line	$P \, \alpha \, T$	$\Delta U = Q$
Isothermal	Temperature	Curved line	$PV \, \alpha \, T$	$\Delta U = 0$
Adiabatic	No heat exchanged	Curved line (jumps to different isotherm)	$PV = nRT$ (only "nR" are constant)	$\Delta U = W$

Thermodynamic work processes, with relationship to Ideal Gas Law (PV = nRT)

Copyright © Sterling Test Prep.

Practice Questions

1. If a stationary gas has a kinetic energy of 500 J at 25 °C, what is its kinetic energy at 50 °C?

   A. 125 J
   B. 450 J
   C. 540 J
   D. 1,120 J

2. What is the purpose of the hollow walls in a closed hollow-walled container that effectively maintains the temperature inside?

   A. Traps air trying to escape from the container, which minimizes convection
   B. Act as an effective insulator, which minimizes convection
   C. Acts as an effective insulator, which minimizes conduction
   D. Provides an additional source of heat for the container

3. How much heat energy (in Joules) is required to heat 21.0 g of copper from 21.0 °C to 68.5 °C? (Use specific heat c of Cu = 0.382 J/g·°C)

   A. 462 J
   B. 381 J
   C. 522 J
   D. 662 J

4. The species in the reaction $KClO_3\,(s) \rightarrow KCl\,(s) + 3/2\,O_2\,(g)$ have the values for standard enthalpies of formation at 25 °C. At constant physical states, assume that the values of $\Delta H°$ and $\Delta S°$ are constant throughout a broad temperature range. Which of the following conditions may apply to the reaction? (Use $KClO_3\,(s)$ with $\Delta H°_f = -391.2$ kJ mol^{-1} and KCl (s) with $\Delta H°_f = -436.8$ kJ mol^{-1})

   A. Nonspontaneous at low temperatures, but spontaneous at high temperatures
   B. Spontaneous at low temperatures but nonspontaneous at high temperatures
   C. Nonspontaneous at all temperatures over a broad temperature range
   D. Spontaneous at all temperatures over a broad temperature range

5. Determine the value of $\Delta E°_{rxn}$ for this reaction, whereby the standard enthalpy of reaction $(\Delta H°_{rxn})$ is -311.5 kJ mol^{-1}:

$$C_2H_2\,(g) + 2\,H_2\,(g) \rightarrow C_2H_6\,(g)$$

   A. -306.5 kJ mol^{-1}
   B. -318.0 kJ mol^{-1}
   C. $+346.0$ kJ mol^{-1}
   D. $+306.5$ kJ mol^{-1}

Any duplication (copies, uploads, PDFs) is illegal.

6. The process of H_2O (g) → H_2O (l) is nonspontaneous under pressure of 760 torr and temperatures of 378 K because:

A. $\Delta H = T\Delta S$

B. $\Delta G < 0$

C. $\Delta H > T\Delta S$

D. $\Delta H < T\Delta S$

7. A fuel cell contains hydrogen and oxygen gas that react explosively, and the energy converts water to steam, which drives a turbine to turn a generator that produces electricity. The fuel cell and the turbine represent which forms of energy, respectively?

A. Electrical and mechanical energy

B. Electrical and heat energy

C. Chemical and mechanical energy

D. Chemical and heat energy

8. The thermodynamic systems with high stability tend to demonstrate:

A. maximum ΔH and maximum ΔS

B. maximum ΔH and minimum ΔS

C. minimum ΔH and maximum ΔS

D. minimum ΔH and minimum ΔS

9. Which is true for the thermodynamic functions G, H and S in $\Delta G = \Delta H - T\Delta S$?

A. G refers to the universe, H to the surroundings, and S to the system

B. G, H, and S refer to the system

C. G and H refers to the surroundings and S to the system

D. G and H refer to the system and S to the surroundings

10. Whether a reaction is endothermic or exothermic is determined by:

A. an energy balance between bond breaking and bond-forming, resulting in a net loss or gain of energy

B. the presence of a catalyst

C. the activation energy

D. the physical state of the reaction system

11. What is the term for a reaction that proceeds by releasing heat energy?

A. Endothermic reaction

B. Isothermal reaction

C. Exothermic reaction

D. Nonspontaneous

Copyright © Sterling Test Prep.

12. The bond dissociation energy is:

    I. useful in estimating the enthalpy change in a reaction

    II. the energy required to break a bond between two gaseous atoms

    III. the energy released when a bond between two gaseous atoms is broken

A. I only

B. II only

C. I and II only

D. I and III only

13. Which of the following statements is true for the following reaction? (Use the change in enthalpy, $\Delta H° = -113.4$ kJ/mol and the change in entropy, $\Delta S° = -145.7$ J/K mol)

$$2\ NO\ (g) + O_2\ (g) \rightarrow 2\ NO_2\ (g)$$

A. The reaction is at equilibrium at 25 °C under standard conditions

B. The reaction is spontaneous at only high temperatures

C. The reaction is spontaneous only at low temperatures

D. The reaction is spontaneous at all temperatures

14. Which law explains the observation that the amount of heat transfer accompanying a change in one direction is equal in magnitude but opposite in sign to the amount of heat transfer in the opposite direction?

A. Law of Conservation of Mass

B. Law of Conservation of Energy

C. Avogadro's Law

D. Boyle's Law

15. Calculate the value of $\Delta H°$ of reaction using provided bond energies.

$$H_2C = CH_2\ (g) + H_2\ (g) \rightarrow H_3C\text{–}CH_3\ (g)$$

C–C: 348 KJ

C≡C: 960 kJ

C=C: 612 kJ

C–H: 412 kJ

H–H: 436 kJ

A. −348 kJ

B. +134 kJ

C. −546 kJ

D. −124 kJ

Any duplication (copies, uploads, PDFs) is illegal.

Notes for active learning

Copyright © Sterling Test Prep.

Detailed Explanations

1. C is correct.

Temperature is a measure of the average kinetic energy of the molecules.

Kinetic energy is proportional to temperature.

In most thermodynamic equations, temperatures are expressed in Kelvin, so convert Celsius to Kelvin:

$$25\ °C + 273.15 = 298.15\ K$$

$$50\ °C + 273.15 = 323.15\ K$$

Calculate the kinetic energy using simple proportions:

$$KE = (323.15\ K\ /\ 298.15\ K) \times 500\ J$$

$$KE = 540\ J$$

2. C is correct.

An *insulator* reduces conduction (i.e., thermal energy transfer through matter).

Air and vacuum are excellent insulators. Storm windows, which have air wedged between two glass panes, work by utilizing this conduction principle.

3. B is correct.

Heat capacity is the amount of heat needed to increase the whole sample's temperature by 1 °C.

Specific heat is the energy required to increase *1 gram* of the sample by 1 °C.

Heat = mass × specific heat × change in temperature:

$$q = m \times c \times \Delta T$$

$$q = 21.0\ g \times 0.382\ J/g \cdot °C \times (68.5\ °C - 21.0\ °C)$$

$$q = 21.0\ g \times 0.382\ J/g \cdot °C \times (47.5\ °C)$$

$$q = 381\ J$$

Any duplication (copies, uploads, PDFs) is illegal.

4. D is correct.

Calculate the value of:

$$\Delta H_f = \Delta H_{f\,product} - \Delta H_{f\,reactant}$$

$$\Delta H_f = -436.8 \text{ kJ mol}^{-1} - (-391.2 \text{ kJ mol}^{-1})$$

$$\Delta H_f = -45.6 \text{ kJ mol}^{-1}$$

To determine spontaneity, calculate the Gibbs free energy:

$$\Delta G = \Delta H^\circ - T\Delta S$$

The reaction is spontaneous if ΔG is negative:

$$\Delta G = -45.6 \text{ kJ} - T\Delta S$$

Typical ΔS values are around 100–200 J.

If the ΔH value is –45.6 kJ (or –45,600 J), the value of ΔG would be negative unless $T\Delta S$ is less than –45,600.

Therefore, the reaction would be spontaneous over a broad range of temperatures.

5. A is correct.

Relationship between enthalpy (ΔH) and internal energy (ΔE):

$$\text{Enthalpy} = \text{Internal energy} + \text{work (for gases, work} = PV)$$

$$\Delta H = \Delta E + \Delta(PV)$$

Solving for ΔE:

$$\Delta E = \Delta H - \Delta(PV)$$

According to ideal gas law:

$$PV = nRT$$

Substitute ideal gas law to the previous equation:

$$\Delta E = \Delta H - \Delta(nRT)$$

R and T are constant, which leaves Δn as the variable.

The reaction is:

$$C_2H_2 \text{ (g)} + 2 H_2 \text{ (g)} \rightarrow C_2H_6 \text{ (g)}.$$

There are three gas molecules on the left and one on the right, which means $\Delta n = 1 - 3 = -2$.

continued...

   Copyright © Sterling Test Prep.

In this problem, the temperature is not provided.

However, the presence of degree symbols ($\Delta G°$, $\Delta H°$, $\Delta S°$) indicates that those are standard values, measured at 25 °C or 298.15 K.

Double-check the units before performing calculations; ΔH is in kilojoules, while the gas constant (R) is 8.314 J/mol K. Convert ΔH to Joules before calculating.

Use the Δn and T values to calculate ΔE:

$$\Delta E = \Delta H - \Delta nRT$$

$$\Delta E = -311,500 \text{ J/mol} - (-2 \times 8.314 \text{ J/mol K} \times 298.15 \text{ K})$$

$$\Delta E = -306,542 \text{ J}$$

$$\Delta E \approx -306.5 \text{ kJ}$$

6. C is correct.

The spontaneity of a reaction is determined by evaluating the Gibbs free energy, or ΔG.

$$\Delta G = \Delta H - T\Delta S$$

A reaction is spontaneous if $\Delta G < 0$ (or ΔG is negative).

For ΔG to be negative, ΔH must be less than $T\Delta S$.

A reaction is nonspontaneous if $\Delta G > 0$ (or ΔG is positive and ΔH is greater than $T\Delta S$).

7. C is correct.

Potential Energy **Stored energy and the energy of position (gravitational)**	**Kinetic Energy** **Energy of motion: motion of waves, electrons, atoms, molecules and substances**
Chemical Energy Chemical energy is the energy stored in the bonds of atoms and molecules. Biomass, petroleum, natural gas, propane, and coal are examples.	**Radiant Energy** Radiant energy is electromagnetic energy that travels in transverse waves. It includes visible light, x-rays, gamma rays, and radio waves.
Nuclear Energy Nuclear energy is the energy stored in the nucleus of an atom, and it is the energy that holds the nucleus together. The nucleus of the uranium atom is an example.	**Thermal Energy** Thermal energy (or heat) is the internal energy in substances, vibration, and the movement of atoms and molecules. Geothermal energy is an example.

<table>
<tr><td>

Stored Mechanical Energy

Stored mechanical energy is energy stored in objects by applying force. Compressed springs and stretched rubber bands are examples.

Gravitational Energy

Gravitational Energy is the energy of place or position. Water in a reservoir behind a hydropower dam is an example of gravitational potential energy. When the water is released to spin the turbines, it becomes kinetic energy.

</td><td>

Motion

The movement of objects or substances from one place to another. Wind and hydropower are examples.

Sound

Sound is energy movement through substances in longitudinal (compression/rarefaction) waves.

Electrical Energy

Electrical energy is the movement of electrons. Lightning and electricity are examples.

</td></tr>
</table>

8. C is correct.

Gibbs free energy:

$$\Delta G = \Delta H - T\Delta S$$

Stable molecules are *non-spontaneous* because they have a negative (or relatively low) ΔG.

ΔG is most negative (most stable) when ΔH is small, and ΔS is large.

9. B is correct.

All thermodynamic functions in $\Delta G = \Delta H - T\Delta S$ refer to the system.

10. A is correct.

ΔH refers to enthalpy (or heat).

Endothermic reactions have heat as a reactant.

Exothermic reactions have heat as a product.

Endothermic reactions absorb energy to break strong bonds to form a less stable state (i.e., positive enthalpy).

Exothermic reactions release energy while forming stronger bonds and a more stable state (i.e., negative enthalpy).

continued...

Copyright © Sterling Test Prep.

The reaction is *nonspontaneous* (i.e., endergonic) when products are *less stable* than reactants and ΔG is *positive.*

The reaction is *spontaneous* (i.e., exergonic) when products are *more stable* than reactants and ΔG is *negative.*

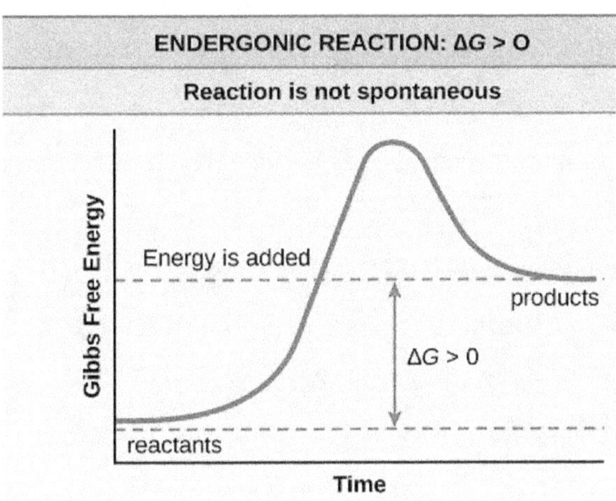

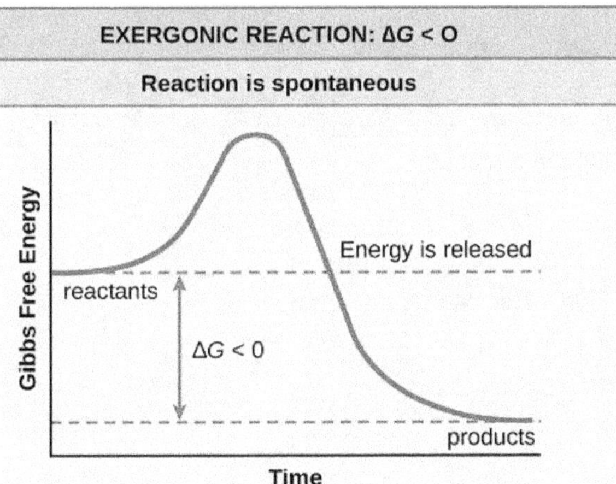

Exothermic reactions *release heat* and cause the surroundings' temperature to rise (i.e., a net loss of energy).

Endothermic reactions *absorb heat* and cool the surroundings (i.e., a net gain of energy).

11. C is correct.

ΔH refers to enthalpy (or heat).

Endothermic reactions have *heat as a reactant*.

Exothermic reactions have *heat as a product*.

Exothermic reactions *release heat* and cause the surroundings' temperature to rise (i.e., a net loss of energy).

Exothermic reactions release energy while forming stronger bonds forming a more stable state (i.e., negative enthalpy).

Endothermic processes *absorb heat* and cool the surroundings (i.e., a net gain of energy).

Endothermic reactions absorb energy to break strong bonds to form a less stable state (i.e., positive enthalpy).

The reaction is nonspontaneous (i.e., endergonic) if products are less stable than reactants and ΔG is positive.

The reaction is spontaneous (i.e., exergonic) if products are more stable than reactants and ΔG is negative.

12. C is correct.

Bond dissociation energy is the energy required to break a bond between two gaseous items and is useful in estimating the enthalpy change in a reaction.

13. C is correct.

To predict spontaneity of reaction, use the Gibbs free energy equation:

$$\Delta G = \Delta H° - T\Delta S$$

The reaction is spontaneous if ΔG is negative.

Substitute the given values to the equation:

$$\Delta G = -113.4 \text{ kJ/mol} - [T \times (-145.7 \text{ J/K mol})]$$

$$\Delta G = -113.4 \text{ kJ/mol} + (T \times 145.7 \text{ J/K mol})$$

Important: note that the units are not identical; $\Delta H°$ is in kJ, and $\Delta S°$ is in J.

Convert kJ to J (1 kJ = 1,000 J):

$$\Delta G = -113,400 \text{ J/mol} + (T \times 145.7 \text{ J/K mol})$$

It can be predicted that the value of ΔG would be negative if the value of T is small.

If T increases, ΔG approaches a positive value, and the reaction is nonspontaneous.

Copyright © Sterling Test Prep.

14. B is correct.

The *Law of Conservation of Energy* states that an isolated system's total energy (e.g., potential or kinetic) remains constant and conserved.

Energy can be neither created nor destroyed but is transformed from one form to another.

For instance, chemical energy is converted to kinetic energy in a firecracker's explosion.

The *Law of Conservation of Mass* states that for any system closed to transfers of matter and energy, the mass of the system must remain constant over time.

The *Law of Definite Proportions* (Proust's Law) states that a chemical compound contains the same proportion of elements by mass.

The law of definite proportions forms the basis of stoichiometry (i.e., from the known amounts of separate reactants, the amount of the product can be calculated).

Avogadro's Law is an experimental gas law relating the volume of a gas to the amount of substance (i.e., moles) of gas present.

Avogadro's Law states that equal volumes of gases have the same number of molecules at the same temperature and pressure.

Boyle's Law is an experimental gas law that describes how the pressure of a gas tends to increase as the volume of a gas decreases. It states that the absolute pressure exerted by a given mass of an ideal gas is inversely proportional to the volume it occupies if the temperature and amount of gas remain unchanged within a closed system.

15. D is correct.

Reactant bonds are broken in a chemical reaction, and new bonds form to create products.

Therefore, in bond dissociation problems,

$$\Delta H \text{ reaction} = \text{sum of bond energy in reactants} - \text{sum of bond energy in products}$$

For $H_2C{=}CH_2 + H_2 \rightarrow CH_3{-}CH_3$:

$$\Delta H_{\text{reaction}} = \text{sum of bond energy in reactants} - \text{sum of bond energy in products}$$

$$\Delta H_{\text{reaction}} = [(C{=}C) + 4(C{-}H) + (H{-}H)] - [(C{-}C) + 6(C{-}H)]$$

$$\Delta H_{\text{reaction}} = [612 \text{ kJ} + (4 \times 412 \text{ kJ}) + 436 \text{ kJ}] - [348 \text{ kJ} + (6 \times 412 \text{ kJ})]$$

$$\Delta H_{\text{reaction}} = -124 \text{ kJ}$$

Remember that this is the opposite of ΔH_f problems,

where: $\Delta H_{\text{reaction}} = (\text{sum of } \Delta H_f \text{ products}) - (\text{sum of } \Delta H_f \text{ reactants})$

Any duplication (copies, uploads, PDFs) is illegal.

Notes for active learning

Copyright © Sterling Test Prep.

CHAPTER 3

Fluids & Solids

- States of Matter

- Properties of Water

- Surface Tension

- Density and Specific Gravity

- Pressure

- Viscosity of Liquids

- Flow of Liquids

- Strain and Stress on Solids

- Elasticity: Elementary Properties

- Practice Questions & Detailed Explanations

Page intentionally left blank

Copyright © Sterling Test Prep.

States of Matter

Four states of matter

Three standard states of matter are solid, liquid, and gas.

Solids have a definite shape and volume.

Liquid has a fixed volume but can be any shape.

Gas can be any shape and can be compressed (variable volume).

Since liquids and gases can "flow," they are fluids.

Plasma is the fourth state of matter, distinct from the other states. Like liquids boiling, changing into a gas when energy is added, heating a gas will form plasma.

Plasma is a mixture of positively charged (ions) and negatively charged particles (electrons). Moving charged particles generate electric currents, and any movement of a charged plasma particle affects and is affected by the fields created by the other charges.

Describing a low-density plasma as merely an *ionized gas* is incorrect despite being similar to the gas phase in that neither assumes a definite shape or volume.

State transitions

Matter changes states when matter loses or absorbs energy (i.e., heat or thermal energy).

When a substance absorbs energy, the atoms and molecules move more rapidly, and this increased kinetic energy pushes particles far enough that they change form. This energy is usually heat or thermal energy.

Usually, changes in the state of matter go in a specific order.

Adding thermal energy to a solid transforms it into a liquid and then a gas; when removing enough thermal energy from a gas, it changes back into a liquid and, finally, a solid.

Phase change also depends on factors like pressure and temperature.

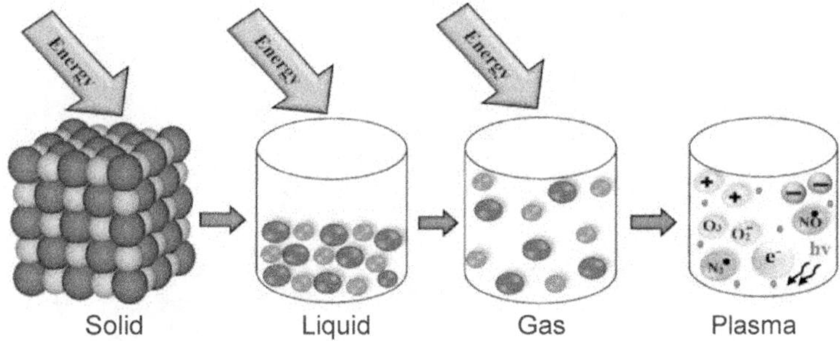

States of matter transition when energy is added

Notes for active learning

Copyright © Sterling Test Prep.

Properties of Water

Molecular formula of water

Water is vital for life on Earth but is an inorganic molecule due to its lack of carbon.

Water molecules contain one oxygen atom covalently bonded to two hydrogen atoms in a bent conformation because of the two lone pairs of electrons on the oxygen atom.

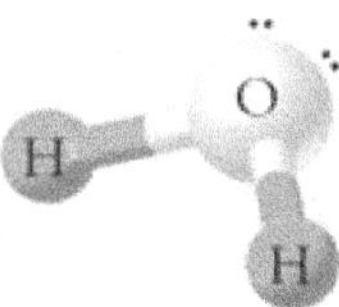

Due to oxygen's electronegativity, bonding electrons are concentrated near this central atom, giving oxygen a partial negative charge on oxygen and a partial positive charge on hydrogen.

Unequal sharing of electrons makes water a *polar* molecule.

Oxygen of water (H_2O, shown below) has two bonding and two non-bonding domains. The water molecule experiences electron repulsion from the two lone pairs on its bonds.

Water molecules adopt a bent shape to minimize van der Waals repulsion of the negatively charged electrons.

A water molecule (H_2O) is bent due to the two lone pairs of e^- on oxygen

Significance of water for living systems

The first cells evolved in water and incorporated water into their living systems; today, organisms are 70–90% water.

Water is liquid between 0 °C and 100 °C, which is critical since it serves as the solvent of living organisms.

Polarity of water allows it to dissolve ionic, polar, or semi-polar molecules.

This solubility allows substances such as nutrients to be carried in the blood and is a suitable medium for metabolic reactions.

Evaporation of water cools plants and animals. These organisms allow heat (i.e., energy) to break hydrogen bonds as water molecules evaporate and release energy into the air.

Any duplication (copies, uploads, PDFs) is illegal.

Hydrogen bonding between water molecules

Water can form four hydrogen bonds: one at each hydrogen atom and two at lone pairs of electrons on the oxygen atom.

When water molecules form hydrogen bonds, they create strong intermolecular forces that give water characteristic properties.

One hydrogen bond (dash line) between water molecules. Two hydrogen bond acceptors from each lone pair of electrons on partial negative oxygen and two hydrogen bond donors from each partial positive hydrogen

Physical properties of water

Physical properties are quantitative measures of a substance that can be determined without permanently altering the substance.

For example, the boiling point of water is a physical property because the water condenses (i.e., the reverse process of boiling) to form a liquid.

Water has a high *heat capacity*, which is the degree to which a substance changes its temperature in response to the gain or loss of heat.

Temperature of a large body of water is stable in response to temperature (average kinetic energy of a molecule) changes in the surrounding air.

Since water can hold more heat, its temperature falls slower than other liquids; this moderates Earth's surface temperature and protects organisms from rapid temperature changes.

Heat capacity is reflected in the *heat of vaporization,* the energy required to boil a substance (change from liquid to a gaseous state).

Water has an unusually high boiling point due to many strong intermolecular hydrogen bonds.

Water boils at 100 °C at standard pressure because high heat is needed to break the hydrogen bonds. The high boiling point of water is vital for life on Earth. If the water boiled at a lower temperature, the water in organisms would boil, and they would not survive.

Water has a high *heat of fusion*, the energy required to melt from its solid state. This makes Earth's ice caps resistant to sudden and frequent melting.

Solid (i.e., ice) water exhibits a lower density than liquid water, unlike most other substances.

    Copyright © Sterling Test Prep.

As water freezes, its hydrogen bonds become rigid and pushed apart, expanding the water and adopting a crystal pattern at 0 °C.

The large air spaces between molecules lower the density of ice and allow it to float in liquid water. Because of this property, bodies of water freeze from the top down.

If ice were denser than water, it would sink, and ponds would freeze solidly.

Water becoming less dense as it freezes is beneficial to organisms; as ice forms at the surface, it insulates the water underneath, maintaining a hospitable environment for aquatic organisms. The ice does not sink and remain deep in the body of water, which, over time, would pose an issue for organisms as the ambient temperature decreases.

Chemical properties of water

Chemical properties are not observable without altering a substance in a chemical reaction.

Water is a *universal solvent* that dissolves many solutes. This is possible because of water's polarity, causing it to attract charged (or polar substances), associating around the individual molecules and breaking the aggregation of the solute.

Ionized or polar molecules attracted to water are *hydrophilic.*

Non-ionized and nonpolar molecules that cannot attract water are *hydrophobic.*

Water is a reactant and product in many chemical reactions outside and within living systems.

Water is an *amphoteric* substance; it can function as an acid or a base.

Water molecules function as a base by accepting a hydrogen ion; a second water molecule acts as an acid.

Hydrogen bonds are responsible for water's cohesive and adhesive properties.

Cohesion is water's tendency to stick due to hydrogen bonding. Strong cohesion between water molecules produces high *surface tension,* measured by disrupting a liquid's surface. Water's high surface tension helps cell membranes from collapsing and allows insects to walk on water.

Adhesion is water's ability to attract other polar molecules.

Adhesion and cohesion work together so water can "climb" against gravity up a thin glass tube, the vascular tissues of a plant, or an animal's blood vessels.

Capillary action of adhesion is an essential process of life.

Any duplication (copies, uploads, PDFs) is illegal.

Notes for active learning

Copyright © Sterling Test Prep.

Surface Tension

Hydrogen bonding produces surface tension

Small objects like paper clips can be placed lightly on the water's surface and float. However, if pushed down, they readily sink. A small object placed gently on top of the water may not weigh enough to break the surface tension. Therefore, it floats despite its higher density than water. Hydrogen bonding attractions between liquid water molecules produce *surface tension*.

Surface tension is due to the *cohesive forces* within water molecules. Cohesion occurs because water molecules are electrostatically attracted and stick to neighboring water molecules.

Some insects, for example, can walk on water, not because they are less dense than the water but because of the resistive force of surface tension.

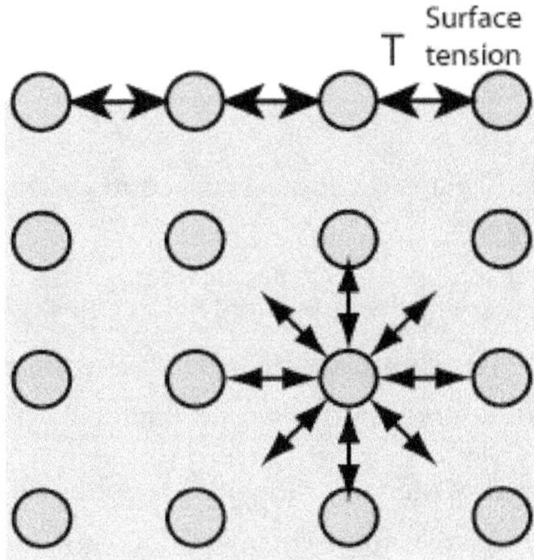

Surface tension as cohesive forces join adjacent water molecules by hydrogen bonding

Adhesive and cohesive forces

Adhesion is the property of a substance to be attracted to a different substance.

Adhesion, unlike cohesion, is caused by several mechanisms and may or may not be due to electrostatic forces, as discussed for water molecules.

Cohesive and *adhesive forces* of the liquid determine the behavior of a liquid in a container.

Often, the surface of a liquid in a container is not flat when in a small container (e.g., graduated cylinder), where the liquid curves up or down as it contacts the container walls. This curved surface for the liquid is the *meniscus* and results from surface tension and adhesion.

If the adhesive force between the liquid and the container is greater than the cohesive force between the liquid molecules, the liquid curves up to meet the edges of the container, creating a *concave meniscus*.

Any duplication (copies, uploads, PDFs) is illegal.

If the cohesive force exceeds the adhesive force, the liquid curves downward, creating a *convex meniscus*. For example, water molecules are more attracted to glass than to each other (adhesion > cohesion), so the water's surface curves upward toward the walls of its container.

Mercury has adhesion < cohesion, so its surface curves downward against the container walls.

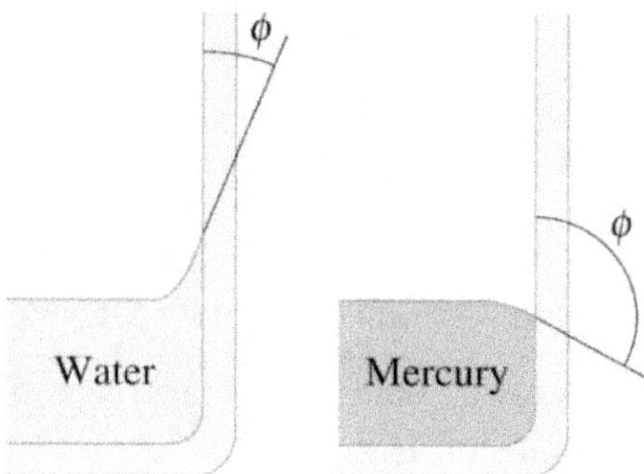

Cohesive and adhesive forces affect the liquid's position relative to the container's walls

Capillary action of liquids

Fluids exhibit capillary action if a narrow tube is placed in the fluid. This phenomenon is due to cohesive and adhesive forces.

If a narrow tube is placed in water (adhesion > cohesion), the water travels up the tube some distance (figure below on the left). This movement is because the adhesive forces pull the water along the edge, and the cohesive forces pull water up the tube.

Suppose the same tube is placed in mercury (adhesion < cohesion); then the cohesive forces of the mercury keep the mercury together rather than climb the tube wall. Mercury becomes lower in the capillary tube than the height of the mercury level around it (figure below on the right).

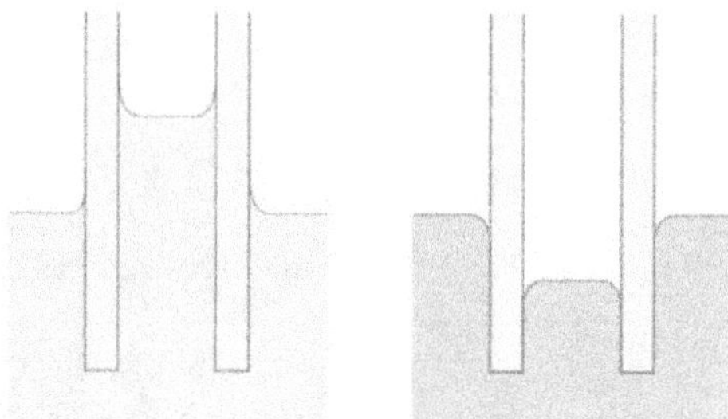

Central capillary tube in water (on the left) compared to in mercury (on the right)

Copyright © Sterling Test Prep.

Density and Specific Gravity

Determining density

Density (ρ) of an object is its mass per unit volume, expressed as:

$$\rho = \frac{m}{V}$$

where ρ is density, m is mass, and V is volume.

Density is expressed in SI units of kg/m^3; it may also be given in g/cm^3.

Density is the weight per unit volume of material at $23°C$ and is expressed as D $23°C$ in g/cm^3.

To convert g/cm^3 to kg/m^3, use dimensional analysis.

However, it is easier to multiply by 1,000 (cm^3kg/m^3g):

$$\left(\frac{g}{cm^3}\right) \cdot \left(\frac{100 \text{ cm}}{1 \text{ m}}\right)^3 \cdot \left(\frac{1 \text{ kg}}{1{,}000 \text{ g}}\right) = \left(\frac{g}{cm^3}\right) \cdot \left(\frac{1{,}000 \text{ cm}^3\text{kg}}{m^3g}\right)$$

$$\left(\frac{g}{cm^3}\right) \cdot \left(\frac{1{,}000 \text{ cm}^3\text{kg}}{m^3g}\right) = \frac{kg}{m^3}$$

Specific gravity

Specific gravity of a substance is the ratio of its density to the density of the reference substance.

Specific gravity is referenced against water. Thus, the specific gravity of water equals 1.

Specific gravity is the ratio of the weight of a given volume of material at 73 °F (23 °C) to that of an equal volume of water at the same temperature, expressed as specific gravity, 23/23°C.

Equation for specific gravity:

$$\text{specific gravity} = \frac{\rho_{\text{object}}}{\rho_{\text{water}}}$$

Any duplication (copies, uploads, PDFs) is illegal.

Notes for active learning

Copyright © Sterling Test Prep.

Pressure

Calculating pressure

Pressure is the force exerted on the area over which the force acts.

Equation for pressure is:

$$P = \frac{F}{A}$$

where P is pressure (N/m^2), F is a force (N), and A is area (m^2).

Unit of pressure is N/m^2, which is the Pascal (Pa).

$$1 \text{ Pa} = 1 \text{ N/m}^2$$

Anything that exerts a force can exert pressure; therefore, the concept of pressure applies to solid objects, liquids, and gases.

When discussing fluids, the pressure exerted by the atmosphere and measurements of pressure relative to it are of concern.

At sea level, atmospheric pressure is about 1.013×10^5 N/m^2; this is one atmosphere (1 atm).

Another unit of pressure is the bar (1 bar = 1.00×10^5 N/m^2). Standard atmospheric pressure is slightly more than 1 bar.

Atmospheric pressure

Atmospheric pressure does not crush organisms because their cells maintain an internal pressure that balances it.

The pressure is measured using two common types.

Absolute pressure is pressure zero-referenced against a perfect vacuum.

Gauge pressure

Gauge pressure is pressure zero-referenced against atmospheric pressure.

Most gauges measure pressure above the atmospheric pressure. Hence, most systems measure gauge pressure.

For example, what would the reading be if there were two different pressure dials (with one displaying absolute pressure and the other displaying gauge pressure) at sea level?

Absolute pressure dial reads 1.013×10^5 Pa, and the gauge pressure dial reads 0 Pa. This difference is because the atmosphere exerts 1.013×10^5 Pa of pressure over a perfect vacuum; thus, the absolute pressure gives this value.

Any duplication (copies, uploads, PDFs) is illegal.

Gauge pressure reads 0 Pa because it references atmospheric pressure and requires pressures above this value to give a reading.

Gauge pressure is the difference between absolute and atmospheric pressure.

$$P_{gauge} = P_{abs} - P_{atm}$$

Absolute pressure equation:

$$P_{abs} = P_{atm} + P_{gauge}$$

Pressure versus depth ($P = pgh$)

Below the surface of a liquid, pressure increases linearly with depth. Liquid pressure on an object is found only using the liquid's density and the object's distance below the surface.

The pressure on an object is not determined by using the object's density in the liquid or the shape of the container.

Liquid pressure is expressed as:

$$P = \frac{F}{A} = \frac{\rho A h g}{A}$$

Simplifies to:

$$P_{liquid} = \rho g h$$

where P_{liquid} is the pressure exerted by the liquid (N/m^2 or Pa), g is the acceleration of gravity, ρ is the density of the liquid, and h is the distance below the surface of the liquid (m).

Liquid pressure is useful in the operation of some types of pressure gauges.

For example, the gauge pictured below is an open-tube manometer. The pressure in the open end is open to atmospheric pressure, and the pressure that is being measured causes the fluid to rise until the pressures on both sides of the tube are equal.

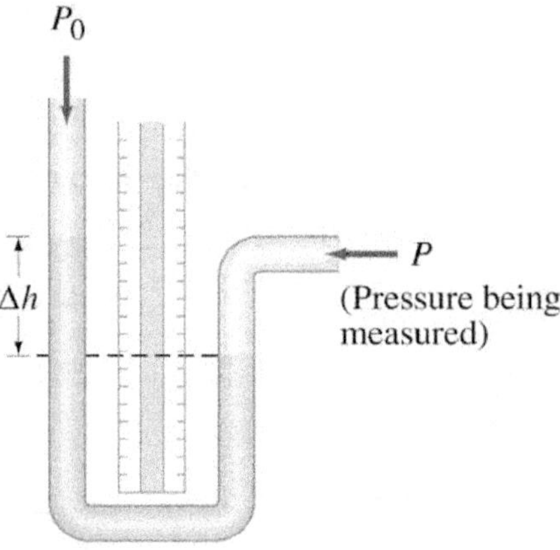

Open tube manometer includes atmospheric pressure

Copyright © Sterling Test Prep.

From the liquid pressure equation, the pressure being measured is calculated as:

$$P_0 = P_{atm}$$

$$P = P_{atm} + P_{liquid}$$

$$P = P_{atm} + \rho g \Delta h$$

The measured pressure is absolute in this example because the equation includes atmospheric pressure. This is an important distinction because if the pressure on an object is evaluated at depth, the gauge pressure only gives the liquid pressure on the object.

In contrast, the absolute pressure gives the gauge pressure plus atmospheric pressure so that it would be more accurate.

Calculating gauge and atmospheric pressure

The figure below shows two containers with different shapes holding water at equal levels. If points 1 and 2 are equal height, what are the gauge and absolute pressure at points 1 and 2?

The containers are different shapes; the two po   are at equal depths within the water. Therefore, they have the same gauge and absolute pressures.

The shape of the container does not influence the liquid pressure; only the depth below the liquid contributes to liquid pressure. Gauge and absolute pressure can be calculated as:

$$P_1 = P_2$$

$$P_{gauge} = \rho g \Delta h$$

$$P_{absolute} = P_{atm} + \rho g \Delta h$$

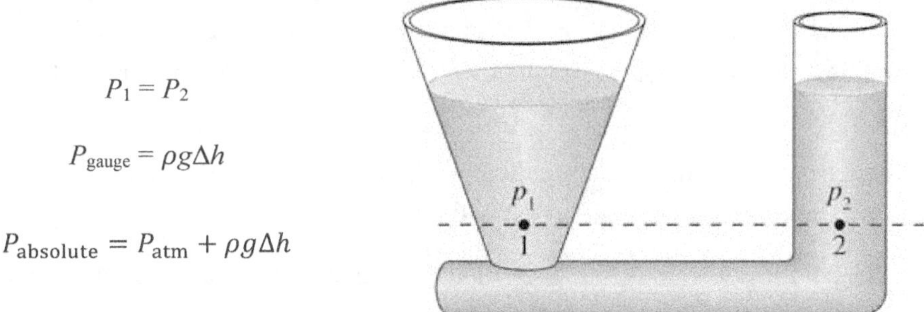

Two points are at equal depths, so they have the same gauge and absolute pressures

Pascal's Law

Pascal's Law (1647-1648) states that *a pressure exerted on an incompressible liquid transmits equally to all liquid parts.*

This Principle has important implications because it explains why pressure differentials are not created within a closed volume of liquid.

A small cube is within a beaker of water. For this example, consider the sides small enough that liquid pressure due to depth is equal on the top and bottom of the cube.

The pressure on all sides of the object is equal because the pressure is the same in every direction in a fluid at a given depth.

If this were not true, the fluid would flow because of the natural tendency of particle movement from high pressure to low pressure.

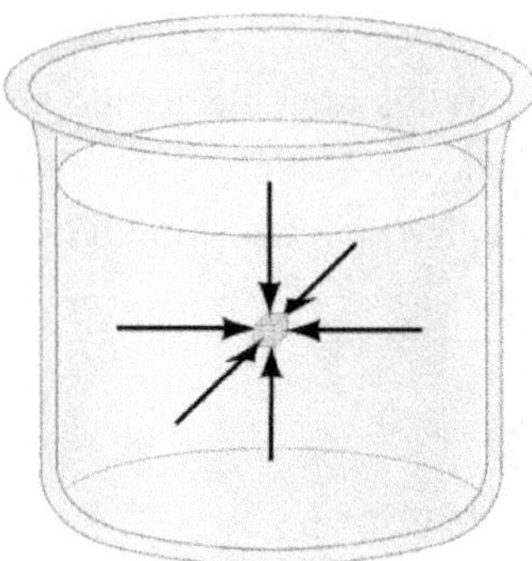

Pressure on all sides of the object is equal because the pressure is the
same in every direction in a fluid at a given depth

Although the pressures across an incompressible liquid are the same, the forces vary depending on the area they are applied over (because of $P = F / A$). This observation is useful because it allows the development of systems that impart mechanical advantage.

If the external pressure is applied to a confined fluid, the pressure at every point within the fluid increases by that amount. Pascal's Principle is used in hydraulic lifts and hydraulic brakes. The fluid is confined, so when pressure is applied, the fluid is compressed, which raises the pressure throughout the system.

In the system below, a small force is applied over a small area (F_1 over A_1). The output force (F_2) at the location with the larger area (A_2) will be higher.

Energy must be conserved; thus, the work done on one end is the same as on the other. The volume of liquid displaced must be conserved to equal the volume displaced at both the input and output.

Work equivalence:

$$W_1 = W_2$$

$$F_1 d_1 = F_2 d_2$$

Volume equivalence:

$$V_1 = V_2$$

$$A_1 d_1 = A_2 d_2$$

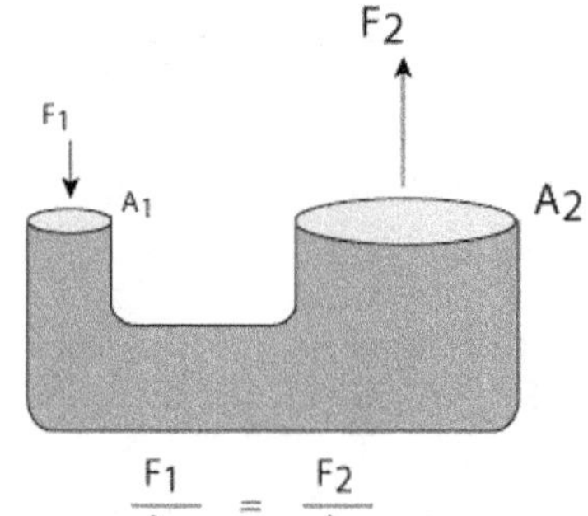

$$\frac{F_1}{A_1} = \frac{F_2}{A_2}$$

Pressure applied compresses the fluid, which raises the pressure throughout the system

Copyright © Sterling Test Prep.

Viscosity of Liquids

Viscosity

When water poured out of a cup is compared to honey poured out of a jar, the flow pattern is substantially different. Although water and honey are both liquids, they behave differently when flowing due to their *viscosity*.

Viscosity of a fluid is a measure of the internal frictional force. All fluids have some measure of viscosity. Viscosity may be high (e.g., honey) or low (e.g., water).

The characteristic viscosity of a fluid is a known value and is represented by the symbol η with units of (Pa·s).

Flow patterns

When highly viscous fluids flow through a pipe, the behavior is markedly different than when a fluid with lower viscosity flows through the pipe.

The flow from highly viscous fluids forms a front shaped like a *parabola bulging outward.*

A parabola shape is observed in the fluid because the frictional forces on the fluid from the sides of the pipe are much higher than the frictional forces in the center of the fluid.

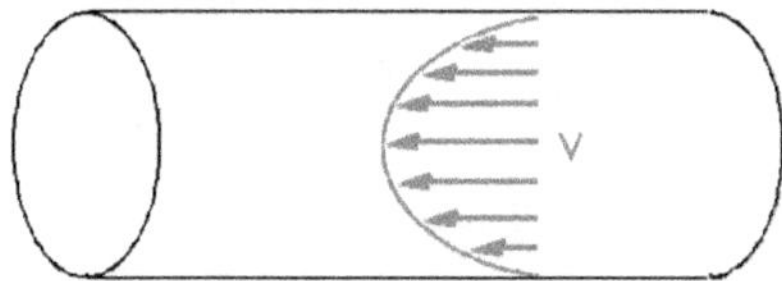

Parabola shape observed for high viscous fluids

Along with the parabola flow shape, the *flow rate* of a fluid through a pipe must account for the viscosity of the liquid.

Poiseuille's Law

Poiseuille's Law is expressed as:

$$\dot{V} = \frac{\Delta P \pi r^4}{8 \eta L}$$

where $\dot{V}$ is the volumetric flow rate (m³/s), ΔP is pressure differential across pipe (Pa), r is the radius of the pipe (m), η is the viscosity of the pipe (Pa·s), and L is the length of the pipe (m).

Poiseuille's Law (1838) has significant consequences for understanding blood flow.

Any duplication (copies, uploads, PDFs) is illegal.

Hagen–Poiseuille equation determines vascular resistance and flow rate of intravenous (IV) fluids achieved by using various sizes of peripheral and central *cannulas* (i.e., tubes inserted into the body).

Flow rate is proportional to the radius to the *fourth power*, meaning that a slight increase in the internal diameter of the cannula yields a significant increase in the flow rate of IV fluids.

For example, if the radius of an artery is halved by cardiovascular occlusion (i.e., blockage), pressure must increase by a factor of 16 to maintain the same blood flow. Usually, the human heart cannot sustain 16 times the output.

Blood pressure increases (i.e., *hypertension*) as the heart pumps (cardiac stress) to satisfy this demand for typical blood flow with blockage.

Copyright © Sterling Test Prep.

Flow of Liquids

Conservation of volume

When liquids flow through pipes, the flow is considered incompressible, and the volume of the liquid remains constant regardless of pressure. As such, the volume of liquid flowing through a pipe in a given time must *always be equal*.

This principle is the conservation of volume and can be expressed by the *continuity equation*, which describes the volume flow rate of an incompressible liquid.

Continuity equation for calculating flow rate

Continuity equation (or *transport equation*) describes the transport of some quantity.

It is usually applied to a conserved quantity but can be generalized for any extensive quantity (e.g., mass, energy, momentum, electric charge, and other conserved quantities).

Often, the continuity equation is used to calculate the speed of the fluid flowing through pipes and is given as:

$$A_1 v_1 = A_2 v_2$$

where A is the area of the pipe's cross-section (m^2), and v is the velocity of the fluid through that point (m/s).

Imagine an incompressible liquid flowing through a pipe that has a diameter reduction. The volume of fluid passing through the constriction over time cannot change.

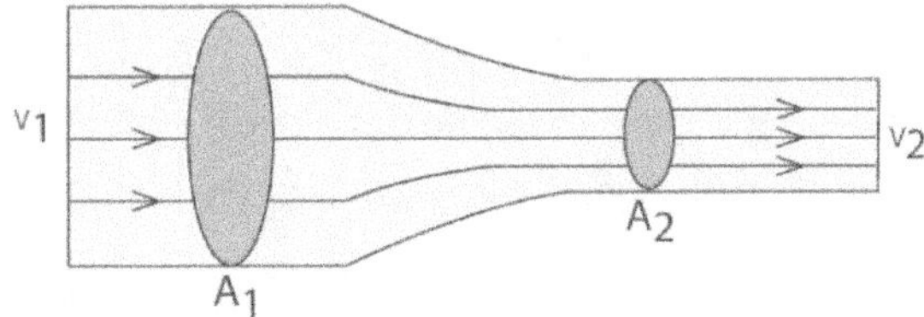

Continuity equation means flow rate through A₁ equals A₂

Continuity equation can solve for the resulting velocity of the liquid:

$$A_1 v_1 = A_2 v_2$$

$$v_2 = \frac{A_1 v_1}{A_2}$$

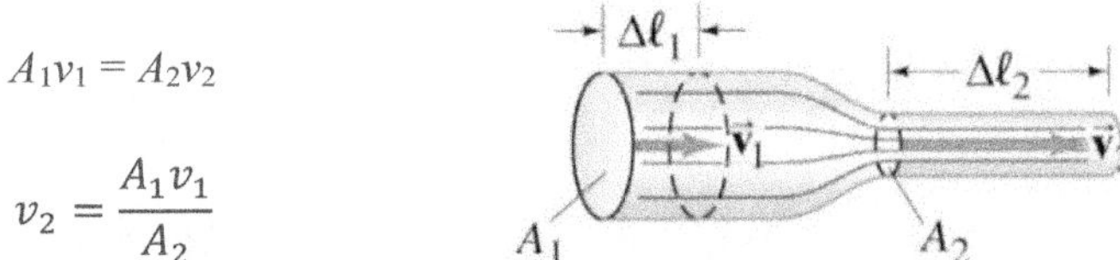

Bernoulli principle

Bernoulli equation (1738) relates energy conservation within flowing fluids. It states that *a flowing fluid's pressure, kinetic, and potential energy is conserved.*

Bernoulli equation is:

$$P_1 + \rho g h_1 + \frac{1}{2}\rho v_1^2 = P_2 + \rho g h_2 + \frac{1}{2}\rho v_2^2$$

where P is pressure (Pa), ρ is the density of the liquid, g is the acceleration due to gravity, h is the height of the fluid (m), and v is the velocity of the fluid through that point (m/s).

Bernoulli equation analyzes fluid flow across pressure or height differentials and the energy needed to transport fluid volumes.

In the diagram below, water flows into the left side at a pressure of P_1, a velocity of v_1, through an area of A_1, and at the height of y_1.

As fluid moves to the right, the pipe rises in height and decreases in area (i.e., circumference).

Given a few conditions, Bernoulli's Principle can be used to find the other values.

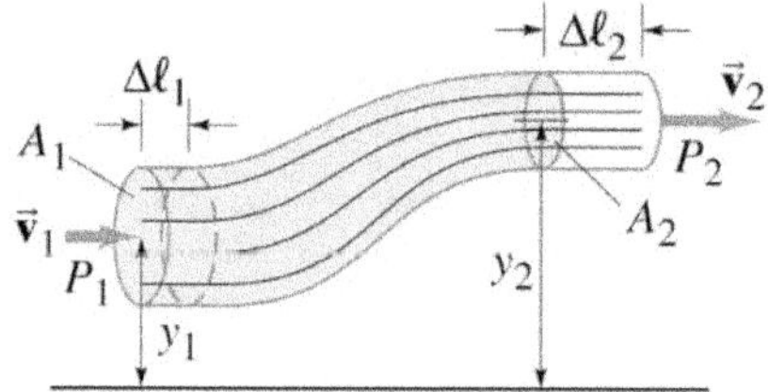

Bernoulli's Principle analyzes fluid flow across pressure or height differentials

Bernoulli's principle solved problem example

Bernoulli's principle correlates the speed of a fluid through an opening to the height of the fluid above the opening. As shown below, suppose a spigot is at the bottom of an open water tank.

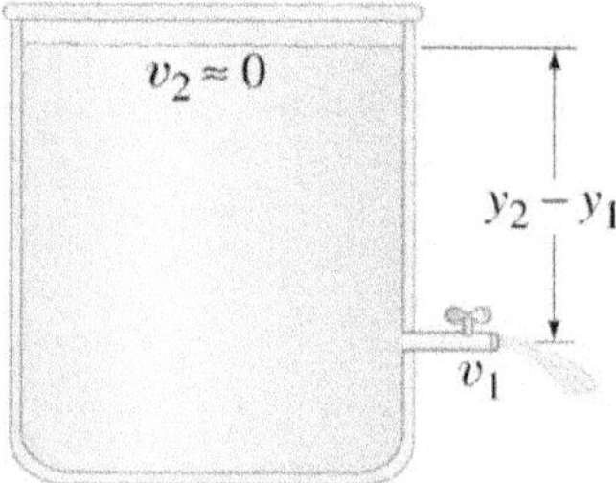

Bernoulli's equation solves for the speed of fluid through bottom spigot

Copyright © Sterling Test Prep.

Speed of the fluid can be found using Bernoulli's Principle and making a few assumptions.

$$P_1 + \rho g y_1 + \frac{1}{2}\rho v_1^2 = P_2 + \rho g y_2 + \frac{1}{2}\rho v_2^2$$

Assume that the top of the container is much larger than the opening of the spigot:

$$A_{\text{top}} \gg A_{\text{spigot}}$$

According to the *continuity equation*, fluid flow velocity through the top is much smaller than that through the spigot. Flow through the spigot can be set to zero:

$$v_{\text{top}} \ll v_{\text{spigot}} \qquad \text{so} \qquad v_{\text{top}} \approx 0$$

Atmospheric pressure acts on both the top of the container and at the end of the spigot. This pressure is equal on both sides of the equation. It cancels, leaving:

$$g y_1 + \frac{1}{2} v_1^2 = g y_2$$

Assume the height of the spigot as the reference height and set it equal to zero. Then, solve for the velocity of fluid flow out of the spigot:

$$\frac{1}{2} v_1^2 = g y_2 \quad \text{or} \quad v = \sqrt{2 g y_2}_{\,1}$$

Same expression as for the conservation of mechanical energy into kinetic energy.

Measuring fluid flow through constricted pipe

Measuring fluid flow velocity through a pipe uses a constriction placed in the pipe. Two pressure meters are attached, one before the constriction and the other at the constriction.

If the areas of the pipe and the constriction are known, the velocity of the fluid flow through the pipe is solved by both the continuity and Bernoulli equation.

$$A_1 v_1 = A_2 v_2 \qquad\qquad \textit{continuity equation}$$

$$P_1 + \frac{1}{2}\rho v_1^2 = P_2 + \frac{1}{2}\rho v_2^2 \qquad \textit{Bernoulli equation}$$

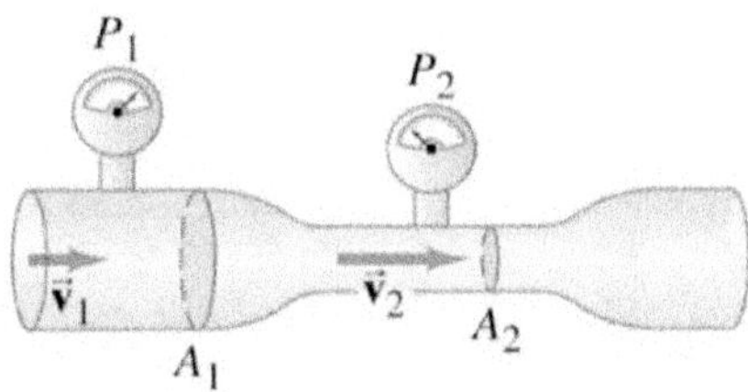

Flow through a constriction increases the velocity of the flow to conserve volume

Any duplication (copies, uploads, PDFs) is illegal.

Venturi effect and pivot tube

From the Bernoulli principle, higher velocity gas experiences a pressure drop. This is the Venturi effect, which measures fluid flow by measuring pressure differences.

Venturi effect (1797) describes the drop in fluid pressure that occurs when a fluid flows through a constriction in a pipe.

From the continuity principle, flow through a constriction in a tube increases the velocity of the flow to conserve volume.

Pitot tube is a flow measurement device that uses the Venturi effect to determine fluid flow velocity. It is commonly used to calculate an aircraft's airspeed.

Copyright © Sterling Test Prep.

Strain and Stress in Solids

Deformation of solids

Solids (e.g., ceramics, metals, and composites) undergo the process of *stress* and *strain*.

When a solid is subjected to an external force, it experiences *stress* (or deformation), measured in terms of *strain*. When a solid material has a force applied to it, it resists deformation (up to a maximum applied force) and retains its shape.

Deformation resistance is from material's stress, which applies a restoring force per unit area.

Stress has the form:

stress = force / area

Normal, shearing and volumetric strain

Strain (or *deformation*) of a solid material generates an internal elastic stress analogous to the reaction force of a spring, which tends to restore the material to its original non-deformed state.

Normal strain (or longitudinal strain) is the ratio of the change in length with respect to its original length. It is a longitudinal strain since it is concerned with the change in the length of the body in relation to its original length.

$$S_1 = \Delta V / V$$

Shearing strain is the angular distortion made by the body, so it is equal to the displacement of the body surface in relation to the body height, but keep in mind that the body surface is the one that bears the strain.

$$S_2 = l / H$$

Volumetric strain (or bulk strain) is the ratio is the volume change of a body compared to its original volume. It is a bulk strain because it considers the body's entire volume.

$$S_1 = \Delta V / V$$

Solids experience stress

Solid materials undergo three basic types of stress: *tension, compression,* and *shear stress*.

Stress and strain are *linearly proportional* in many materials when the stress is small.

The material obeys Hooke's Law and the ratio of stress to strain is the *elastic modulus*.

Hooke's Law only holds for a range of stresses, referred to as the elastic region.

These types of stress are important because they dictate how structures are built, the loads they can tolerate, and how the failure of materials may occur.

Any duplication (copies, uploads, PDFs) is illegal.

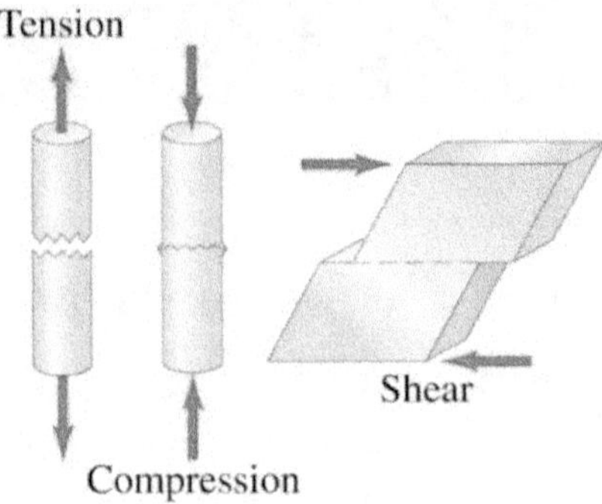

Applied stress includes tension, compression, and shear force

Tension

Tension force (or *tensile stress*) is transmitted through an object (e.g., string, rope, or wire) when pulled by forces acting from opposite ends. The tension force is directed along the length and pulls equally on *opposite ends* of the object.

Tensile stress is expressed as:

$$\sigma = F / A$$

where σ is the stress an object experiences (N/m^2), F is the force exerted on an object (N), and A is the area over which it is applied (m^2).

Compressive stress

Compressive stress results when a force acts to compress the material. The same formula expresses compressive stress as for tensile stress:

$$\sigma = F / A$$

where σ is the stress an object experiences (N/m^2), F is the force exerted on an object (N), and A is the area over which it is applied (m^2).

Shear stress

Shear stress occurs when two forces act opposite such that the material slides on a plane between the applied forces.

Shear stress is:

$$\tau = F/A$$

where τ is the stress an object experiences (N/m^2), F is the force exerted on an object (N), and A is the area over which it is applied (m^2).

Copyright © Sterling Test Prep.

Elasticity: Elementary Properties

Elastic properties influence shape distortions

Elasticity is an essential property in tensile and compressive stress.

When a material undergoes tensile or compressive deformation, elasticity measures its ability to return to its *original shape after distortion*.

For example, rubber bands have high elasticity because, after moderate stretching, a rubber band returns to its original shape.

Conversely, if a low-elasticity material is deformed, such as a piece of ceramic pottery, it experiences slight deformation before snapping.

Elastic properties determine the behaviors of these two materials.

Stretching and compression

When materials undergo *tension*, they stretch according to the force acting upon them. As mentioned earlier, all materials have some degree of elasticity.

After exceeding this inherent value, they stretch according to the tensile stress applied.

When a solid material undergoes tension, its elasticity is described as *ductile* or *brittle*.

Material is *ductile* if it tends to deform (stretch) under tensile stress.

However, if the material undergoes slight deformation before breaking, the material is *brittle*.

Compression is the inverse of tension and occurs when force points inward on a material, compacting it. Similarly, if the applied force is high enough, the material deforms.

Malleability is the propensity for a material to deform under compression.

Young's Modulus measures elasticity

Material can be compressed and deformed easily if it is highly *malleable*.

Non-malleable materials break rather than deform.

The two properties that indicate the elasticity of material in tension and compression are strain and Young's modulus.

Strain, given by the lowercase Greek letter epsilon (ε), is the change in length of an object in the direction of the applied force, divided by the original length.

$$\varepsilon = \frac{\Delta l}{l_0}$$

where ε is the strain, Δl is the change in length (m), and l_0 is the original length (m).

Young's modulus (elastic modulus) is a measure of the force per unit needed to stretch material.

Young's modulus is expressed as the ratio of stress divided by strain:

$$E = \frac{\sigma}{\varepsilon} = \frac{F l_0}{A \Delta l}$$

Most structural materials are considered somewhat elastic, and the behavior of these materials under *tensile* or *compressive stress* is important.

Strength and breaking point

A general force *vs.* elongation graph of an elastic material undergoes *tension* or *compression*:

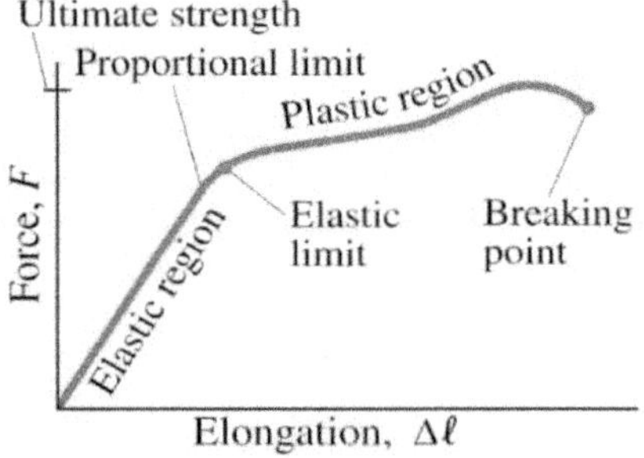

Force vs. elongation (Δl) graph shows an elastic and plastic region

Elastic materials under stress display a pattern of force *vs.* elongation (or compression), as seen above. When the maximum amount of stress is applied to the material, it no longer deforms *elastically* but deforms *plastically* (i.e., permanent change in shape without fracture).

Copyright © Sterling Test Prep.

Elastic limit and shear modulus

Elastic limit (yield strength) is the maximum stress a material can sustain before all subsequent deformation becomes permanent (i.e., plastic deformation).

After passing the elastic limit, the material continues to deform plastically until the ultimate strength is reached, after which the material breaks apart into two pieces.

Like tension and compression, all materials resist shearing to a certain degree. This resistance is governed by the shear strain and the *shear modulus*, which, like Young's modulus, is used to measure the stiffness of materials.

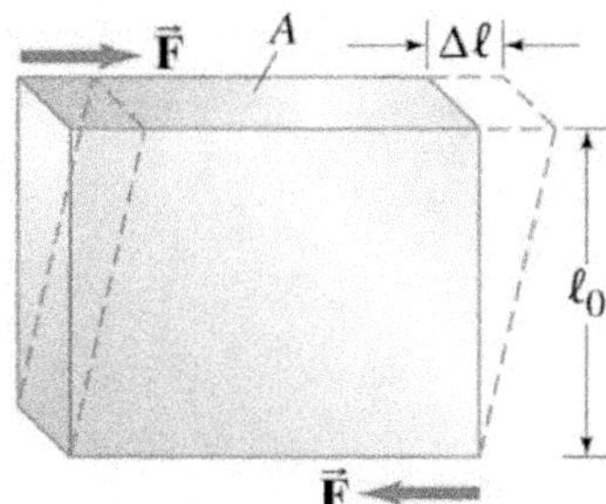

Block undergoing shearing forces and deforming according to shear forces

The shear strain is given by:

$$\varepsilon_{shear} = \frac{\Delta l}{l_0}$$

Shear Modulus can be written as:

$$S = \frac{\tau}{\varepsilon_{shear}}$$

$$\frac{\tau}{\varepsilon_{shear}} = \frac{F l_0}{A \Delta l}$$

$$S = \frac{F l_0}{A \Delta l}$$

When an object undergoes shear forces, the *shear strain* and *shear modulus* are calculated with Δl and l_0, which are perpendicular.

Notes for active learning

Copyright © Sterling Test Prep.

Practice Questions

1. The Bernoulli effect describes the lift force on an airplane wing. Wings must be designed to ensure those air molecules:

   A. move more rapidly past the upper surface of the wing than past the lower surface
   B. flow around wings that are smooth enough for an easy flow of the air
   C. are deflected upward when they hit the wing
   D. are deflected downward when they hit the wing

2. Two horizontal pipes (A and B) are the same length, but pipe B has twice the diameter of pipe A. Water undergoes viscous flow in both pipes, subject to the same pressure difference across the lengths of the pipes. If the flow rate in pipe A is Q, what is the flow rate in pipe B?

   A. 2Q   **C.** 8Q
   B. 4Q   **D.** 16Q

Questions **3-5** are based on the following:

A pressurized cylindrical tank is 5 m in diameter. Water exits from the pipe at point C with a velocity of 13 m/s. Point A is 10 m above point B, and point C is 3 m above point B. The pipe's cross-sectional area at point B is 0.08 m^2, and the pipe narrows to a cross-sectional area of 0.04 m^2 at point C. Assume an ideal fluid in laminar flow. The density of water is 1,000 kg/m^3. (Use the acceleration due to gravity $g = 9.8$ m/s^2)

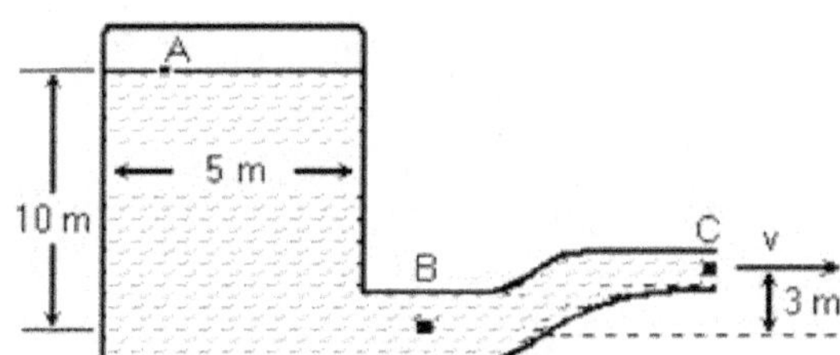

3. What is the mass flow rate in the pipe at point C?

   A. 320 kg/s   **C.** 570 kg/s
   B. 520 kg/s   **D.** 610 kg/s

4. What is the rate at which the water is falling into the tank?

   A. 12 mm/s   **C.** 86 mm/s
   B. 26 mm/s   **D.** 44 mm/s

Any duplication (copies, uploads, PDFs) is illegal.

5. What is the gauge pressure in the pipe at point B?

    A. 82 kPa   **C.** 71 kPa

    B. 167 kPa   **D.** 98 kPa

6. A closed cubical chamber resting on the floor contains oil and a piston. If the piston is pushed down hard enough to increase the pressure just below the piston by an amount ΔP, which of the statements is correct?

    A. The pressure at the top of the oil increases by less than ΔP

    B. The pressure on the sides of the chamber remains the same.

    C. The pressure in the oil increases by less than ΔP

    D. The increase in the force on the top of the chamber equals the increase in the force on the bottom of the chamber

7. What is the gauge pressure in the water at the deepest point of the Pacific Ocean, 11,030 m? (Use the density of seawater $\rho = 1{,}025$ kg/m^3 and the acceleration due to gravity $g = 9.8$ m/s^2)

    A. 1.1×10^8 Pa   **C.** 4.2×10^7 Pa

    B. 3.1×10^8 Pa   **D.** 7.6×10^7 Pa

8. A cubical box with 25 cm sides is immersed in a fluid. The pressure at the top surface of the box is 108 kPa, and the pressure on the bottom surface is 114 kPa. What is the density of the fluid? (Use acceleration due to gravity $g = 9.8$ m/s^2)

    A. 980 kg/m^3 **C.** 2,452 kg/m^3

    B. 1,736 kg/m^3 **D.** 2,794 kg/m^3

9. A polar bear of mass 240 kg stands on floating ice 100 cm thick. What is the minimum area of the ice that will support the bear? (Use specific gravity of ice = 0.98 and saltwater = 1.03)

    A. 2.6 m^2 **C.** 4.8 m^2

    B. 4.9 m^2 **D.** 11.2 m^2

Copyright © Sterling Test Prep.

10. A 14,000 N car is raised using a hydraulic lift. The lift consists of a U-tube with arms of unequal areas, initially at the same level. The lift is filled with oil with a 750 kg/m3 density with tight-fitting pistons at each end. The narrower arm has a radius of 6 cm, while the broader arm of the U-tube has a radius of 16 cm. The car rests on the piston on the broader arm of the U-tube. What force must be applied to the smaller piston to lift the car after it has been raised 1.5 m? (Ignore the weight of the pistons and use the acceleration due to gravity g = 9.8 m/s^2)

A. 4,568 N	**C.** 2,094 N
B. 3,832 N	**D.** 1,379 N

11. An object has a volume of 4.2 m^3 and weighs 41,800 N. What is its apparent weight in water? (Use acceleration due to gravity g = 9.8 m/s^2 and density of water ρ = 1,000 kg/m^3)

A. 1,140 N	**B.** 230 N	**C.** 800 N	**D.** 640 N

12. A solid sphere of mass 9.2 kg, made of metal whose density is 3,650 kg/m^3, hangs by a cord. When the sphere is immersed in a liquid of unknown density, the tension in the cord is 42 N. What is the density of the liquid? (Use acceleration due to gravity g = 9.8 m/s^2)

A. 1,612 kg/m^3	**C.** 1,950 kg/m^3
B. 1,468 kg/m^3	**D.** 1,742 kg/m^3

13. The tensile strength of a steel wire is $2,860 \times 10^6$ N/m^2. What is the maximum load that can be applied to a wire made of this steel if its diameter is 3.5 mm?

A. 46.4 kN	**C.** 27.5 kN
B. 19.7 kN	**D.** 58.5 kN

14. A grade of motor oil with a viscosity of 0.3 N·s/m2 flows through a 1 m long tube with a radius of 3.2 mm. What is the average speed of the oil if the drop in pressure over the length of the tube is 225 kPa?

A. 0.82 m/s	**C.** 1.2 m/s
B. 0.96 m/s	**D.** 1.4 m/s

15. Ice has a lower density than water because ice:

A. molecules vibrate at lower rates than water molecules
B. is made of open-structured, hexagonal crystals
C. is denser and, therefore, sinks when in liquid water
D. molecules are more compact than when in the solid state

Any duplication (copies, uploads, PDFs) is illegal.

Notes for active learning

Copyright © Sterling Test Prep.

Detailed Explanations

1. A is correct.

According to the Bernoulli effect, air moving with a higher velocity exerts less pressure against a surface than air with a lower velocity.

Thus, to produce lift, the pressure on the underside of a wing should be higher, and the air should be slower on the bottom surface of the wing compared to air on the top surface.

2. D is correct.

Poiseuille's Law:

$$Q = \pi \Delta P r^4 / 8\eta L$$

$$D_B = 2D_A$$

$$r_B = 2r_A$$

$$Q_A = \pi \Delta P r_A{}^4 / 8\eta L$$

$$Q_B = \pi \Delta P (2r_A)^4 / 8\eta L$$

$$Q_B = 16(\pi \Delta P r_A{}^4 / 8\eta L)$$

$$Q_B = 16 Q_A$$

3. C is correct.

Mass flow rate $\dot{m}$ is:

$$\dot{m} = \rho v A_C$$

where ρ = density of the fluid, v = velocity of flow, A_C = cross-sectional area

Dimensions of the tank are irrelevant to this answer, so $\dot{m}$ is calculated as:

$$\dot{m} = (1{,}000 \text{ kg/m}^3){\cdot}(13 \text{ m/s}){\cdot}(0.04 \text{ m}^2)$$

$$\dot{m} = 520 \text{ kg/s}$$

4. B is correct.

Relationship between area and velocity of both exists as:

$$A_1 v_1 = A_2 v_2$$

$$v_1 = A_2 v_2 / A_1$$

$$v_1 = [(0.04 \text{ m}^2){\cdot}(13 \text{ m/s})] / [(\pi / 4){\cdot}(5 \text{ m/s})]$$

$$v_1 = 0.26 \text{ m/s} = 26 \text{ mm/s}$$

Any duplication (copies, uploads, PDFs) is illegal.

5. D is correct.

Gauge pressure is the pressure reference against the surrounding air pressure. Because of this, it is valid to ignore both the pressure in the tank and atmospheric pressure.

$$P = \rho g h$$

$$P = (10^3 \text{ kg/m}) \cdot (9.8 \text{ m/s}^2) \cdot (10 \text{ m})$$

$$P = 98,000 \text{ Pa} = 98 \text{ kPa}$$

6. D is correct.

According to Pascal's Law, the pressure is transmitted undiminished in an enclosed static fluid. Thus, the pressure increases by ΔP everywhere in the oil. If the chamber is cubic, the top and bottom sides have the same area and experience the same increase in force:

$$A_{top} = A_{bottom}$$

$$\Delta P_{top} = \Delta P_{bottom}$$

$$P = F / A$$

Thus, force is directly proportional to pressure, and if the area of the top equals the bottom:

$$\Delta F_{top} = \Delta F_{bottom}$$

7. A is correct.

Gauge pressure is referenced relative to atmospheric pressure and is calculated by:

$$P_{gauge} = \rho g h$$

$$P_{gauge} = (1{,}025 \text{ kg/m}^3) \cdot (9.8 \text{ m/s}^2) \cdot (11{,}030 \text{ m})$$

$$P_{gauge} = 1.1 \times 10^8 \text{ Pa}$$

8. C is correct.

$$P_{top} = \rho g h$$

$$P_{top} = 108 \times 10^3 \text{ Pa}$$

$$h = (1 / \rho) \cdot (108 \times 10^3 \text{ Pa} / 9.8 \text{ m/s}^2)$$

$$h = (1 / \rho) \cdot (11{,}020 \text{ kg/m}^2)$$

$$P_{bottom} = \rho g (h + 25 \text{ cm})$$

$$\rho g (h + 25 \text{ cm}) = 114 \times 10^3 \text{ Pa}$$

$$h = (1 / \rho) \cdot (114 \times 10^3 \text{ Pa} / 9.8 \text{ m/s}^2) - 0.25 \text{ m}$$

$$h = (1 / \rho) \cdot (11{,}633) - 0.25 \text{ m} \qquad \qquad \textit{continued...}$$

Copyright © Sterling Test Prep.

Set equal and solve for ρ:

$$(1 / \rho) \cdot (11{,}020 \text{ kg/m}^2) = (1 / \rho) \cdot (11{,}633 \text{ kg/m}^2) - 0.25 \text{ m}$$

$$(11{,}020 \text{ kg/m}^2) = (11{,}633 \text{ kg/m}^2) - 0.25 \text{ m}(\rho)$$

$$0.25 \text{ m}(\rho) = 613 \text{ kg/m}^2$$

$$\rho = (613 \text{ kg/m}^2) / (0.25 \text{ m})$$

$$\rho = 2{,}452 \text{ kg/m}^3$$

9. C is correct.

Solve for density of ice and saltwater:

$$SG = \rho_{\text{substance}} / \rho_{\text{water}}$$

$$SG_{\text{ice}} = 0.98 = \rho_{\text{ice}} / 10^3 \text{ kg/m}^3$$

$$\rho_{\text{ice}} = 980 \text{ kg/m}^3$$

$$SG_{\text{saltwater}} = 1.03 = \rho_{\text{saltwater}} / 10^3 \text{ kg/m}^3$$

$$\rho_{\text{saltwater}} = 1{,}030 \text{ kg/m}^3$$

Solve for volume of ice:

$$F_B = F_{\text{bear}} + F_{\text{ice}}$$

$$\rho_{\text{saltwater}} V_{\text{ice}} g = m_{\text{bear}} g + \rho_{\text{ice}} V_{\text{ice}} g, \text{ cancel } g \text{ from all terms}$$

$$V_{\text{ice}} = m_{\text{bear}} / (\rho_{\text{saltwater}} + \rho_{\text{ice}})$$

$$V_{\text{ice}} = (240 \text{ kg}) / (1{,}030 \text{ kg/m}^3 - 980 \text{ kg/m}^3)$$

$$V_{\text{ice}} = 4.8 \text{ m}^3$$

Solve for area of ice:

$$A = V / h$$

$$A = (4.8 \text{ m}^3) / (1 \text{ m})$$

$$A = 4.8 \text{ m}^2$$

Any duplication (copies, uploads, PDFs) is illegal.

10. C is correct.

$$P_1 = P_2 + \rho g h$$

$$F_1 / A_1 = F_2 / A_2 + \rho g h$$

$$F_1 = A_1(F_2 / A_2 + \rho g h)$$

$$F_1 = \pi(0.06 \text{ m})^2 \cdot [(14{,}000 \text{ N}) / \pi(0.16 \text{ m})^2 + (750 \text{ kg/m}^3) \cdot (9.8 \text{ m/s}^2) \cdot (1.5 \text{ m})]$$

$$F_1 = \pi(0.0036 \text{ m}^2) \cdot [(14{,}000 \text{ N}) / \pi(0.0256 \text{ m}^2) + (750 \text{ kg/m}^3) \cdot (9.8 \text{ m/s}^2) \cdot (1.5 \text{ m})]$$

$$F_1 = (0.0036 \text{ m}^2) \cdot [(14{,}000 \text{ N}) / (0.0256 \text{ m}^2) + (11{,}025 \text{ N})\pi]$$

$$F_1 = 1{,}969 \text{ N} + 125 \text{ N}$$

$$F_1 = 2{,}094 \text{ N}$$

11. D is correct.

$$F_B = \rho V g$$

$$F_{net} = F_{object} - F_B$$

$$F_{net} = (41{,}800 \text{ N}) - (1{,}000 \text{ kg/m}^3)(4.2 \text{ m}^3)(9.8 \text{ m/s}^2)$$

$$F_{net} = (41{,}800 \text{ N}) - (41{,}160 \text{ N})$$

$$F_{net} = 640 \text{ N}$$

12. C is correct.

$$F_{net} = F_{object} - F_B$$

$$F_{net} = mg - \rho_{fluid} V g$$

$$\rho_{fluid} = -(F_{net} - mg) / V_{sphere} g$$

$$\rho_{fluid} = -[42 \text{ N} - (9.2 \text{ kg}) \cdot (9.8 \text{ m/s}^2)] / [(9.2 \text{ kg} / 3{,}650 \text{ kg/m3}) \cdot (9.8 \text{ m/s}^2)]$$

$$\rho_{fluid} = 1{,}950 \text{ kg/m}^3$$

13. C is correct.

Tensile strength $= F_{max} / A$

$$F_{max} = A(\text{tensile strength})$$

$$F_{max} = (\pi / 4) \cdot (0.0035 \text{ m})^2 \cdot (2{,}860 \times 10^6 \text{ N/m}^2)$$

$$F_{max} = 27{,}516 \text{ N} = 27.5 \text{ kN}$$

Copyright © Sterling Test Prep.

14. B is correct.

By Poiseuille's Law, the volumetric flow rate of a fluid is given by:

$$V = \Delta P A r^2 / 8\eta L$$

The volumetric flow rate is the volume of fluid that passes a point per unit time:

$$V = Av$$

where v is the speed of the fluid.

Therefore:

$$Av = \Delta P A r^2 / 8\eta L$$

$$v = \Delta P r^2 / 8\eta L$$

$$v = (225 \times 10^3 \text{ Pa}) \cdot (0.0032 \text{ m})^2 / [8 \, (0.3 \text{ Ns/m}^2) \cdot (1 \text{ m})]$$

$$v = 0.96 \text{ m/s}$$

15. B is correct.

For most substances, the solid form is denser than the liquid phase. Therefore, a block of most solids sinks in the liquid. For pure water, a block of ice (solid phase) floats in liquid water because ice is less dense than liquid water.

Like other substances, when liquid water is cooled from room temperature, it becomes increasingly dense. However, at approximately 4 °C (39 °F), water reaches its maximum density, and as it is cooled further, it expands the distance between molecules and becomes less dense.

This phenomenon is negative thermal expansion and is attributed to strong intermolecular interactions that are orientation-dependent.

Density of water is about 1 g/cm^3 and depends on the temperature. When frozen, the density of water is decreased by about 9%. This decreased density of water is due to the decrease in intermolecular vibrations, which allows water molecules to form stable hydrogen bonds with other water molecules around.

As hydrogen bonds form, molecules lock into positions as hexagonal structures.

Even though hydrogen bonds are shorter in the crystal than in liquid, this position locking decreases the average coordination number of water molecules as the liquid reaches solid phase.

Any duplication (copies, uploads, PDFs) is illegal.

Notes for active learning

Copyright © Sterling Test Prep.

CHAPTER 4

Chemical Bonding

- Electrostatic Forces Between Ions

- Electron Pair Sharing

- *Sigma* and *Pi* Bonds

- Hybridization

- Valence Bond (VB) Theory

- Molecular Orbital (MO) Theory

- Formal Charge

- Resonance

- Partial Ionic Character

- Intermolecular Forces

- Polarity

- Practice Questions & Detailed Explanations

Copyright © Sterling Test Prep.

Electrostatic Forces Between Ions

Ionic bond formation

An *ionic bond* forms when electrons transfer from one atom to another. This electron transfer results in oppositely charged species that attract each other via electrostatic interaction.

Atoms gain and lose electrons and become charged, known as *ions*.

Nonmetals (i.e., right side of the periodic table) tend to gain electrons and form negatively charged anions.

Metals (left side of the periodic table) exhibit weaker nuclear forces on valence electrons and lose electrons forming positively charged cations.

Ionic bonds form between a metal cation and a nonmetal anion so that each atom obtains a complete *valence shell*.

Ionic bonds *transfer an electron* from the electropositive to the electronegative element.

An ionic bond occurs if the *electronegativity difference* (> 1.6 D) is significant (table below).

Ionic bonds generally occur between a metal (left two columns; groups 1 and 2) and a nonmetal (right side; generally, groups 16 and 17).

If a metal has low ionization energy and a nonmetal has a high electron affinity (discussed elsewhere), ionic bonding is more likely.

A familiar ionic compound is sodium chloride (NaCl), or table salt.

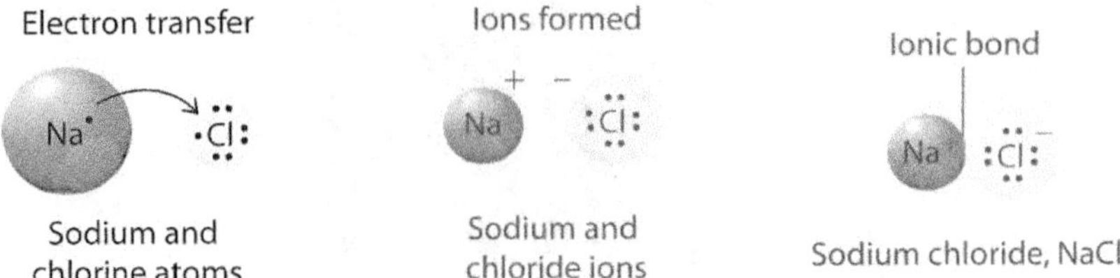

The cations and anions in an ionic compound are like a magnet's North and South poles, which attract.

Similarly, two cations (or anions) repel.

Any duplication (copies, uploads, PDFs) is illegal.

Crystalline structures

Therefore, the molecules arrange into rigid crystalline structures.

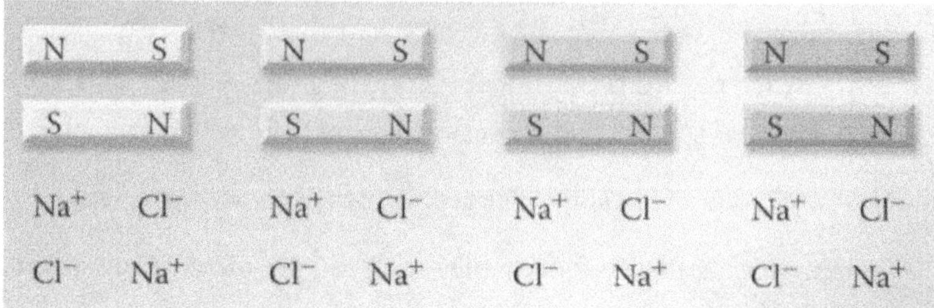

Crystalline Na$^+$ and Cl$^-$ with alternating positive and negative charges

Due to the strong, polarized, intermolecular forces, ionic compounds generally possess high melting and boiling points. Therefore, they are often solid at room temperature.

Ionic crystals are three-dimensional arrangements of ions in an ionic compound, sometimes called a *crystal lattice* (e.g., solid crystalline structure of sodium chloride).

Sodium chloride (NaCl) has a high melting point (800 °C) and dissolves in water to give a conducting solution. Like ionic compounds, sodium chloride is a strong electrolyte, dissociating entirely in water.

Solid ionic compounds are non-conductive; however, molten compounds (i.e., compounds liquefied by heat) or compounds dissolved in water conduct electricity.

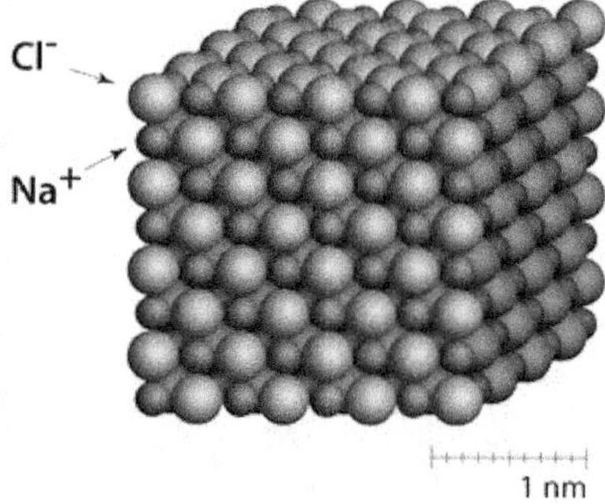

The crystalline lattice structure of table salt (NaCl)

Like all types of bonding, the valence (outermost) electrons are involved in ionic bonding.

Transferring the lone $3s$ electron of a sodium atom to the half-filled $3p$ orbital of a chlorine atom generates a sodium cation and a chloride anion, forming the compound sodium chloride.

Initially, the sodium and chlorine atoms each had a net charge of 0. After the transfer of one electron took place, they had a charge of +1 and –1, respectively, due to the atom's imbalance between protons and electrons.

Copyright © Sterling Test Prep.

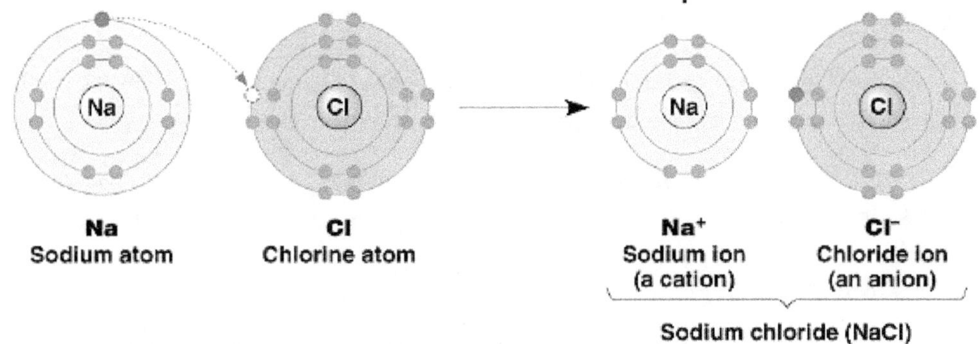

Ionic bonds form by the transfer of a valence electron

Electrostatic potential energy $\propto q_1 q_2 / r$

The strength of an ionic bond is related to the amount of *electrostatic potential energy* between opposing charges.

The *electrostatic potential energy* is:

$$U_E = k q_1 q_2 / r$$

where U_E is electrostatic potential energy, k is Coulomb's constant (9×10^9 m/F), q_1 and q_2 are the two charges, and r is the distance between them

Electrostatic energy is negative in ionic bonds because q_1 and q_2 have opposing charges.

If q_1 and q_2 did not have opposing charges (i.e., both positive or negative), they repel, and no ionic bond forms.

The negative sign is often dropped, and the magnitude of the electrostatic energy is used.

The higher the magnitude of the electrostatic potential energy, the stronger the ionic bond.

Strong ionic bonds are formed by high charge magnitudes (q values) close together (small r value).

Ions that form strong ionic bonds have a high charge density; the charge-to-size ratio is high.

A cation's high charge density can distort the anion's electron cloud to promote electron sharing.

Any duplication (copies, uploads, PDFs) is illegal.

Lattice energy

Lattice energy is the energy required to break an ionic bond. The energy released by the reaction correlates to the ionic bond strength, proportional to electrostatic energy.

Conversely, it is the energy released when oppositely charged; gaseous ions join to form an ionic crystalline solid.

Electrostatic force $\propto q_1 q_2 / r^2$

The electrostatic force that holds charged particles is defined as *electrostatic energy*.

Coulomb's Law defines electrostatic force F with electrostatic charges and distance.

$$F = kq_1q_2 / r^2$$

where F is the electrostatic force, k is Coulomb's constant (9×10^9 m/F), q_1 and q_2 are the two charges, and r is the distance between the charges

From the equation, the electrostatic force is directly proportional to the product of opposite charges that attract (negative F) or the same charges that repel (positive F).

The electrostatic force is inversely proportional to the square of the distance between them.

Therefore, larger charge magnitudes closer exhibit a greater electrostatic force.

Coulomb's Law is analogous to the *universal law of gravitation*:

$$F = Gm_1m_2 / r^2$$

where G (gravitational constant) is analogous to k, and m (mass) to q

The major difference is that G is tiny compared to coulomb's constant k (stationary electrically charged particles) because the gravitational force is weak compared to the stronger electrostatic force between charged particles.

There is another way to write the formula for an electrostatic force specific to ionic compounds.

The equation describes the force of attraction between the cation n^+ and the anion n^- at a distance d apart.

$$F = R(n^+ \times e) \cdot (n^- \times e) / d^2$$

where R is coulomb's constant (usually written as k)

$n^+ \times e$ = charge of cation in coulombs = positive charge (n^+) × coulombs per electron (e)

$n^- \times e$ = charge of anion in coulombs = negative charge (n^-) × coulombs per electron (e)

Copyright © Sterling Test Prep.

Electron Pair Sharing

Covalent bond formation

A *covalent bond* forms when two atoms share an electron pair.

These electron pairs are shared, or bonding pairs and covalent bonds result from the overlap of electron orbitals.

Covalent bonds typically form between nonmetal elements.

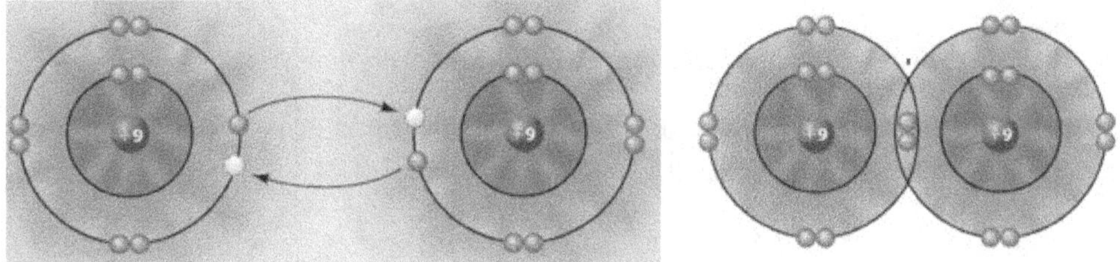

A covalent bond forms between two fluorine atoms. The left image shows the valence electron from each fluorine atom that becomes shared to form the covalent bond in the F_2 molecule

Most atoms with similar electronegativity share electrons to achieve a full valence shell of eight electrons (i.e., octet rule for $2s$ and $6p$ electrons).

Illustrations of covalent bonds following the octet rule are shown below.

Electron dot formulas

Simple atomic notation with dots designating valence electrons:

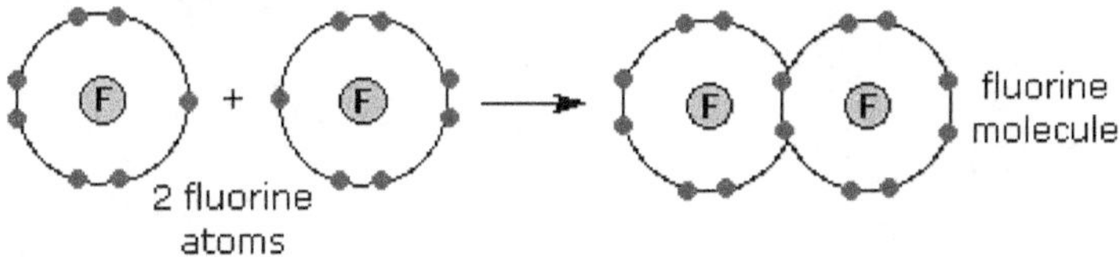

Covalent bond forms between 2 fluorine atoms that share a valence electron

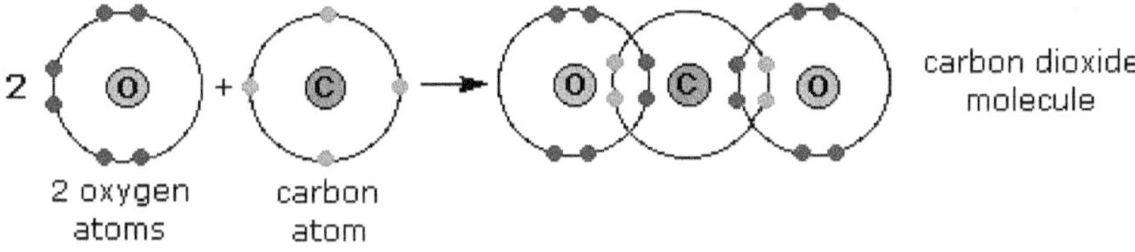

Covalent bond forms between 2 O atoms and one carbon atom that share two pairs of valence electrons from the carbon atom to form CO_2

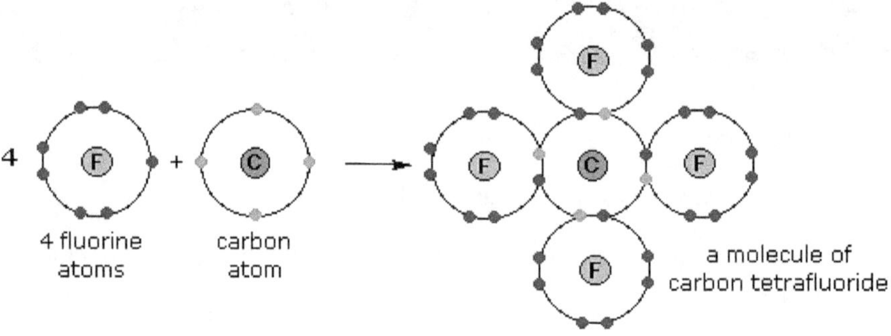

Covalent bond forms between 4 fluorine atoms and one carbon atom

Each atom contributes one valence electron to form carbon tetrafluoride (CF_4)

The hydrogen atom is an exception to the octet rule because it is at its lowest energy when it has two $1s$ electrons in its valence shell.

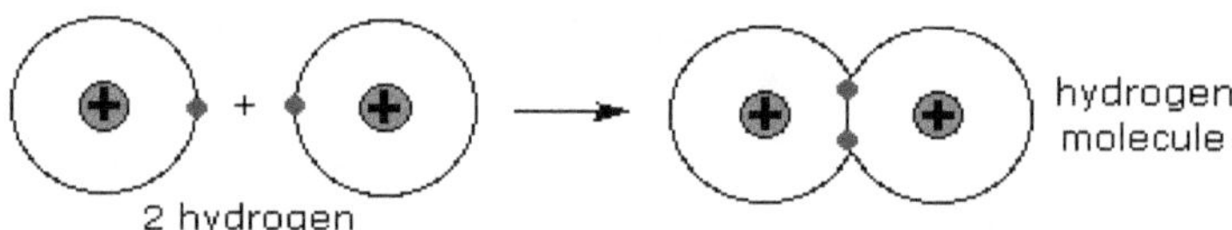

Atoms form covalent bonds to obtain the ideal number of electrons.

Non-bonding lone pairs

The *non-bonding lone pairs* are electrons not involved in bonding.

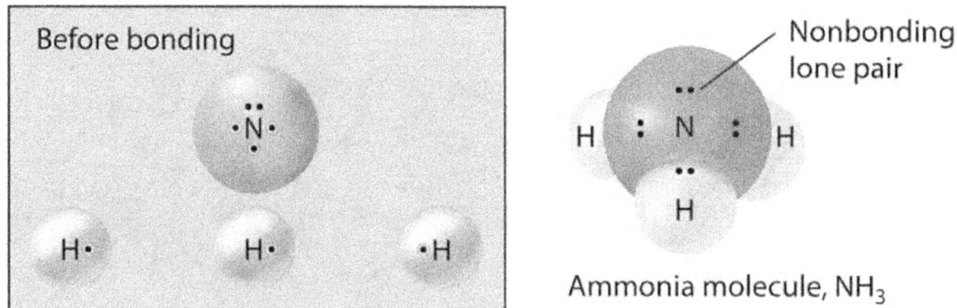

Nitrogen has 5 valence electrons. Nitrogen shares one valence electron with each of three
hydrogen atoms. Each hydrogen contributes 1 valence electron. With three covalent bonds,
nitrogen has a complete octet, and hydrogen satisfies a complete 1s subshell

---

   Copyright © Sterling Test Prep.

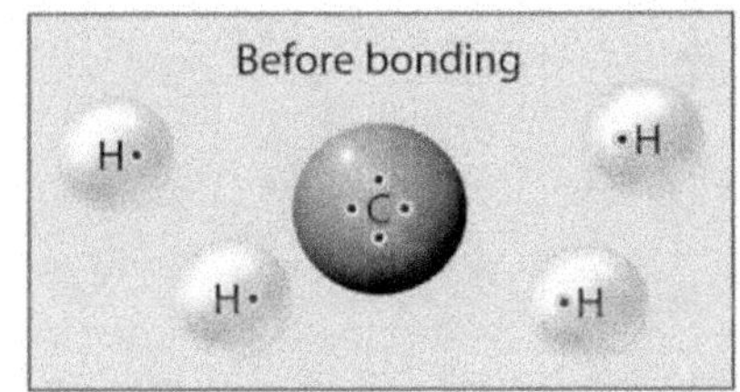

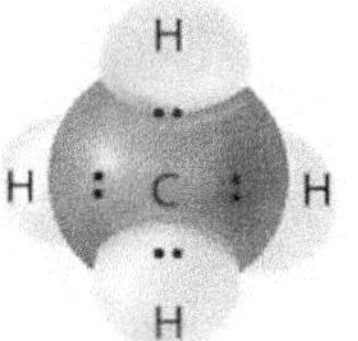

Methane molecule, CH_4

Carbon has 4 valence electrons and shares one electron with four H atoms.

Each H contributes 1 valence electron.

With four covalent bonds, carbon has a complete octet, and H satisfies a complete 1s subshell

Multiple bonds

Covalent bonds in double- and triple-bonded nonmetal atoms involve more than one pair of electrons.

In double bonds, *two pairs* of electrons (four electrons) are shared.

In triple bonds, *three pairs* of electrons (six electrons) are shared.

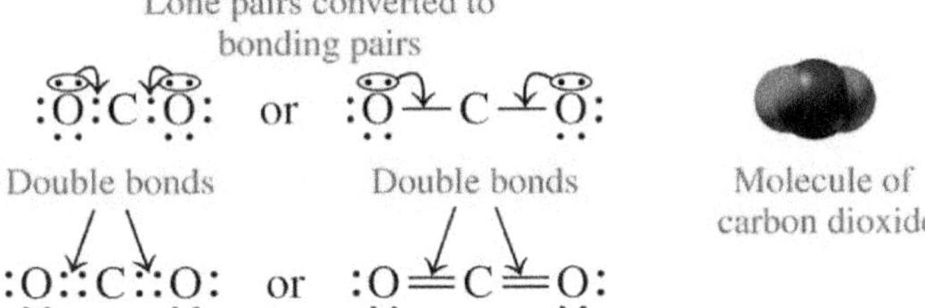

Carbon dioxide forms from two double bonds on the central carbon. Each double bond results from the oxygen and the carbon contributing a pair of electrons

$$·\ddot{N}:\ddot{N}· \longrightarrow :N::N: \quad :N\equiv N:$$

Three shared pairs   *Triple bond*   *N_2 molecule*

Coordinate covalent bonds

Coordinate covalent bonds result from one atom donating an electron pair to another atom; both electrons come from one donor atom.

Ozone, O_3, is a molecule with a coordinate covalent bond.

$$:\ddot{O}::\ddot{O}: + \ddot{O}: \longrightarrow :\ddot{O}::\ddot{O}:\ddot{O}:$$

nonbonding electron pairs   *coordinate covalent bond: O=O–O*

Any duplication (copies, uploads, PDFs) is illegal.

Notes for active learning

Copyright © Sterling Test Prep.

Sigma and *Pi* Bonds

Single bonds

The electron density of a chemical bond can be divided into regions.

Sigma (σ) *bonds* form when one orbital from each atom overlaps and creates a new orbital.

Typically, a *sigma* bond is a single bond and is the first bond involved in double and triple bonds. The sigma bonds lie along the imaginary line joining two nuclei, the *internuclear axis*.

In σ bonds, the electron density is between the two atoms' nuclei and is symmetrical about the internuclear axis.

Atoms on each side of a σ single bond are free to rotate with the bond as their axis.

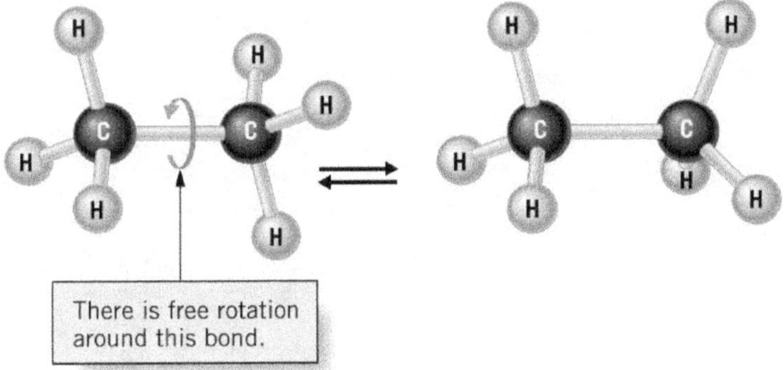

S and *p* orbitals

In the image below, the σ bonds of *s*–*s* and *p*–*p* orbitals are shown.

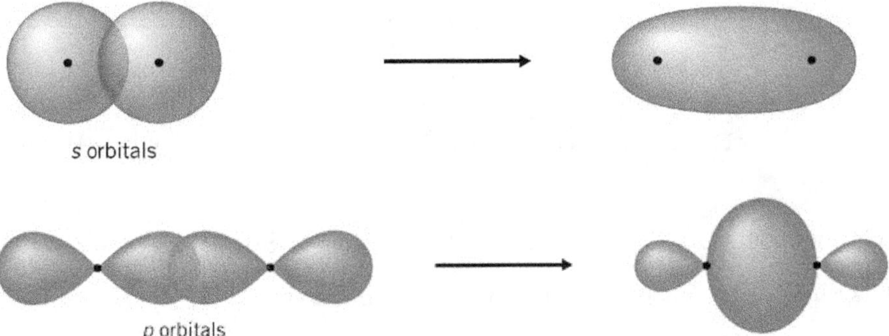

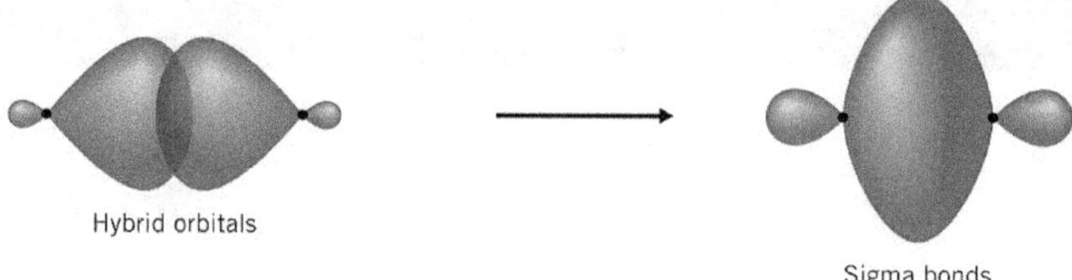

Double bonds

Pi (π) bonds form when unhybridized *p* orbitals on the top and bottom of the internuclear axis overlap.

A single π bond consists of regions above and below the intermolecular axis.

There is no electron density along the π bond intermolecular axis.

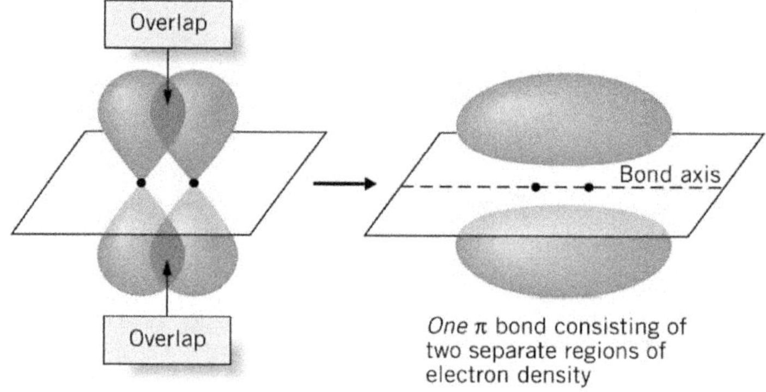

When electron density above and below the σ bonds overlap, a multiple bond (i.e., double or triple) forms. Thus, π bonds form after a σ bond has connected two atoms.

A double bond: one σ bond and one π bond.

A triple bond: one σ bond and two π bonds.

Nitrogen (N_2) gas is connected by a triple bond ($N\equiv N$). The first is a σ bond, and the remaining two are π bonds.

The π bond introduces rigidity to molecules; π bonds do not allow for the rotation of atoms.

Molecules that involve π bonds require more energy to break.

Copyright © Sterling Test Prep.

Hybridization

Hybrid orbitals

Hybrid orbitals are produced by the hybridization (mixing) of existing electron orbitals.

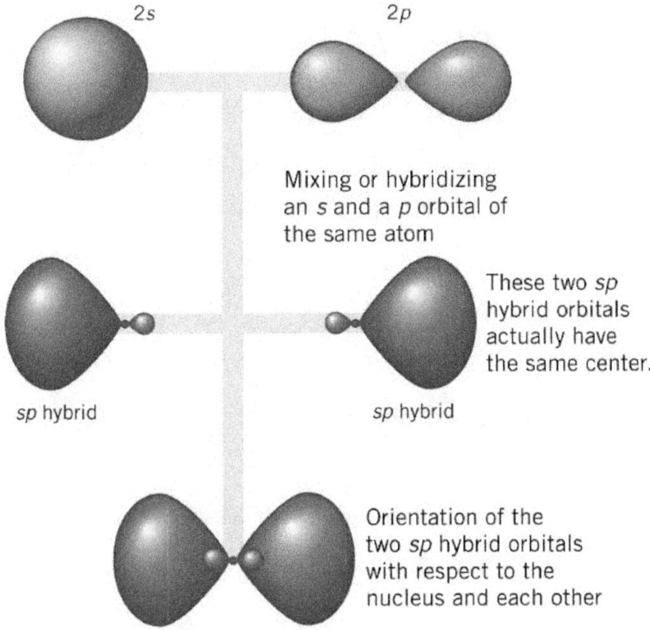

Hybridization and molecular geometry

The possible hybridizations of the carbon atom are as follows.

- sp^3: a hybrid between one s with three p orbitals.

  Tetrahedral geometry.

  Contains single bonds only.

- sp^2: a hybrid between one s with two p orbitals.

  Trigonal planar geometry.

  Contains a double bond.

- sp: a hybrid between one s with one p orbital.

  Linear geometry.

  Contains a triple bond.

Any duplication (copies, uploads, PDFs) is illegal.

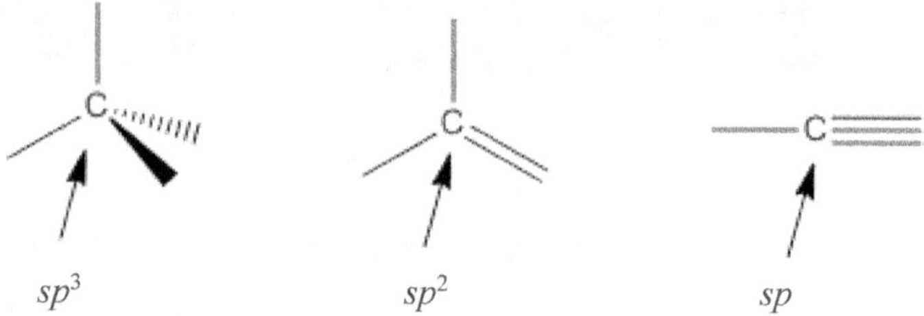

Hybridization of the carbon atom for 4 single bonds, 1 double or 1 triple bond

Valence shell electron-pair repulsion (VSEPR) theory predicts molecular shapes

Valence Shell Electron Pair Repulsion (VSEPR) theory states that electron pairs surrounding an atom repel each other and have the greatest bond angles separating the bonds and lone pairs around a central atom.

In VSEPR, *electron domains* are electron pairs in the molecule.

Molecules have *bonding* and *non-bonding domains*.

When atoms bond to create molecules, the atoms arrange as far apart to minimize the same-charge repulsion between electrons.

VSEPR theory predicts the *molecular geometry* based on the *number of electron pairs*.

The *electron pair domain geometry* (EDG) indicates bonding and non-bonding electron pairs around the central atom.

The *molecular shape geometry* (MG) indicates atoms' arrangements around the central atom after accounting for electron repulsion.

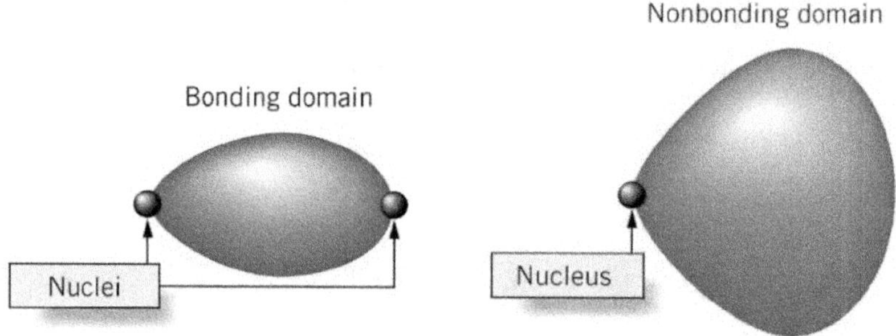

Bonding domains are directed between two adjacent nuclei, while the antibonding domains are directed away from adjacent nuclei

Copyright © Sterling Test Prep.

Shape and bond angle

Specific configurations (i.e., specific shapes) maximize the distance between the bonding electron pairs and substituents attached *via* the bond.

Molecules with *two* attached atoms (or substituents) are *linear* (a straight line, with one atom on each side of the central atom).

Bond angle = 180°

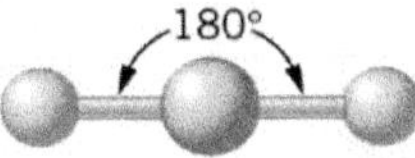

Molecules with *three* attached atoms are *trigonal planar*.

Bond angle = 120°

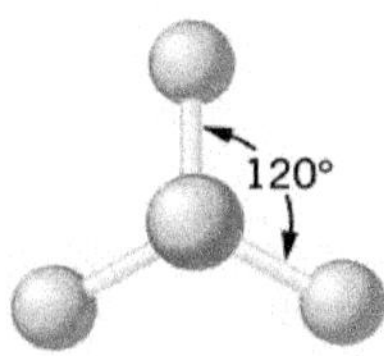

Molecules with *four* attached atoms are a *tetrahedron*.

Bond angle = 109.5°

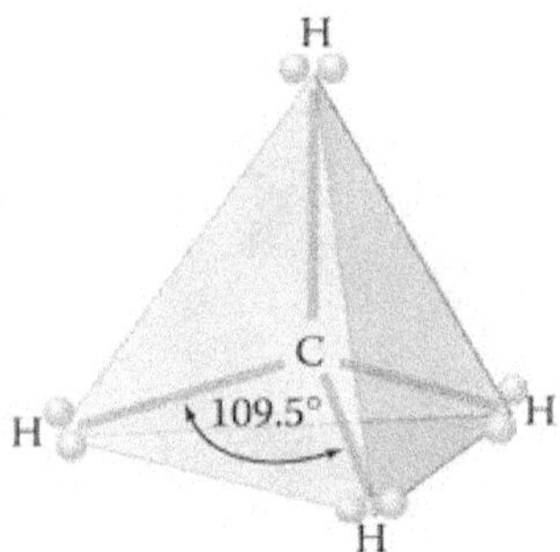

Molecules with *five* attached atoms are *trigonal bipyramidal*.

Bond angles = 120° and 90°

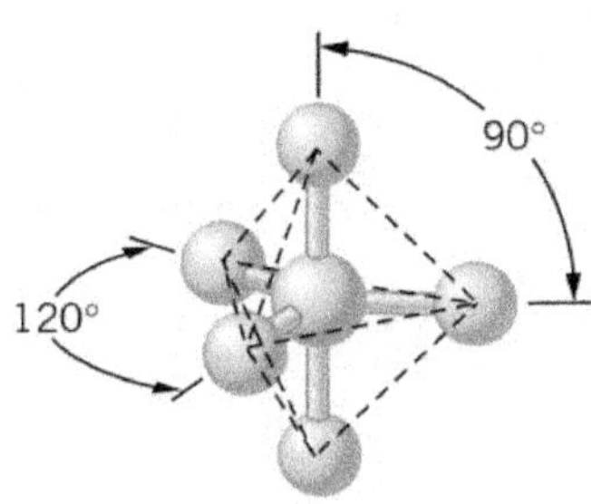

A molecule with *six* attached atoms is an *octahedron*.

Bond angle = 90°

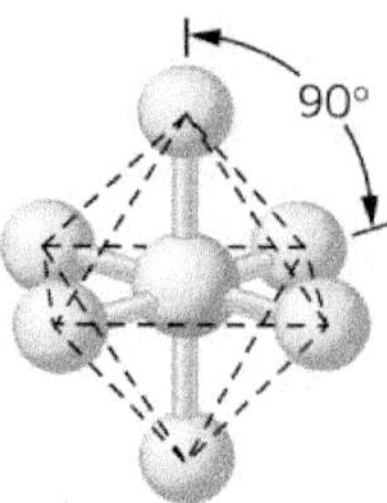

The molecular shapes above apply to molecules containing only bonding domains (i.e., central atoms do not have non-bonding domains or lone pairs of electrons).

Non-bonding domains

When the central atom has non-bonding domains, the lone pair occupies the space where a bond usually forms.

These non-bonding electron domains are relatively dispersed in space and exert electrostatic repulsions that cause bonding electrons to bend away.

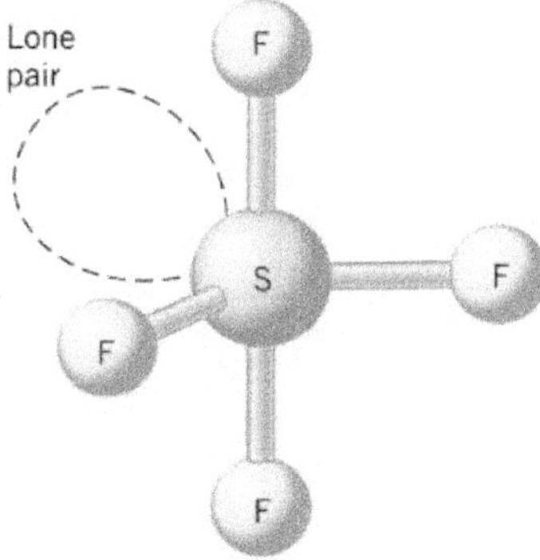

Lone pairs must be closer to the atom whose orbital they occupy than the bonding pairs shared along the two bonding atoms' internuclear axis.

Thus, lone pairs are not restricted to the internucleus axis occupy more available space within the orbital (i.e., more electrostatic repulsion).

Copyright © Sterling Test Prep.

An example is SF$_4$, which has one lone pair:

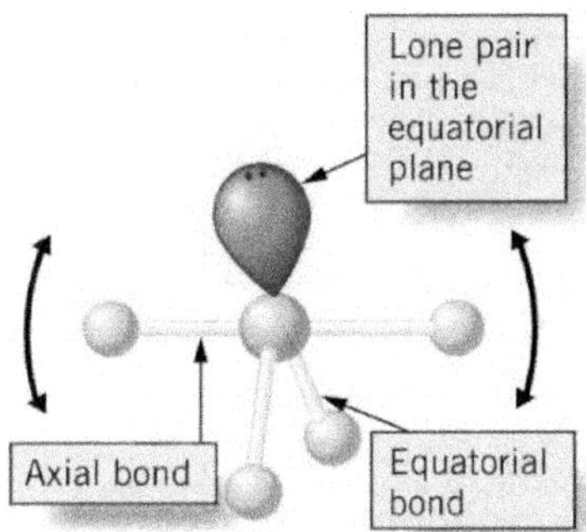

Electrons in sigma bonds occupy a restricted space while nonbonding lone pairs are more dispersed and therefore contribute electrostatic repulsion to the bonding electrons

Hybridization, bond angles, and molecular geometry rules

Hybridization	Electron geometry	Electron groups	Bonding groups	Lone pairs	Approx. bond angles	Molecular geometry
sp	Linear	2	2	0	180°	Linear
sp²	Trigonal planar	3	3	0	120°	Trigonal planar
		3	2	1	< 120°	Bent
sp³	Tetrahedral	4	4	0	109.5°	Tetrahedral
		4	3	1	< 109.5°	Trigonal pyramidal
		4	2	2	<< 109.5°	Bent
sp³d	Trigonal bipyramidal	5	5	0	120° (equatorial) 90° (axial)	Trigonal bipyramidal
		5	4	1	< 120° (equatorial) < 90° (axial)	Seesaw
		5	3	2	< 90°	T-shaped
		5	2	3	180°	Linear
sp³d²	Octahedral	6	6	0	90°	Octahedral
		6	5	1	< 90°	Square pyramidal
		6	4	2	90°	Square planar

Copyright © Sterling Test Prep.

Determining the molecule's geometric shape

1. Start by counting the domains (bonding and non-bonding) around the central atom.

   The number of domains determines the molecule's general shape (discussed above).

   For every non-bonding domain, remove a surrounding atom.

   For example, a four-domain molecule has a tetrahedral shape.

   If one of the domains is non-bonding, one of the surrounding atoms is removed, leaving a central atom with three surrounding atoms; the molecule has a trigonal pyramid shape.

   If there are two non-bonding domains, remove two surrounding atoms; the molecule is a bent shape.

2. For molecules with five domains, start by removing atoms above and below the pyramid before the 3 atoms surrounding it.

   The molecule is more stable when the atoms adopt a configuration with the least repulsion.

   The angles between the top or bottom atoms and the three side atoms are $90°$.

   The angles between the side atoms are $120°$.

   Thus, the molecule sheds the more repulsive atoms first (i.e., ones with smaller angles).

   The same approach applies to molecules with six domains.

   The domains above and below are removed before the central atom domains.

Non-bonding domains affect the angle of bonds in the molecule due to occupying more significant space (increased electrostatic repulsion) than bonded electrons.

The following diagram summarizes the shapes of atoms based on the number of bonding and non-bonding electron pairs (also called nonbonded electron pairs):

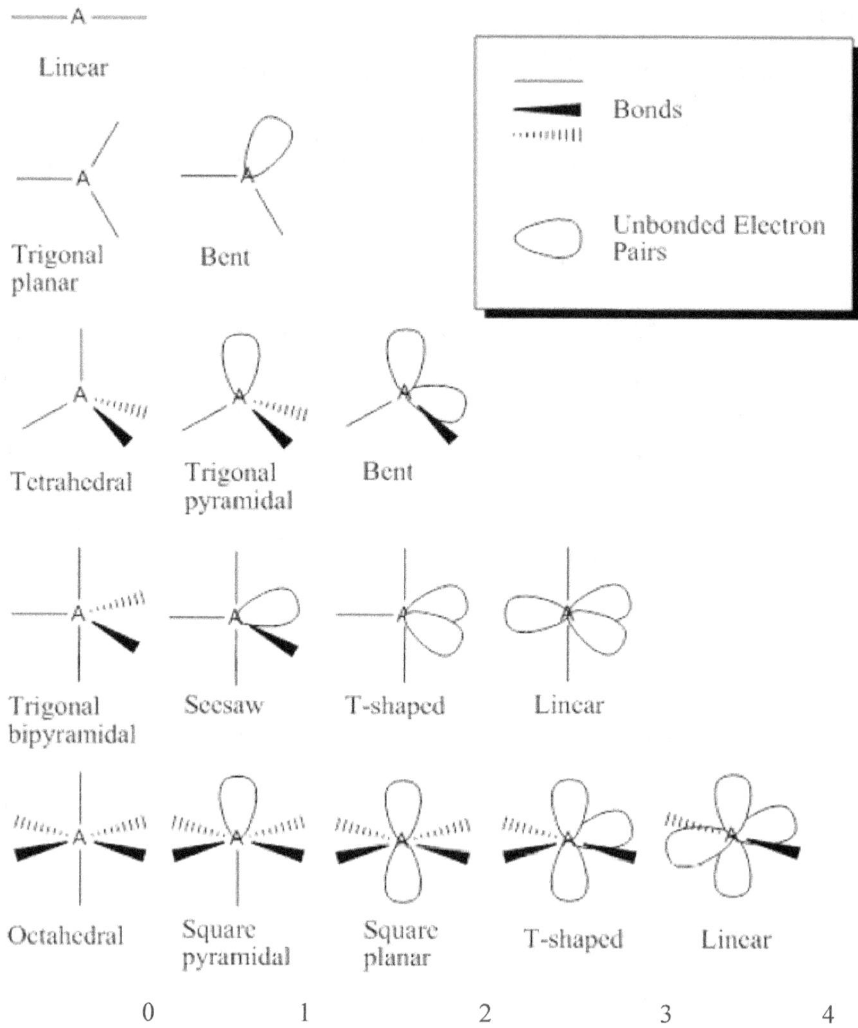

Numbers along the bottom indicate the # of nonbonded electron pairs

The oxygen of water (H_2O, shown below) has two bonding and two non-bonding domains. The water molecule experiences electron repulsion from the two lone pairs on its bonds.

The water molecule adopts a bent shape (see the previous chart) to minimize the van der Waals repulsion of the negatively charged electrons.

Water's bent shape comes from a tetrahedral configuration, with two vertices removed.

Bonds in a tetrahedron are 109.5°, which means bonds in a bent molecule should be 109.5°.

However, because of the repulsion between lone electron pairs and the bonding electrons, the reported bond angle is slightly decreased to approximately 105°.

Copyright © Sterling Test Prep.

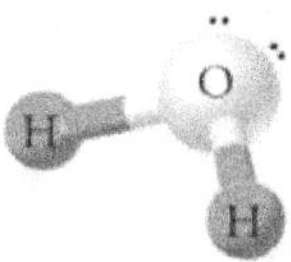

A water molecule (H₂O) is bent due to the two lone pairs of e⁻ on oxygen

VSEPR geometric shapes classify double bonds as a single domain.

Formaldehyde (CH$_2$O) has one double and two single bonds, adding to three bonding domains.

The molecular geometry of formaldehyde is trigonal planar (i.e., flat shape with ~120 bonds).

Trigonal plane geometry for formaldehyde (CH₂O) with 1 double (π) and 2 single (σ) bonds

Two major theories explain how molecules bond: valence bond theory and molecular orbital theory. Both theories are used to predict the structure of a molecule.

Notes for active learning

Copyright © Sterling Test Prep.

Valence Bond (VB) Theory

Atomic orbital (AO) theory

The *Valence bond theory* (VB theory) is primarily based on valence electrons and states that each atom has an orbital, and *orbitals overlap to form bonds*.

The extent of overlap of atomic orbitals (AO) is related to bond strength; more overlap results in a stronger bond.

The VB theory explains the bonding of F_2. F_2 bonds form from atomic valence orbitals overlap (e.g., $2p$ orbitals), one from each fluorine.

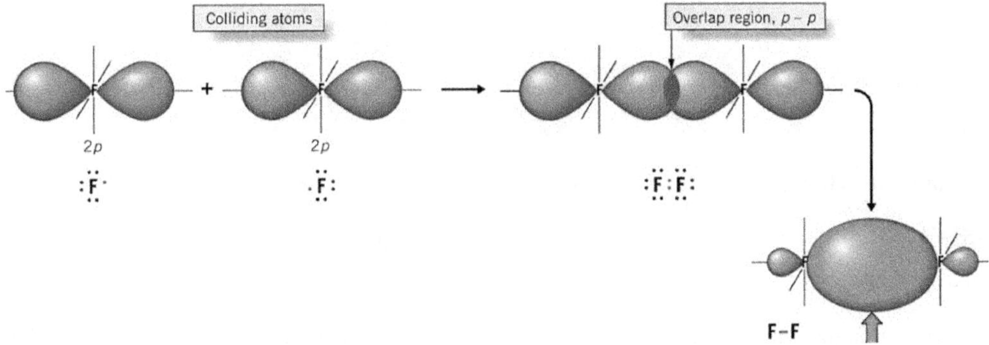

The overlap of p–p orbitals form the F_2 bond

The orbital overlap is not limited to identical orbitals.

Bonding in HF molecules uses the overlap between the $1s$ orbital of H and the $2p$ orbital of F.

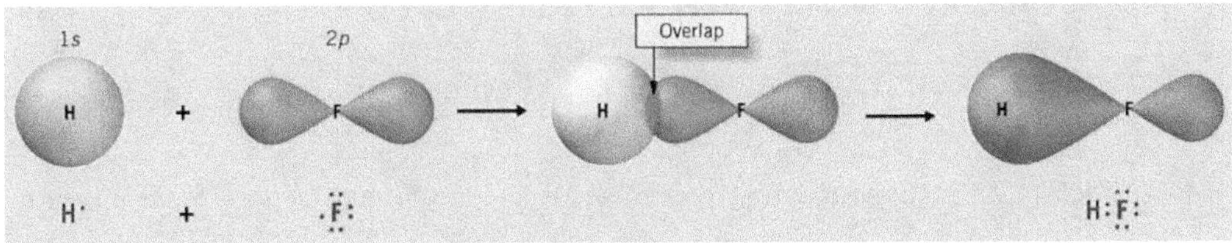

The bonding of H_2S can be described according to VB theory. Sulfur has one filled orbital (contains two electrons) and two partially filled orbitals (contains one electron).

The partially filled p orbitals overlap with hydrogen's s orbital.

The predicted 90° bond angle is close to the measured value of 92°.

Any duplication (copies, uploads, PDFs) is illegal.

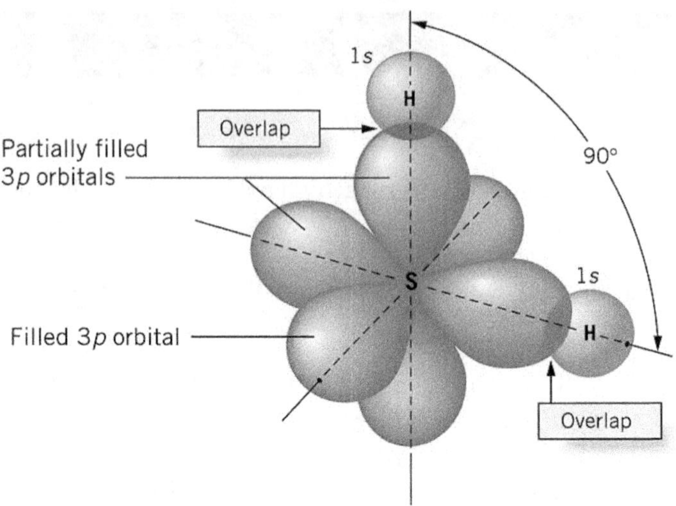

VB theory describes many molecules' bonding, but it cannot explain existing molecular geometries.

Methane (CH_4) has a tetrahedral shape, seen in the image below.

Carbon's electron configuration ($1s^2 2s^2 2p^2$) indicates that its electrons are paired, except for two electrons that occupy p orbitals.

VB theory predicts that carbon can only bond with two hydrogen molecules, resulting in a CH_2 molecule and that the bond angle is close to $90°$ (like H_2S above).

Observations of methane reveal four equal bonds in the shape of a tetrahedral with bond angles of $109.5°$.

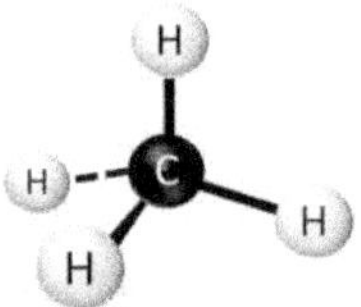

Methane (CH_4) adopts a tetrahedral shape to maximize repulsion from the 4 H substituents

Hybridization mixes atomic orbitals

The *hybridization* concept improved VB theory to resolve theoretical and experimental values differences.

Hybridization is mixing *atomic orbitals* (AO) to form bonds with unexpected bond angles.

Atomic orbitals have the highest probabilities of finding electrons, and *hybridization* is a rearrangement of orbitals.

     Copyright © Sterling Test Prep.

Hybrid orbitals have new shapes, directional properties, and combined constituent orbitals (shown below).

Hybrid orbitals combine the symbols of the initial orbitals.

The sum of exponents in hybrid orbital notation equals the number of atomic orbitals used.

- One *s* and one *p* orbital form two *sp* hybrid orbitals

- One *s* and two *p* orbitals form three *sp²* hybrid orbitals

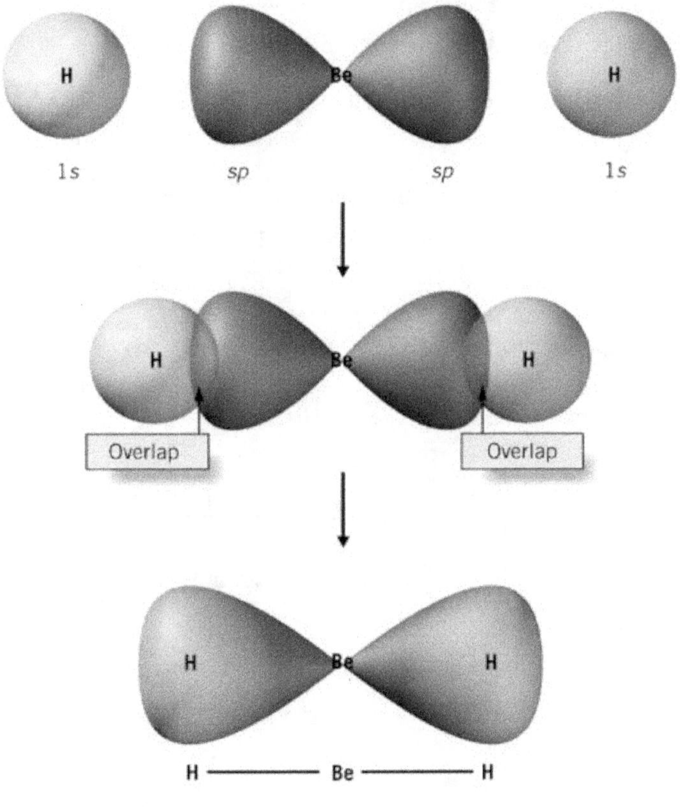

Bonding of BeH$_2$ Be: $1s^2 2s^2$ H: $1s^1$

Beryllium (Be) has two electrons in the $2s$ orbital. When bonding with two hydrogens, it splits the electrons into two orbitals. The Be atom uses the next available orbital ($2p$) combined with $2s$ orbital, resulting in two *sp* orbitals.

Each of those *sp* orbitals overlaps with hydrogen's $1s$ orbital, creating a covalent bond.

VSEPR theory predicts this molecule has a linear shape confirmed by experimental observations.

Most atoms form up to four bonds (i.e., octet rule); four orbitals are needed (1 *s* and 3 *p*).

Any duplication (copies, uploads, PDFs) is illegal.

Bonding and non-bonding pairs

For molecules with more than four bonds/lone pairs around them, *d* orbitals are included in the hybridization to accommodate the extra bonds, known as *expanded octet hybridization.*

Bond angle experiments suggest that NH_3 and H_2O use sp^3 hybrid orbitals in bonding (a total of four orbitals) even though each has two bonds.

This evidence for sp^3 hybrid orbitals concluded that hybrid orbitals are not exclusively used for bonding; they accommodate the non-bonding electron pairs.

Bonding and non-bonding pairs contribute to the molecule's geometry.

For example, the orbital diagram of hybridized nitrogen in NH_3:

Hybridization accommodates 3 bonding and 1 nonbonding electron pairs

Diagram showing the hybridization of oxygen in H_2O:

Hybridization accommodates 2 bonding and 2 nonbonding electron pairs

Copyright © Sterling Test Prep.

Molecular Orbital (MO) Theory

Wave interference

Molecular orbital theory (MO theory), the second dominant theory, views molecules as a collection of positively charged nuclei having a set of molecular orbitals filled with electrons (like filling atomic orbitals with electrons).

It does not focus on how individual atoms join to form molecules.

The MO theory is more difficult to visualize than the VB theory.

MO theory is advantageous because it accurately predicts a molecules' *magnetic properties* and other characteristics.

The energies of molecular orbitals (MO) are determined by combining *atomic orbitals'* (AO) *electron waves*.

For H_2 (below), two $1s$ wave functions, one from each atom, combine to make two MO wave functions.

Two MO produce *constructive interference* of the waves.

Bonding and antibonding molecular orbitals

For H_2, the bonding MO's energy is lower than the atomic orbitals' energy (AO), and the molecule is collectively more stable than the individual atoms (atoms forming the molecule).

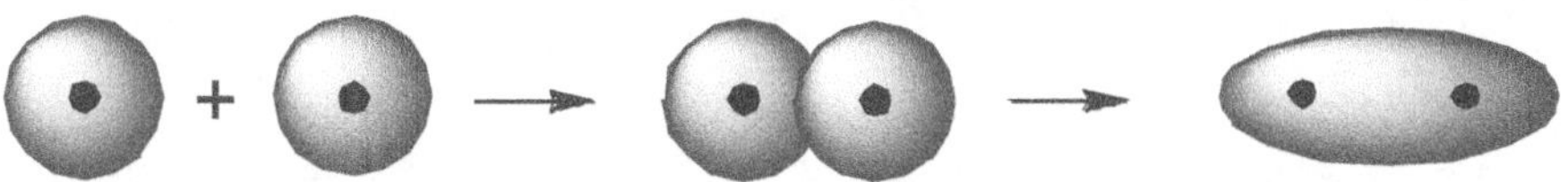

As there are σ orbitals, there are corresponding σ^* antibonding MOs.

The possible combination of two $1s$ orbitals involves destructive interference of the $1s$ waves.

In the example below, the antibonding MO's energy is higher than the energy of parent atomic orbitals. The molecule *does not form*, and the atoms remain separate.

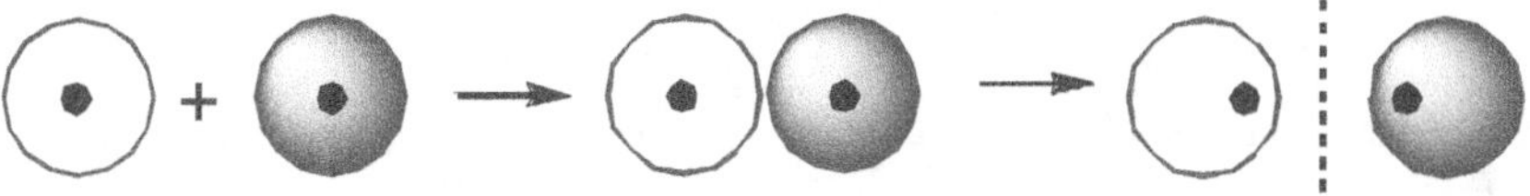

Antibonding MOs are high in energy and do not contribute to bond formation

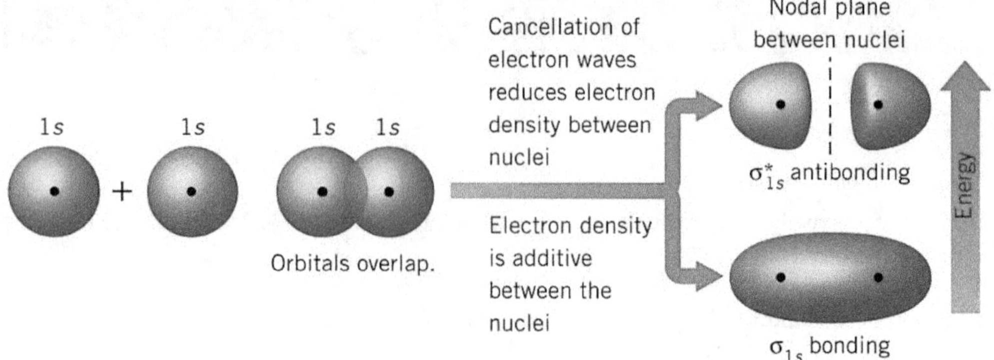

Bonding and antibonding MOs with the relative energy of each

In *bonding MOs*, electron density is between nuclei; the electrons tend to stabilize the molecule.

In the *antibonding MO*, the cancellation of electron waves reduces the nuclei's electron density.

The electrons in antibonding MOs tend to destabilize the molecule.

Molecular orbital theory diagrams

MO energy diagrams (shown below) represent the interactions between atomic orbitals.

A MO energy diagram displays the orbitals arranged vertically, from highest to lowest energy.

The atomic orbitals (AO) for the atoms are on the left and right sides of the diagram.

The molecular orbitals (MO) are in the central column of the diagram.

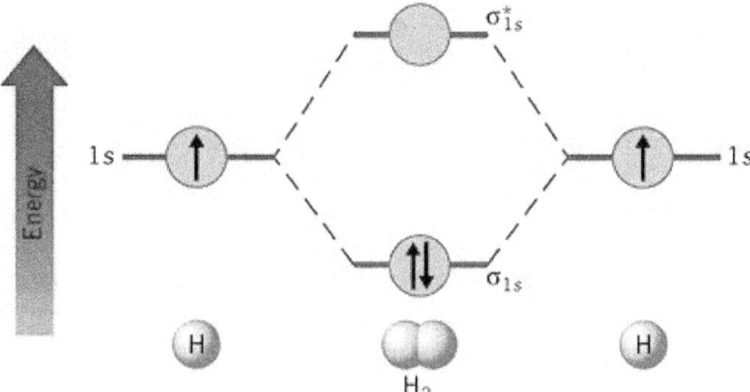

The MO energy diagram for H₂: atomic orbitals are designated 1s and have single arrows up, while the lower energy bonding molecular orbital has 2 electrons of opposite spin.

The high-energy antibonding MO is *vacant*. The molecule forms from separate atoms because the bonded molecule is *more stable* than the separate atoms (relative stability with energy on the y-axis).

   Copyright © Sterling Test Prep.

Three rules for filling MO energy diagrams:

1. Electrons fill the lowest-energy orbital (i.e., Aufbau's principle).

2. No more than two electrons (with opposite spins) can occupy any orbital (i.e., Pauli exclusion principle).

3. Electrons with unpaired spins spread over the same energy orbitals (i.e., Hund's rule).

Bond order

Bond order indicates the number of electron pairs shared between atoms (i.e., number of bonds between two atoms).

A bond order of one corresponds to a single bond.

For example, the H_2 bond order and the C−H bond order are each one, while the N≡N bond order in diatomic nitrogen is three.

The bond order is calculated by:

$$\text{Bond order} = \frac{(\text{number of bonding } e^-) - (\text{number of antibonding } e^-)}{2 \text{ electrons / bond}}$$

Valence bonding and molecular orbital theory limitations

Neither the valence bonding (VB) nor the molecular orbital theory (MO theory) is entirely correct because neither theory explains all aspects of bonding; each theory has its strengths and deficiencies.

VB theory is based on Lewis structures (discussed later) and the related geometric shapes.

MO theory correctly predicts the unpaired electrons in O_2, while the Lewis structures do not. MO theory is complicated because even simple molecules have complex energy level diagrams, and molecules with three or more atoms require extensive calculations.

VB theory uses three-dimensional structures based on electron domains without extensive calculations.

Simple hybrid orbitals are invoked where experimental evidence shows the need, and the integer bond orders are often correct.

Structural formulas for heteroatom molecules

Lewis structures (explained later) are used to represent atoms and molecules visually.

The Lewis structures for elements in the same column (group or family) are similar.

For example, sulfur can be substituted for oxygen in Lewis structures of oxygen.

Below are structural formulas for some important molecules.

- **Hydrogen** Lewis structures

    hydrogen proton: H^+

    hydride ion: H^-

- **Boron** – Group 13 Lewis structures

Since group 13 elements have three valence electrons, they often have six total electrons in a molecule (an exception to the octet rule).

Borane:

Borohydride ion:

Borohydride is expressed as BH_4^- with a formal charge on boron (B), shown with brackets and a negative charge on the top right side.

- **Carbon** – Group 14 Lewis structures

Methane:

Carbocation:

Copyright © Sterling Test Prep.

Carbanion: 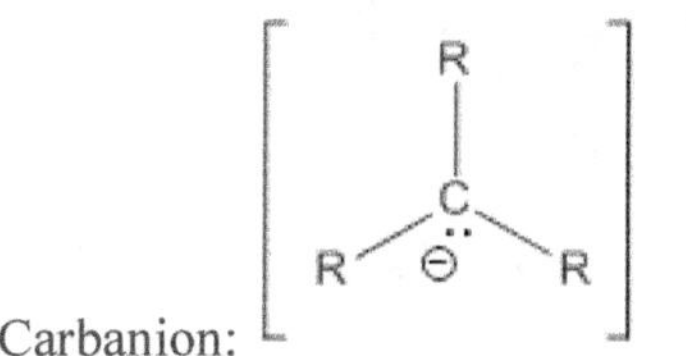

Note: R in the figures is carbon or hydrogen.

- **Nitrogen** – Group 15 Lewis structures

Some group 15 elements have more than eight electrons after bonding, an exception to the octet rule (e.g., PCl_5).

The presence of *d* orbitals permits the atom to accommodate these additional (> than 4) bonds.

Amine / Ammonia:

Ammonium:

Imine:

- **Oxygen** – Group 16 Lewis structures

Some group 16 elements have more than eight electrons after bonding, an exception to the octet rule (e.g., SF_6).

Molecular oxygen:

Water, alcohol, and ethers:

Ozone:

Any duplication (copies, uploads, PDFs) is illegal.

- **Halogen** – Group 17 Lewis structures

Hydrogen fluoride: H—$\ddot{\underset{\cdot\cdot}{F}}$:

Chloromethane: $H-\underset{\underset{H}{|}}{\overset{\overset{H}{|}}{C}}-\ddot{\underset{\cdot\cdot}{Cl}}:$

Bromide ion: $[:\ddot{\underset{\cdot\cdot}{Br}}:]^{-}$

Copyright © Sterling Test Prep.

Formal Charge

Formal charge for resonance structures

The *formal charge* is assigned to each atom in a Lewis structure from equal sharing of bonded electron pairs.

Formal charges are used in drawing resonance structures (multiple Lewis structures that collectively describe a single molecule) of covalently bonded molecules.

This technique describes and compares resonance structures, like how oxidation numbers balance chemical equations in oxidation-reduction reactions.

Formal charge is calculated by:

$$\text{Formal charge} = \text{valence } e^- \text{ in neutral atom} - (\text{unshared valence } e^- + \tfrac{1}{2} \text{ shared } e^-)$$

where e^- represents electrons

Every atom in a molecule has a formal charge.

In a neutral molecule, the sum of atomic charges is zero. That does not mean the atoms are not charged; they cancel to create a neutral molecule.

The charge is *not evenly distributed* in charged molecules, and formal charges can help determine where the charge rests.

Calculating formal charge

$$\text{Formal charge} = \text{valence } e^- \text{ in a neutral atom} - (\text{unshared valence } e^- + \tfrac{1}{2} \text{ shared } e^-)$$

For Lewis structures:

$$\text{Formal charge} = [\# \text{ of valence } e^-] - [\text{dots around atom} + \text{lines connected to atom}]$$

The number of valence electrons is generally equal to the atom's group number on the periodic table (e.g., N has five valence electrons, O has six valence electrons, and F has seven valence electrons).

Dots around the atom represent electrons held entirely by the atom.

The lines connecting the atom represent bonding electron pairs; the atom gets one of the two electrons.

Label the atom with the formal charge for formal charges other than zero.

Examples of atoms with formal charges:

Oxygen with a single bond: -1

Oxygen with no bonds but with an octet: -2

Carbon with three bonds:

$+1$ as a carbocation

-1 as a carbanion

Nitrogen with four bonds: $+1$

Nitrogen with three bonds: -1

Halogen with no bonds but has an octet: -1

Boron with four bonds: -1. (e.g., $^-BH_4$)

Carbon:

$=C=$ $\quad$ $-C\equiv$ $\quad$ $>C=$ $\quad$ $-C-$ $\qquad$ 4 covalent bonds: Formal charge $= 0$

Nitrogen:

$=\overset{\oplus}{N}=$ $\quad$ $-\overset{\oplus}{N}\equiv$ $\quad$ $>\overset{\oplus}{N}=$ $\quad$ $-\overset{\oplus}{N}-$ $\qquad$ 4 covalent bonds: Formal charge $= +1$

$-\ddot{N}=$ $\quad$ $:N\equiv$ $\quad$ $-\ddot{N}-$ $\qquad$ 3 covalent bonds, 1 lone pair: Formal charge $= 0$

$-\ddot{\underset{\cdot\cdot}{N}}{}^{\ominus}$ $\quad$ $\overset{\ominus}{:N}=$ $\qquad$ 2 covalent bonds, 2 lone pairs: Formal charge $= -1$

Oxygen:

$-\overset{\oplus}{\underset{\cdot\cdot}{O}}=$ $\quad$ $:\overset{\oplus}{O}\equiv$ $\quad$ $-\overset{\oplus}{O}-$ $\qquad$ 3 covalent bonds, 1 lone pair: Formal charge $= +1$

$-\ddot{O}-$ $\quad$ $\ddot{\ddot{O}}=$ $\qquad$ 2 covalent bonds, 2 lone pairs: Formal charge $= 0$

$^{\ominus}:\ddot{\underset{\cdot\cdot}{O}}-$ $\qquad$ 1 covalent bond, 3 lone pairs: Formal charge $= -1$

 Copyright © Sterling Test Prep.

The valence number is the number of outer shell (valence) electrons an atom needs to gain or lose to achieve a complete octet.

In covalent compounds, the number of bonds characteristically formed by an atom equals the valence number.

Atom	H	C	N	O	F	Cl	Br	I
Valence	1	4	3	2	1	1	1	1

Determining formal charge

The valences in the chart above represent the most common form of these elements in organic compounds.

Many elements (e.g., chlorine, bromine, iodine) have several valences in different inorganic compounds.

If the number of covalent bonds to an atom is higher than its typical valence, it carries a positive formal charge.

If the number of covalent bonds to an atom is lower than its typical valence, it carries a negative formal charge.

A *step-by-step approach* to determine the formal charges of each atom in a molecule of nitric acid (HNO_3), using the structure displayed below.

$$\overset{\displaystyle O}{\underset{HO \diagdown \; \overset{\textstyle +}{N} \diagup O^-}{\|}}$$

The formal charge for nitric acid (HNO_3) with formal charge shown on the atoms

Formal charge on **H**

- Hydrogen shares two electrons with oxygen.

- Assign one electron to H and one to O.

- Hydrogen has one valence electron, and $1 - 1 = 0$.

- Therefore, the formal charge of H in nitric acid is zero.

Formal charge on the **O** bonded to N and H

- Oxygen has four electrons in covalent bonds (2 with N, 2 with H).

- Assign two of these four electrons to O.

- Oxygen has two unshared pairs; assign four electrons to O.

- The total number of electrons assigned to O is 2 + 4 = 6.

- Oxygen has six valence electrons, and 6 – 6 = 0.

- Formal charge of O bonded to H and N in nitric acid is zero.

Formal charge on the **double-bonded O**

- Oxygen has four electrons in covalent bonds with N.

- Assign two of these four electrons to O.

- Oxygen has two unshared pairs; assign four electrons to O.

- The total number of electrons assigned to O is 2 + 4 = 6.

- Oxygen has six valence electrons, and 6 – 6 = 0.

- Formal charge of the double-bonded O in nitric acid is 0.

Formal charge on the **single-bonded O**

- O has two electrons in a covalent bond.

- Assign one of those electrons to O.

- O has three unshared pairs. Assign six electrons to O.

- The total number of electrons assigned to O is 1 + 6 = 7.

- Oxygen has six valence electrons, 6 – 7 = –1.

- Typical charge of single-bonded O in nitric acid is –1.

Formal charge of **N**

- N has eight electrons in covalent bonds.

- Assign four of those electrons to N.

- Nitrogen has five valence electrons, 5 – 4 = 1.

- Therefore, the formal charge of the N in nitric acid is +1.

$$6 - 1 - 6 = -1$$

$$5 - 4 - 0 = +1$$

$$6 - 2 - 6 = 0$$

Copyright © Sterling Test Prep.

If atoms in a molecule have a formal charge, it does not necessarily mean that the entire molecule has a charge.

In the ozone structure (O_3) below, the central oxygen atom has three bonds and is positively charged.

The right-hand oxygen has a single bond and is negatively charged.

The overall charge of the ozone molecule is, therefore, zero.

ozone   *nitromethane*   *azide ion*

Nitromethane (CH_3NO_2) has positively charged nitrogen and negatively charged oxygen; the total molecular charge is again zero.

The azide (N_3^-) anion has two negatively charged nitrogens (2 bonds each) and one positively charged nitrogen (four bonds), and the total charge is minus one.

Notes for active learning

Copyright © Sterling Test Prep.

Resonance

Resonance structures

When more than one relatively stable structure exists, it forms *resonance structures.*

Resonance is the averaging of electron distribution over two or more hypothetical contributing structures to produce a hybrid electronic structure.

Resonance structures describe molecules that may contain fractional bonds and charges.

The resonance structures must have the same number of paired and unpaired electrons.

No atoms change position within the common structural framework, not breaking existing chemical bonds.

Electrons (not atoms, as is the case for isomers) move. Visualize the molecule as quickly "shifting" between its resonance structures and intermittently existing in several configurations.

However, the molecule spends more time in the most stable resonance form.

More accurately, the molecule structure is a "combination" (or a weighted average) of its resonance structures, favoring the most stable resonance structures.

In a molecule with a single-and a double-bond resonance structure, the bond length is between a single- and a double-bond length.

Resonance structures for SO_3^{2-} (single-bonded O with negative charge). The images above do not include the formal charge or the double-headed arrows to show the individual contributing resonance structures.

Three contributing resonance structures for CO_3^{2-}

The double-headed arrow seen in the image above for CO_3^{2-} acts as a notation symbol for resonance structures contributing to the hybrid structure. The principle of resonance is useful in rationalizing the chemical behavior of compounds.

Any duplication (copies, uploads, PDFs) is illegal.

The electron delocalization provided by resonance significantly enhances the molecules' stability (lowers PE).

Determine the most stable resonance structure of a molecule when a stable molecule's properties are known.

In stable molecules, the octet is satisfied in each atom (aside from hydrogen and exceptions, like boron).

Neutral molecules are more stable if the formal charges on atoms are spread accordingly (i.e., + charge on a positive atom, – charge on electronegative atoms).

Resonance structure for formaldehyde (CH₂O)

In the example above, the neutral molecule is the most stable.

The preferred charge distribution has a positive charge on the less electronegative atom (carbon) charged species, and a negative charge is held by the more electronegative atom (oxygen).

Therefore, the middle formula represents a more reasonable and stable structure than the right.

Resonance hybrids

The stable structure(s) indicates that the molecule spends most time in resonance structure(s) over the others.

If the double bond is broken heterolytically (both electrons going to one of the bonding atoms), formal charge pairs result, as shown in the other two structures.

Since the middle, the charge-separated contributor has an electron-deficient carbon atom; this explains the tendency of electron donors (nucleophiles) to bond at this site.

The application of resonance to this example requires a weighted average of these structures.

Consequently, if one structure has greater stability than the others, the hybrid closely resembles it electronically and energetically.

The resonance hybrid is exceptionally stable if two or more forms have identical low-energy structures.

Copyright © Sterling Test Prep.

Examples of low-energy molecules stabilized by resonance hybrid structures include sulfur dioxide (SO_2) and nitric acid (HNO_3).

Three resonance forms of sulfur dioxide (SO_2); the central structure is neutral and therefore the greatest contributor (most molecules exist in this state)

Two resonance forms of nitric acid (HNO_3); structures are equal in energy with nitrogen positive, and the negative charge shared by two oxygens

Three considerations for drawing resonance structures

1. The number of covalent bonds in a structure - the greater the bonding, the more essential and stable the contributing structure.

2. Formal charge separation: *charge separation* decreases the contributing structure's stability and importance.

3. Evaluate electronegativity of charge-bearing atoms and charge density.

    The *positive charge* is accommodated on atoms of low electronegativity.

    The *negative charge* is on highly electronegative atoms.

Lewis acids and bases

Molecules can be categorized by the number of free electron pairs:

Lewis *acids accept* electron pairs.

Lewis *bases donate* electron pairs.

Acids have vacant orbitals and do not have lone pairs on the central atom (e.g., BF_3), while Lewis bases have lone pair electrons (e.g., NH_3).

$$A^+ \; + \; ^-B \; \longrightarrow \; A-B$$

Lewis acid (A): electron acceptor   Lewis base (B): electron donor

Any duplication (copies, uploads, PDFs) is illegal.

$$\text{F}-\overset{\displaystyle \text{F}}{\underset{\displaystyle \text{F}}{\text{B}}} \; + \; :\overset{\displaystyle \text{H}}{\underset{\displaystyle \text{H}}{\text{N}}}-\text{H} \; \longrightarrow \; \text{F}-\overset{\displaystyle \text{F}}{\underset{\displaystyle \text{F}}{\text{B}}}-\overset{\displaystyle \text{H}}{\underset{\displaystyle \text{H}}{\text{N}}}-\text{H}$$

Acid   *Base*

The product would have a formal negative charge on boron and a positive charge on nitrogen

Acids and bases are discussed in the acids and bases chapter.

Copyright © Sterling Test Prep.

Partial Ionic Character

Electron sharing

Covalent bonds can be categorized into nonpolar covalent bonds and polar covalent bonds.

Nonpolar covalent bonds occur between atoms of the same element or atoms with similar electronegativity sharing electrons equally to achieve a more stable electron configuration.

Polar covalent bonds form between atoms with differences in electronegativity. These bonds have *partial ionic character* due to the unequal distribution of the electrons. One atom has a stronger attraction for the shared electrons resulting in one end of the bond being partially negative (δ^-), while the other is partially positive (δ^+).

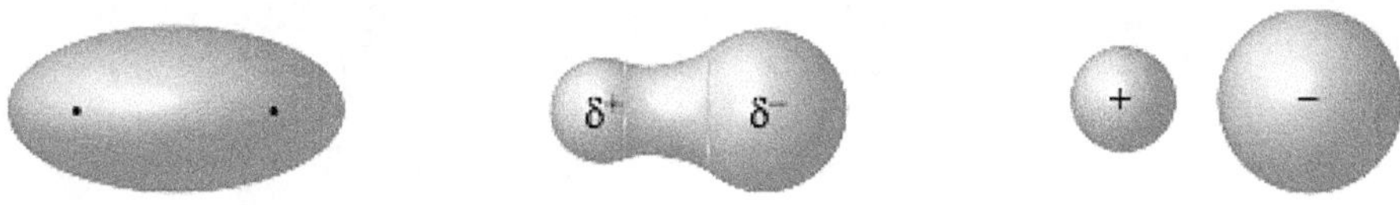

Three types of bonds: covalent (left), polar covalent (center), and ionic (right)

The left structure above shows a covalent bond for identical atoms (i.e., same electronegativity).

The center structure is a polar covalent bond between elements of moderately different electronegativities with the notation for the molecule's resulting partial positive and partial negative ends.

The right figure shows an ionic bond between elements with significant differences in electronegativities, usually a metal and a nonmetal. In ionic bonds electrons from one atom are transferred completely to another, resulting in ions with opposite charges that are attracted to each other by electrostatic forces (e.g., in NaCl sodium donates an electron to chlorine).

Nonpolar covalent bonds involve equal sharing of electrons, polar covalent bonds involve unequal sharing, and ionic bonds result from the complete transfer of electrons from one atom to the other. The difference in electronegativity between the atoms determines the type of bond.

Electronegativity for determining charge distribution

Electronegativity (EN) is an element's ability to attract electrons through bonding due to differing nuclear charges and shielding by inner electron shells; each element has different electronegativity.

The molecule's polarity is the difference in EN values of its atoms.

Linus Pauling (Nobel laureate in chemistry, 1954) elucidated the chemical bond's nature and application to complex molecules' structure. Pauling established a quantitative scale of electronegativity values.

A larger number (e.g., F = 3.98) signifies a higher affinity for electrons.

H 2.20	Electronegativity Values for Some Elements					
Li 0.98	Be 1.57	B 2.04	C 2.55	N 3.04	O 3.44	F 3.98
Na 0.90	Mg 1.31	Al 1.61	Si 1.90	P 2.19	S 2.58	Cl 3.16
K 0.82	Ca 1.00	Ga 1.81	Ge 2.01	As 2.18	Se 2.55	Br 2.96

For the periodic table, EN values increase from left to right, from bottom to top

Higher electronegativity values are located towards the nonmetal side (right side) on the periodic table.

Fluorine has the highest electronegativity (3.98). The heavier alkali metals (e.g., potassium, rubidium, cesium) have lower electronegativity values.

Carbon has an electronegativity of 2.55. This value is mid-range for electronegativity and is slightly more electronegative than hydrogen (2.20).

Of an atom's subatomic particles (protons, neutrons, electrons), only the electrons move within the atom.

When two atoms bond, most electrons congregate between the two nuclei.

The distribution depends on the electronegativity of the atoms.

Shared electrons in a bond are attracted to the more electronegative atom, shifting electron density toward the electronegative atom.

Nonpolar covalent bonds form between identical atoms as *homonuclear diatomic molecules*.

For example, in diatomic hydrogen (H_2), both atoms are identical; each has the identical EN value.

The electrons, therefore, are evenly distributed between the two hydrogens.

$$H : H \qquad H \; H$$

Copyright © Sterling Test Prep.

Polarity based on electronegativity

Hydrogen fluoride (HF) has atoms of different EN values – fluorine (3.98) is more electronegative than hydrogen (2.20). The electrons are more attracted to the fluorine nucleus and are closer to the F than H.

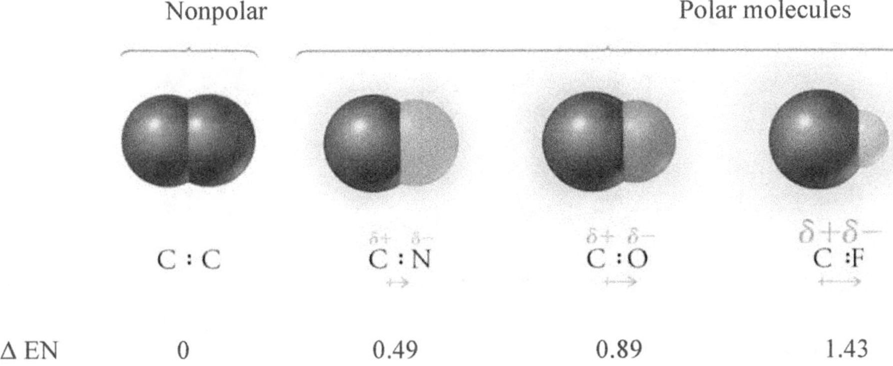

$$H \; : \; F$$

The electron density for the H–F bond with the greatest density on the electronegative F

For example, suppose one atom in a molecule is more electronegative than another. This causes a higher concentration of electrons to locate on one side of the bond, and the molecule is polar (i.e., dipole present).

The molecule's *polarity* is proportional to the difference in the bonded atoms' electronegativity, shown below.

Nonpolar	Polar molecules		
C : C	C : N	C : O	C : F
0	0.49	0.89	1.43

Δ EN

The lowercase Greek letter delta δ denotes partial charges on a polar molecule.

A plus or minus sign (i.e., (δ^+ or δ^-) indicates the partial positive or negative charges.

$$\delta^+ \; H\!-\!Cl \; \delta^-$$

Arrows show the direction of partial charges. The arrow points in the direction of electron movement, originating at the positively charged atom and pointing toward the negatively charged atom.

$$\overset{\delta^-}{C}\!-\!\!-\!\overset{\delta^+}{H} \qquad \overset{\delta^-}{O}\!-\!\!-\!\overset{\delta^+}{H} \qquad \overset{\delta^+}{C}\!-\!\!-\!\overset{\delta^-}{Cl} \qquad \overset{\delta^-}{C}\!-\!\!-\!\overset{\delta^+}{Li}$$

The partial positively δ^+ and negatively δ^- charged ends are *dipoles*.

If the difference in electronegativity is high enough in a direction, the molecule acquires a net dipole moment (geometric polarity).

Any duplication (copies, uploads, PDFs) is illegal.

Dipole moment

The *dipole moment* measures polarity in a molecule; it is the difference in electronegativity values of participating or bonded atoms.

The geometric shape of a molecule affects its overall dipole moment.

A molecule may have polar bonds, but if those bonds arrange to cancel the dipole moments, it has a zero net dipole and is nonpolar.

A molecule is nonpolar if:

    1) It is symmetrical.

    2) It has identical atoms surrounding the central atom.

    3) Non-bonding electron pairs are evenly distributed and cancel (illustrated below).

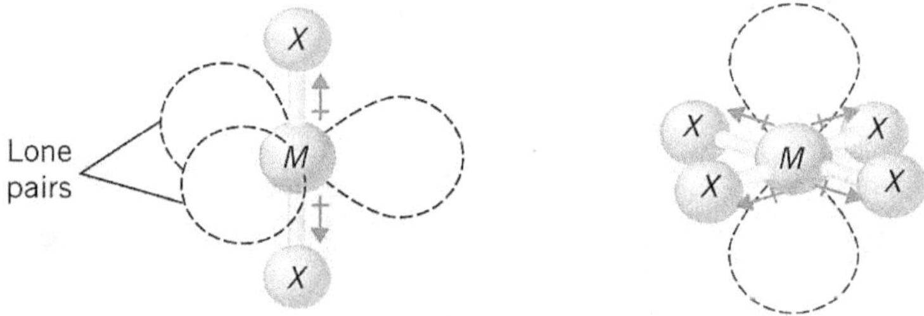

Symmetrical molecules have no net dipole even with individual bond polarity

Dipole based on shape

In the right-hand linear configuration of the molecule below (bond angle 180°), the bond dipoles cancel, and the molecular dipole is zero.

The molecular dipole varies in size for bond angles 90° to 120°.

The molecular dipole is the largest at the 90° configuration.

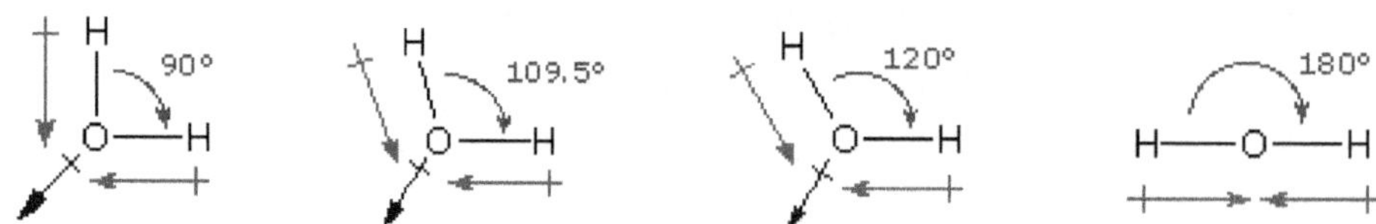

For example, boron trichloride (BCl_3) has three B−Cl bonds. B−Cl is a polar bond, but because the molecule is symmetrical and charges evenly distribute in a single plane, the charges cancel (shown below).

Copyright © Sterling Test Prep.

BCl₃ does not have non-bonding electron pairs, and the symmetrical BCl₃ molecule has a zero dipole moment.

$$\text{Cl} - \text{B} \big(\text{Cl}\big) \text{Cl}$$

Boron trichloride BCl₃ is a symmetric molecule with a zero dipole moment

The configurations of methane (CH_4) and carbon dioxide (CO_2) may be deduced from their zero dipole moments.

Since the dipoles cancel, these molecules' configurations must be tetrahedral and linear, respectively.

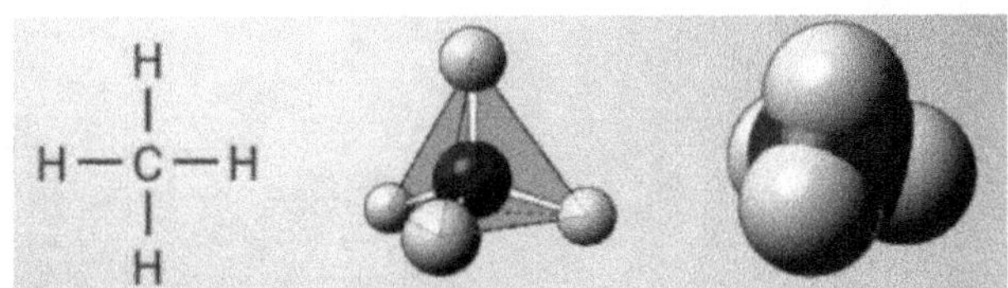

Methane (l to r) as simple drawing, ball-and-stick, and space-filling models

$$O = C = O$$

Carbon dioxide (CO_2) is a linear molecule and exhibits a net dipole of zero

Any duplication (copies, uploads, PDFs) is illegal.

Notes for active learning

Copyright © Sterling Test Prep.

Intermolecular Forces

Forces between molecules

Intermolecular forces are *attractive or repulsive* forces between molecules (e.g., hydrogen bonding, London dispersion).

However, some other intermolecular interactions are not as strong as covalent or ionic bonds but have a significant role in determining substances' properties.

Dipole interactions

Molecules with dipole moments (from differences in electronegativity of bonded atoms) are attracted to each other. The more polar the molecule is, the stronger the dipole attraction.

Four types of dipole attractions:

- *Ion-dipole attractions* form between an ion and a polar molecule. Since a polar molecule has a slight charge (i.e., dipole) and an ion is a charged atom, these particles are attracted to each other.

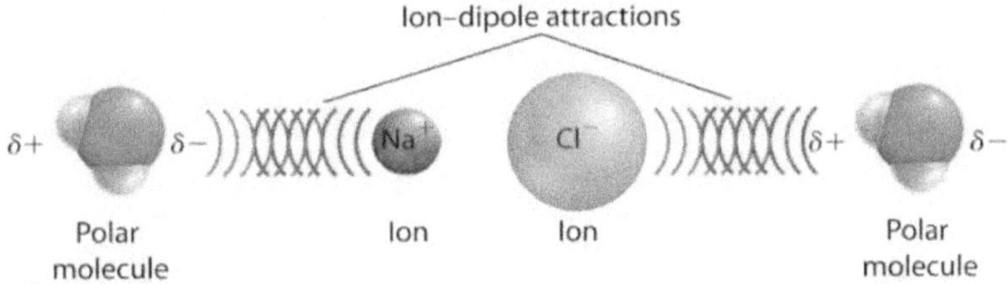

- *Dipole-dipole attraction* forms between two polar molecules. The partially positive side of one polar molecule attracts the partially negative side of another polar molecule.

- *Dipole-induced dipole attraction* forms between partially polar and nonpolar molecules.

  For example, the polar H_2O molecules induce the nonpolar O_2 molecules, shifting electron density and inducing temporary partial positive (δ^+) and negative (δ^-) regions in the nonpolar O_2 molecules.

  positive pole ⟶ negative pole

Any duplication (copies, uploads, PDFs) is illegal.

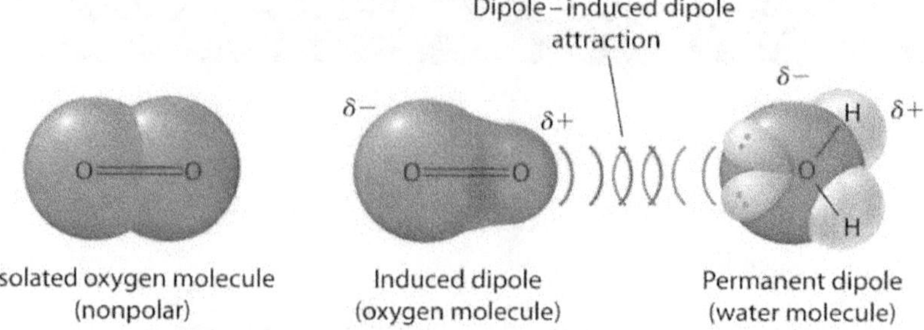

- ***Induced dipole-induced dipole*** *attraction* (or *London dispersion force*) forms between two (or more) nonpolar molecules. These nonpolar molecules become temporarily polar (i.e., momentary flux) due to random, instantaneous induction by another polar molecule that results in the probability of finding the negative electron in a specific region within its orbital.

Longer molecules have more *surface contact* and experience higher London dispersion forces.

For example, octane (C_8H_{18}) has a stronger London dispersion force than methane (CH_4).

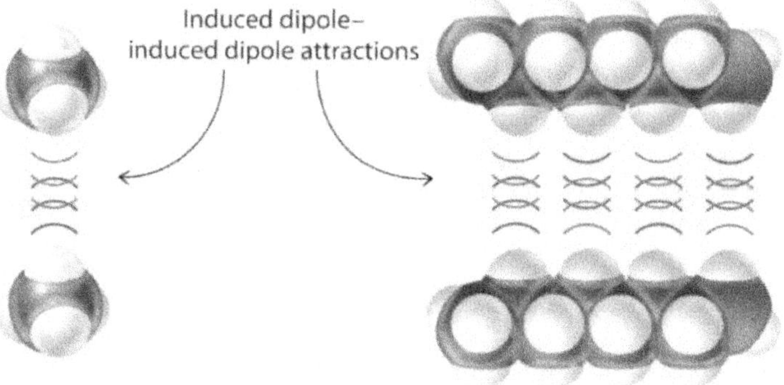

Molecular attractions involving dipoles

Attraction	Relative Strength
Ion–dipole	Strongest
Dipole–dipole	
Dipole–induced dipole	
Induced dipole–induced dipole	Weakest

Copyright © Sterling Test Prep.

Hydrogen bonding

Hydrogen bonding is a specific type of bond that, based on its strength, is categorized as a *dipole-dipole* attraction.

Hydrogen (hydrogen bond donor with one valence electron) bonding occurs when hydrogen is covalently bonded directly to the electronegative atom of N, O, or F (hydrogen bond donors).

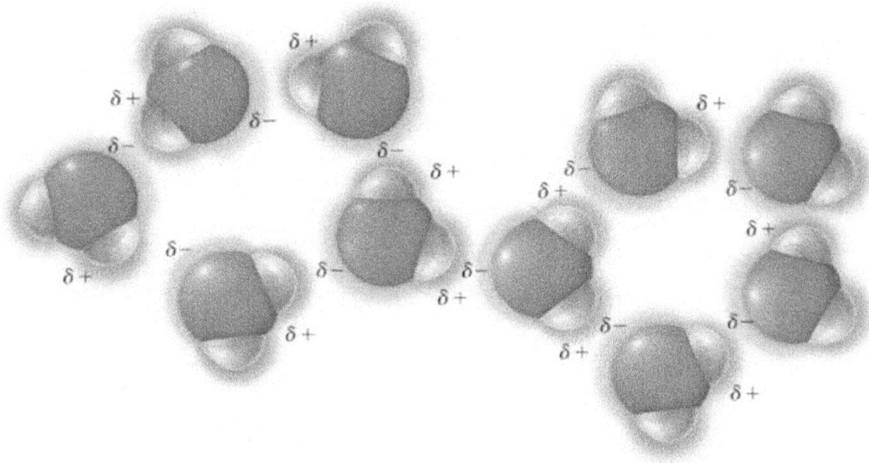

Hydrogen bonding produces differences in physical properties

The bonding electrons are closer to the more electronegative N, O, or F atom.

Due to these factors, the hydrogen nucleus is exposed.

The highly positive hydrogen nucleus attracts the electronegative atom from a neighboring molecule. The neighboring molecule needs a lone electron pair to form a hydrogen bond. This creates a relatively powerful intermolecular force.

The more polar a bond is, the stronger the hydrogen bond.

The H−F bond is most polar, then H−O bonds, and then H−N bonds.

Intermolecular forces in polar molecules

Hydrogen bonding significantly increases the boiling point of compounds.

For example, methanol (CH_3OH) and methane (CH_4) have similar molecular structures.

However, methanol is liquid at room temperature, while methane is a gas.

Methanol is liquid at room temperature because of hydrogen bonding between methanol molecules, which is *much stronger* than the London dispersion (weak intermolecular) forces between the nonpolar methane molecules.

Therefore, methane is a gas at room temperature.

Any duplication (copies, uploads, PDFs) is illegal.

Hydrogen bond donors and acceptors

The molecule bonding (by lone pair electrons) to the hydrogen (bonded to F, O, N) is the hydrogen bond *acceptor*.

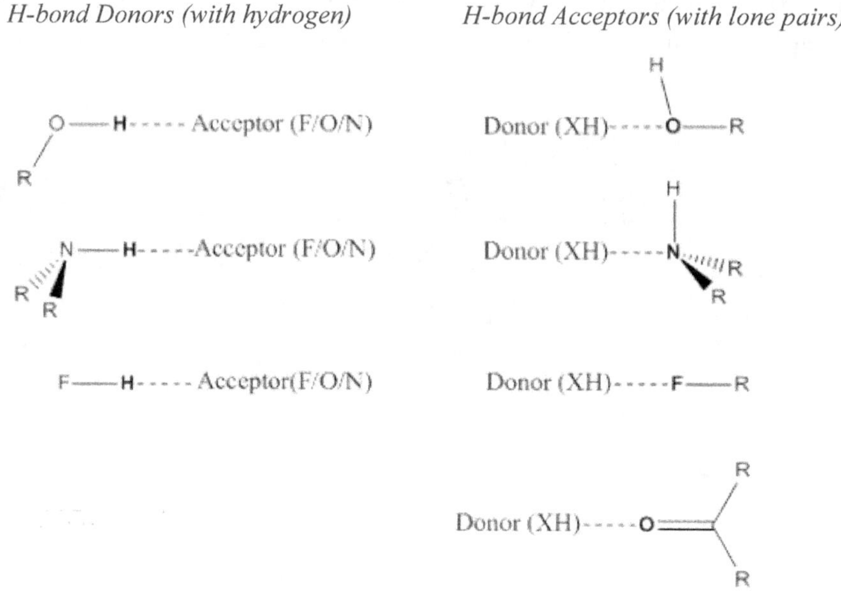

The hydrogen bond donor has a δ^+ H bonded to the electronegative F, O, N.

The hydrogen bond *acceptor* has a *lone pair of electrons* to bond to the δ^+ hydrogen of the donor.

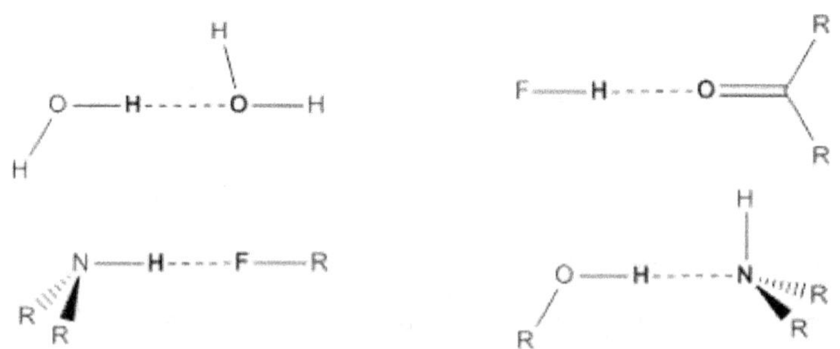

Examples of hydrogen-bonded molecules

Copyright © Sterling Test Prep.

Van der Waals' forces

Atoms and molecules exhibit weak intermolecular attractions, known as *van der Waals forces*. This results from a molecule's nuclei's weak attraction to another molecule's valence electrons.

The van der Waals equation (described previously) accounts for these *attractive intermolecular forces*.

The van der Waals force affects the *boiling point* because the boiling point reflects the kinetic energy needed to release a molecule from the liquid's cooperative attractions.

A larger *molecular mass* reduces the kinetic energy ($KE = \frac{1}{2}mv^2$) for the same energy input.

London dispersion forces

Nonpolar molecules have fluctuating dipoles that align with other molecules from one instant to the next.

The *momentary flux of electron density* is attracted to nearby nuclei's positive charge, and this attraction decreases dramatically as the distance between the molecules increases.

London dispersion forces are a subset of Van der Waals' attractive forces resulting from temporarily fluctuating molecules ("induced dipole or induced dipole attraction").

London dispersion forces exist for many molecules but are a significant source of intermolecular forces for nonpolar molecules that do not have dipole-dipole, dipole-induced dipole, or hydrogen bonding.

London dispersion forces are weak compared to the forces generated between polar molecules.

For polar molecules, dipole forces are strong, influential, and predominant.

In general, larger nonpolar molecules have higher boiling points than smaller nonpolar molecules due to increased London dispersion forces.

Branched molecules (i.e., hydrocarbons) have weaker London dispersion forces.

Any duplication (copies, uploads, PDFs) is illegal.

Notes for active learning

Notes for active learning

Copyright © Sterling Test Prep.

Polarity

Molecular polarity

The methane CH_4 molecule (below) provides insight confirming its tetrahedral configuration.

Substituting one hydrogen by a chlorine atom gives the methyl chloride CH_3Cl molecule below.

Since the tetrahedral, square-planar, and square-pyramidal configurations have structurally equivalent hydrogen atoms, each give a single substitution product.

One chlorine atom at the apex has a different electronegativity in the trigonal-pyramidal configuration than the three hydrogens at the pyramid base.

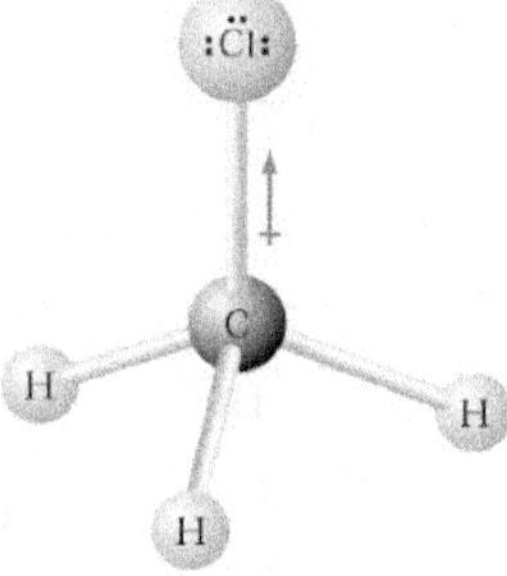

Methyl chloride (CH$_3$Cl; or chloromethane) with dipole pointing towards the single chlorine

Substitution should give two CH_3Cl compounds when hydrogen reacts.

The tetrahedral configuration of methane leads to a single dichloromethane CH_3Cl product below.

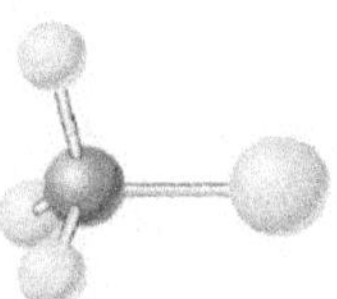

Methyl chloride CH_3Cl (above) is a refrigerant and used to manufacture synthetic rubber and silicones.

The *dipole moment* points toward the single electronegative chlorine atom.

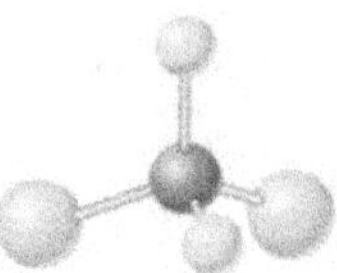

Dichloromethane (methyl chloride; above) CH_2Cl_2 is an industrial solvent used to remove paint and varnish.

Any duplication (copies, uploads, PDFs) is illegal.

The *dipole moment* points toward the two electronegative chlorine atoms.

The vector is resolved between the two chlorines and away from the two hydrogens.

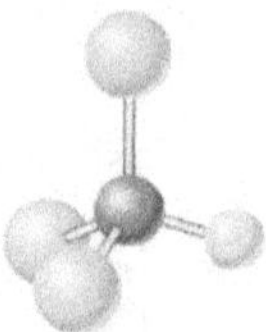

Trichloromethane (chloroform; above) $CHCl_3$ is a solvent to remove resins and adhesives.

The *dipole moment* points toward the three electronegative chlorine atoms.

The vector is resolved mathematically between the three Cl and away from H.

Most covalent compounds show some degree of local charge separation, resulting from differences in electronegativity between the atoms.

If a molecule has polar bonds arranged so that the bonds are in a specific direction, the molecule has a net dipole in that same direction. Some examples of polar molecules are PCl_3 and HCN.

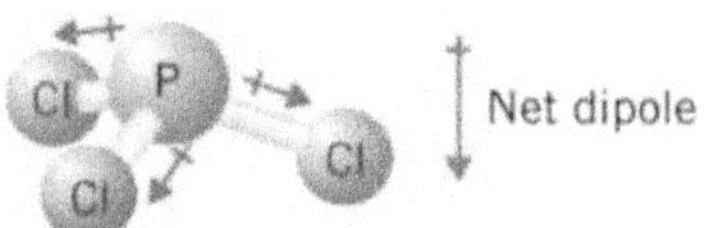

Phosphorus trichloride (PCl_3) has a trigonal pyramidal molecular shape

The phosphorus trichloride (PCl_3) molecule has three chlorine atoms positioned on one side.

Since chlorine is very electronegative, the electrons are attracted to the chlorine side, creating a dipole moment.

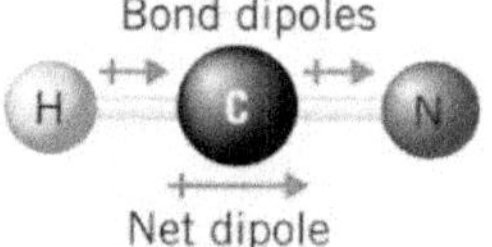

Phosphorus trichloride exhibits a net dipole because of the asymmetric pyramidal shape

Hydrogen cyanide (HCN) is a symmetrical molecule with one atom on each side of carbon.

Nitrogen is more electronegative than hydrogen; electrons concentrate near the nitrogen.

Copyright © Sterling Test Prep.

Polarity affects physical properties

Some physical properties (e.g., melting, boiling) are affected by molecular polarity.

Polar molecules have stronger intermolecular forces, resulting in higher boiling and melting points (below).

Substance	Boiling Point (°C)
Polar	
Hydrogen fluoride, HF	20
Water, H_2O	100
Ammonia, NH_3	−33
Nonpolar	
Hydrogen, H_2	−253
Oxygen, O_2	−183
Nitrogen, N_2	−196
Boron trifluoride, BF_3	−100
Carbon dioxide, CO_2	−79

Boiling points of polar (high BP) and nonpolar (low BP) substances

A critical example of polarity is the water (H_2O) molecule (shown below).

The center oxygen of water has two pairs of non-bonding electrons, two hydrogens (one on each side), and is theoretically symmetrical if the two lone pairs are ignored.

VSEPR theory predicts water has a bent shape, and the electron pairs are grouped (below).

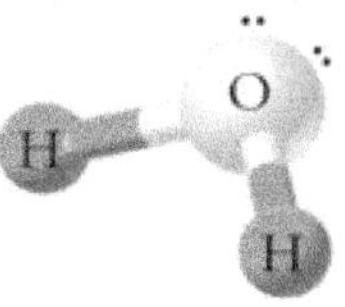

Water with the two lone pairs. The dipole moment is indicated in the central structure.

Any duplication (copies, uploads, PDFs) is illegal.

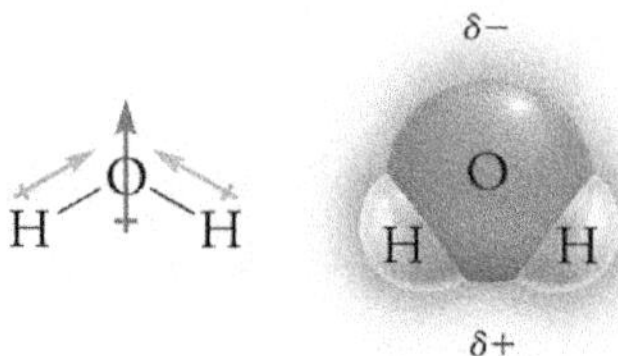

The space-filling water model with the molecule's partial negative and partial positive ends.

For water, the electron pairs are not evenly distributed; they are shifted downwards and concentrated on one side of the oxygen atom. That side has a partial negative charge, while the opposing side has a partial positive charge.

Therefore, the water molecule has a net dipole moment and is a polar molecule.

The polar water molecules exert strong intermolecular forces, known as *hydrogen bonds*.

Hydrogen bonds are a particular class of strong dipole-dipole interactions when a hydrogen atom is bonded *directly* to one of the three most electronegative atoms (F, Cl, N).

Copyright © Sterling Test Prep.

Practice Questions

1. Unhybridized p orbitals participate in π bonds as double and triple bonds. How many distinct and degenerate p orbitals exist in the second electron shell, where n = 2?

    A. 3　　　　**B.** 2　　　　**C.** 1　　　　**D.** 0

2. Which type of attractive forces occurs in molecules regardless of the atoms they possess?

    A. Dipole–ion interactions　　　　**C.** Dipole–dipole attractions
    B. London dispersion forces　　　　**D.** Hydrogen bonding

3. Ignoring other factors, it is possible to approximate the energy released when an ionic bond forms by considering how much the electrostatic potential energy changes while forming the bond (i.e., bringing the two ions together from infinity). Using approximation, what is the amount of energy released when 1 mole of NaCl forms?

For two ions with charges q_1 and q_2 separated by distance r, the electrostatic potential energy is given by $U_e = k \cdot q_1 q_2 / r$, where electrostatic constant $k = 9 \times 10^9$ J·m / C^2. (The charge on Na$^+$ = +e, the charge on Cl$^-$ = –e, the charge of electron $e = 1.602 \times 10^{-19}$ C, and the separation distance $r = 282$ pm)

    A. –493 kJ/mol　　　　**C.** +91.2 kJ/mol
    B. –288 kJ/mol　　　　**D.** +421 kJ/mol

4. During strenuous exercise, why does perspiration on a person's skin forms droplets?

    A. The ability of H_2O to dissipate heat　　　　**C.** Adhesive properties of H_2O
    B. High specific heat of H_2O　　　　**D.** Cohesive properties of H_2O

5. Based on the Lewis structure, how many polar and nonpolar bonds are present in H_2CO?

    A. 1 polar bond and 2 nonpolar bonds　　　　**C.** 3 polar bonds and 0 nonpolar bonds
    B. 2 polar bonds and 1 nonpolar bond　　　　**D.** 0 polar bonds and 3 nonpolar bonds

6. Which of the following occur(s) naturally as nonpolar diatomic molecules?

    I. sulfur　　　　II. chlorine　　　　III. argon

    A. I only　　　　**C.** I and III only
    B. II only　　　　**D.** I and II only

Any duplication (copies, uploads, PDFs) is illegal.　　　　

7. In the nitrogen monoxide molecule, the dipole moment is 0.16 D, and the bond length is 115 pm. What is the sign and magnitude of the charge on the oxygen atom? (Use the conversion factor of 1 D = 3.34×10^{-30} C·m and the charge of 1 electron = 1.602×10^{-19} C)

   A. $-0.098\ e$ **B.** $-0.71\ e$ **C.** $-1.3\ e$ **D.** $-0.029\ e$

8. From the electronegativity, which single covalent bond is the most polar?

Element:	H	C	N	O
Electronegativity	2.1	2.5	3.0	3.5

   A. O–C   **C.** N–C

   B. O–N   **D.** C–H

9. Under what conditions is graphite converted to diamond?

   A. low temperature, high-pressure   **C.** high temperature, high-pressure

   B. high temperature, low-pressure   **D.** low temperature, low-pressure

10. An ion with an atomic number of 34 and 36 electrons has what charge?

   A. +2   **C.** +34

   B. −36   **D.** −2

11. Which of the following pairings of ions is NOT consistent with the formula?

   A. Co_2S_3 (Co^{3+} and S^{2-})   **C.** Na_3P (Na^+ and P^{3-})

   B. K_2O (K^+ and O^-)   **D.** BaF_2 (Ba^{2+} and F^-)

12. The ability of an atom in a molecule to attract electrons to itself is:

   A. ionization energy   **C.** electronegativity

   B. paramagnetism   **D.** electron affinity

13. All of the following are examples of polar molecules, EXCEPT:

   A. H_2O **C.** CH_2Cl_2

   B. CCl_4 **D.** HF

Copyright © Sterling Test Prep.

14. When NaCl dissolves in water, what is the force of attraction between Na^+ and H_2O?

    A. ion–dipole   **C.** ion–ion

    B. hydrogen bonding   **D.** dipole–dipole

15. Which element likely forms a cation with a +2 charge?

    A. Na   **B.** S   **C.** Si   **D.** Mg

Any duplication (copies, uploads, PDFs) is illegal.

Notes for active learning

Notes for active learning

Copyright © Sterling Test Prep.

Detailed Explanations

1. A is correct.

Three degenerate p orbitals exist with an electron configuration in the second shell or higher.

The first shell only has access to s orbitals.

The d orbitals become available from $n = 3$ (third shell).

2. B is correct.

London dispersion forces result from the momentary flux of valence electrons and are present in all compounds; they are the attractive forces that hold molecules.

They are the weakest intermolecular forces, and strength increases with increasing size (i.e., surface area contact) and the molecules' polarity.

3. A is correct.

Calculate the potential energy for each molecule of NaCl:

$$U_e = k\, q_1 q_2\, /\, r$$

$$U_e = [(9 \times 10^9\ \text{J·m·C}^{-2}) \cdot (1.602 \times 10^{-19}\,\text{C}) \cdot (-1.602 \times 10^{-19}\,\text{C})]\, /\, (282 \times 10^{-12}\ \text{m})]$$

$$U_e = -8.19 \times 10^{-19}\ \text{J per molecule of NaCl}$$

Calculate the energy released for one mole of NaCl:

$$[(-8.19 \times 10^{-19}\ \text{J}) \times (6.02 \times 10^{23}\ \text{mol}^{-1})] = -493{,}077\ \text{J}$$

$$-493{,}077\ \text{J} \approx -493\ \text{kJ/mol}$$

This approximation disregards a few important effects.

For example, the crystal lattice orientation affects the energy, and this model does not consider quantum mechanical effects.

4. D is correct.

Water molecules stick to each other (i.e., cohesion) due to hydrogen bonds' collective action between individual polarized water molecules.

These hydrogen bonds are continually breaking and reforming; these bonds hold many molecules.

Water sticks to surfaces (i.e., *adhesion*) because of water's polarity.

continued...

Any duplication (copies, uploads, PDFs) is illegal.

Water may form a thin film on a highly smooth surface (e.g., glass) because the molecular forces between glass and water molecules (adhesive forces) are stronger than the water molecules' *cohesive* forces.

5. A is correct.

The *electronegative oxygen* pulls electron density away from the carbon atom and creates a net dipole towards the oxygen, resulting in a polar covalent bond.

The electronegativity values of carbon and hydrogen are similar and result in a covalent bond (i.e., about equal sharing of bonded electrons).

$$\underset{H \diagdown \;\; \diagup H}{\overset{\textstyle O}{\underset{\textstyle C}{\|}}}$$

6. B is correct.

The *octet rule* states that atoms of main-group elements combine so that each atom has eight electrons in its valence shell.

7. D is correct.

Formula to calculate *dipole moment*:

$$\mu = qr$$

where μ is dipole moment (coulomb meter or C·m), q is the charge (coulomb or C), and r is the radius (meter or m)

Convert the unit of dipole moment from Debye to C·m;

$$0.16 \times 3.34 \times 10^{-30} = 5.34 \times 10^{-31}\,\text{C·m}$$

Convert the unit of radius to meter:

$$115\,\text{pm} \times 1 \times 10^{-12}\,\text{m/pm} = 115 \times 10^{-12}\,\text{m}$$

Rearrange the dipole moment equation to solve for q:

$$q = \mu / r$$

$$q = (5.34 \times 10^{-31}\,\text{C·m}) / (115 \times 10^{-12}\,\text{m})$$

$$q = 4.65 \times 10^{-21}\,\text{C}$$

Express the charge in terms of electron charge (e).

$$(4.65 \times 10^{-21}\,\text{C}) / (1.602 \times 10^{-19}\,\text{C/e}) = 0.029\,\text{e}$$

In this NO molecule, oxygen is the more electronegative atom. Therefore, the charge experienced by the oxygen atom is negative: –0.029e.

Copyright © Sterling Test Prep.

8. A is correct.

The greater the difference in *electronegativity* between two atoms in a compound, the more polar a bond these atoms form.

The atom with the higher electronegativity is the partial (delta) negative end of the dipole.

9. C is correct.

The substantial difference between the two forms of carbon (i.e., graphite and diamonds) is mainly due to their crystal structure, hexagonal for graphite and cubic for diamond.

The conditions to convert graphite into diamond are high pressure and high temperature.

Creating synthetic diamonds is time-consuming, energy-intensive, and expensive since carbon is forced to change its bonding structure.

10. D is correct.

The *valence shell* is an atom's outermost shell (i.e., highest principal quantum number, n).

Valence electrons are those electrons of the outermost electron shell that can participate in a chemical bond.

11. B is correct.

If O's charge is -1, the proper formula of its compound with K ($+1$) should be KO.

12. C is correct.

Electronegativity is a chemical property that describes an atom's tendency to attract electrons.

The most common use of electronegativity pertains to polarity along the *sigma* (single) bond.

The most electronegative atom is F, while the least electronegative atom is Fr.

The trend for increasing electronegativity on the periodic table is up and toward the right (i.e., fluorine).

13. B is correct.

The greater the difference in electronegativity between two atoms, the more polar bond these atoms form.

The atom with the higher electronegativity is the partial (delta) negative end of the dipole.

Although each C–Cl bond is polar, the dipole moments of each of the four bonds in CCl_4 (carbon tetrachloride) cancel because the molecule is a symmetric tetrahedron.

14. A is correct.

When NaCl dissolves in water, ion–dipole is the force of attraction between Na^+ and H_2O.

15. D is correct.

Group IA elements (e.g., Li, Na, and K) *lose* 1 electron to achieve a complete octet as +1 cations.

Group IIA elements (e.g., Mg and Ca) *lose* 2 electrons to achieve a complete octet as +2 cations.

Group VIIA elements (halogens such as F, Cl, Br, and I) *gain* 1 electron for a complete octet as a −1 anion.

Copyright © Sterling Test Prep.

Notes for active learning

Notes for active learning

Copyright © Sterling Test Prep.

CHAPTER 5

Stoichiometry

Copyright © Sterling Test Prep.

Properties of Molecular Compounds

Molecular compounds

A *molecular compound* is an electrically neutral particle consisting of two or more nonmetals covalently bonded.

A covalent bond shares electrons between two atoms during chemical bonding.

Groups of atoms within a molecule behave like single particles or discrete units.

Based on the composition and type of atoms, molecules have a specific molecular weight (MW).

For example, a water molecule consists of two hydrogen atoms bonded to one oxygen atom.

Particle diagrams

A particle diagram has elements and compounds represented as distinctive shapes (e.g., balls shaded or unshaded). Particle diagrams can represent elements and compounds and their molecular composition by the types of shapes and how they are connected.

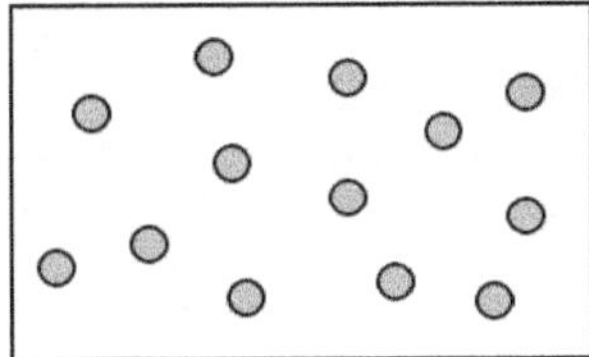 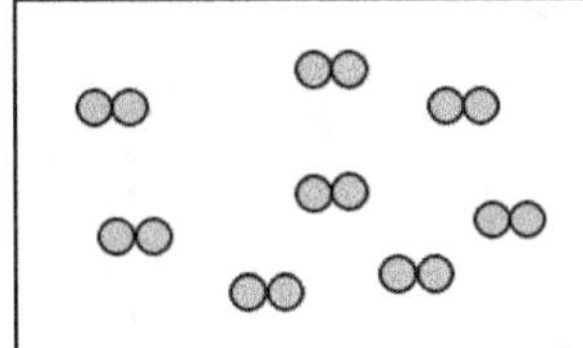

Elements

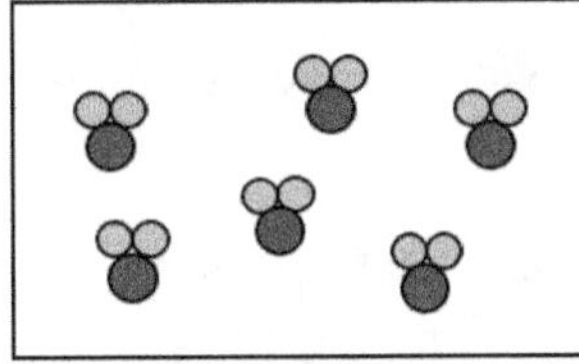 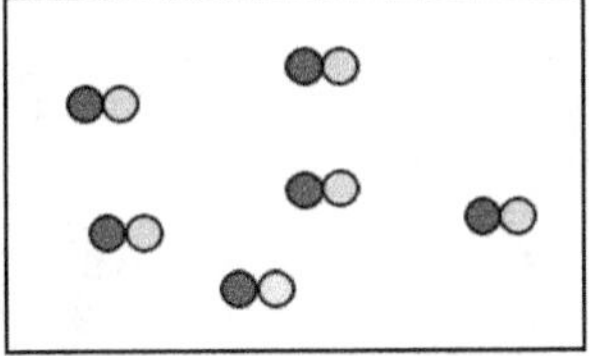

Compounds

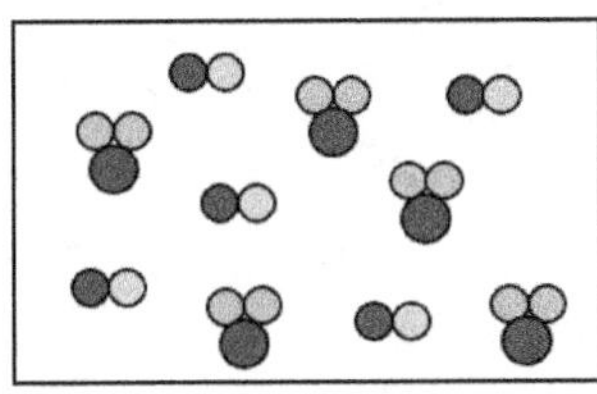 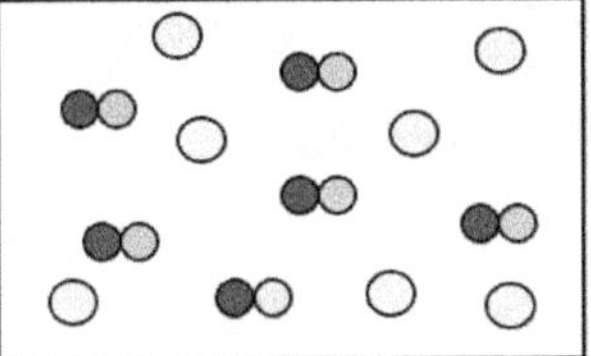 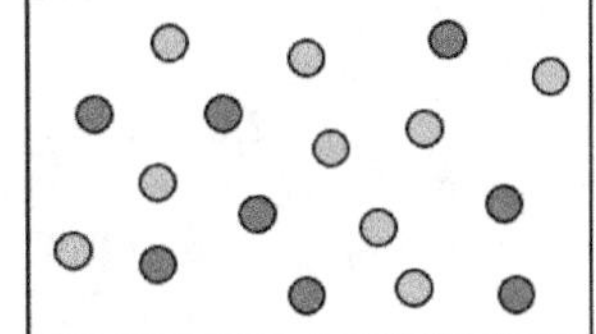

Mixtures

Any duplication (copies, uploads, PDFs) is illegal.

Avogadro's number

Molecular weight (MW) is the weight of 1 mole of molecules, where 1 mole equals 6.02×10^{23} particles.

The 6.02×10^{23} particles value is Avogadro's number (N_A).

The molecular weight of a substance is the sum of the atomic masses of its atoms. Typically, molecular weight is expressed in grams per mole (g/mol).

Molecular weight can be expressed in *atomic mass units* (amu or u), where 1 u = 1 g/mol.

$\qquad$ 1 amu = 1 Dalton (Da).

For example, ^{12}C weighs 12 amu = 12 g/mol.

Atomic mass

The *atomic mass* is the mass of a single atom.

The *molecular weight* is the sum of the weights of the atoms in the molecule.

The *atomic mass* is listed on the periodic table; it is directly below the symbol for the element, as indicated by the arrow for lithium (Li).

Molar mass

The *molar mass* is the mass of a given substance divided by the amount of a substance and is expressed in g/mol.

Atomic mass gives the mass of atoms, while molecular weight gives the mass of molecules.

Formula weight is the sum of weights of the atoms in an empirical formula of a molecule.

For example, the molar mass of methane (molecular formula = CH_4)

$\qquad CH_4$ = 1 carbon (12.0108 g/mol) + 4 hydrogen (1.0079 g/mol)

$\qquad CH_4$ = 12.0108 g/mol + 4(1.0079 g/mol)

$\qquad CH_4$ = 16.0424 g/mol

Copyright © Sterling Test Prep.

Molecular mass

The *molecular mass* (*m*) is the mass of a given molecule, measured in Daltons (Da or u).

The molecular mass is the *ratio of the mass of a molecule to the atomic mass unit*. The molecular mass is more commonly used when referring to the mass of a single molecule.

The molecular mass is often used interchangeably with molecular weight (MW).

Molecules of the same compound have different molecular masses with different isotopes of an element.

Molecular weight

Molecular weight (MW) is the collection of weights of the atoms in a molecular formula.

Molecular weight commonly refers to a weighted average of a sample.

For example, determine the molecular weight of water (H_2O).

Water, from its chemical formula, has two hydrogens and one oxygen.

From the periodic table, the hydrogen (H) atom's mass is 1.008 amu, and oxygen (O) is 15.999 amu.

For example, the equation to determine the MW of water (H_2O):

The chemical formula for water:

$$2\ (H) + 1\ (O)$$

Molecular weight for water:

$$MW = 2\ H\ (1.008\ g/mol) + 1\ O\ (15.999\ g/mol)$$

$$MW = 18.015\ g/mol$$

Another example of atomic mass is seen in the graphic below.

A single carbon atom is approximately 12 amu (12 g/mol).

Oxygen exists as a diatomic molecule.

$$O_2\ (2 \times 15.999\ g/mol = 32\ g/mol)$$

Any duplication (copies, uploads, PDFs) is illegal.

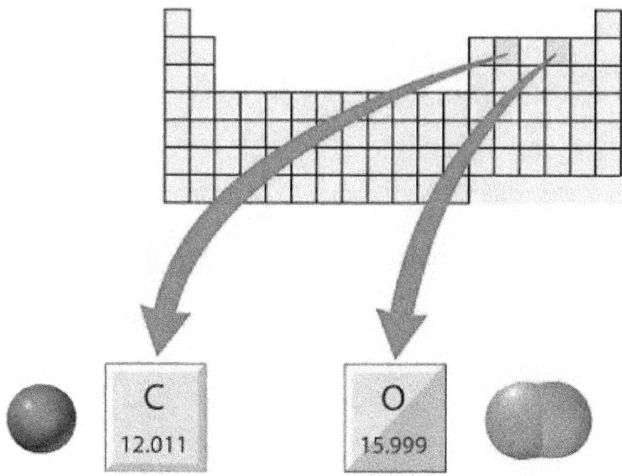

The mass of one carbon atom is approximately 12 amu.

The mass of an O_2 molecule is approximately (16 + 16) 32 amu.

A carbon (C) atom is [(12 g/mol) / (32 g/mol)], or 3/8 as massive as an O_2 molecule.

Copyright © Sterling Test Prep.

Common Metric Units

Seven base units of SI measurements

The metric system is a decimal system of measurement known as the *International System of Units* (commonly abbreviated as SI).

The SI has seven base units:

length (meters)

mass (kilograms)

time (seconds)

electric current (amperes)

temperature (Kelvin)

amount of substance (moles)

luminous intensity (candelas)

Other SI units of measurements, such as volume (liters), density (kg/m^3), and pressure (atmospheres), are derived from the base units.

The SI units are based on 10 (or multiples of 10); conversion between the units is uniform and straightforward. The SI system uses prefixes for the magnitude of a measured quantity; the prefix gives the conversion factor.

Common prefixes, symbols, and powers

Learn the common prefixes shown:

Prefix	Symbol	Power	Prefix	Symbol	Power
mega-	M	10^6	centi-	c	10^{-2}
kilo-	k	10^3	milli-	m	10^{-3}
hecto-	h	10^2	micro-	μ	10^{-6}
deca-	D	10^1	nano-	n	10^{-9}
deci-	d	10^{-1}	pico-	p	10^{-12}

The SI units do not allow for double prefixes (e.g., $1{,}000\ g \neq 1$ hectodecagram); instead, one prefix is used for the quantity of base units (e.g., $1{,}000\ g = 1$ kilogram).

Any duplication (copies, uploads, PDFs) is illegal.

Common conversions

Memorize a few typical metric-imperial conversions:

Length: 2.54 cm = 1 inch

Mass: 454 g = 1 pound

Volume: 0.946 L = 1 quart

Temperature: °C = (°F − 32) / 1.8

Significant figures

Significant figures indicate the number of digits known for a measured (or calculated) quantity within a *degree of uncertainty*.

For example, if a thermometer indicates a boiling point of 36.2 °C and has an uncertainty of ± 0.2 °C, three figures in 36.2 are significant (including the 0.2, which is uncertain).

The notation 36.2 ± 0.2 °C indicates the known quantity and the uncertainty digits (± 0.2 °C).

Converting between units

Conversions between metric units involve adding or subtracting zeros.

For example, suppose a 250 mg aspirin tablet's mass needs to be converted to grams. Start by using the units to set up the problem. If a unit is to be converted (in this case, mg), it is placed in the numerator. That unit must be in the denominator of the conversion factor for it to cancel:

$$(250 \text{ mg}/1) \cdot (1 \times 10^{-3} \text{ g}/1 \text{ mg})$$

$$(250 \text{ mg}/1) \cdot (1 \times 10^{-3} \text{ g}/1 \text{ mg}) = 0.250 \text{ g}$$

Units cancel to give grams.

The conversion factor is shown as a numerator of 1×10^{-3} as it must be entered on most calculators, not 10^{-3}.

The mg is a value of 1, and the prefix "*milli-*" applies to the gram unit. 1 mg = 1×10^{-3} g.

Conversions between English/imperial and metric units work similarly.

The difference is the conversion factors; they are not powers of ten and are different for each unit, making conversions between English/imperial and metric units more complex.

Therefore, the imperial system is not ideal for scientific calculations.

Copyright © Sterling Test Prep.

Converting with multiple units

For example, convert the mass of a 23 lbs. object to kilograms?

Pound units convert to grams, and grams are converted to kilograms.

Use the proper units to set up the problem:

$$23 \text{ lbs.} = (23 \text{ lbs.} / 1) \times (454 \text{ g}/ 1 \text{ lbs.}) \times (1 \text{ kg}/1 \times 10^3 \text{ g})$$

$$23 \text{ lbs.} = (23 \text{ \sout{lbs.}} / 1) \times (454 \text{ \sout{g}}/ 1 \text{ \sout{lbs.}}) \times (1 \text{ kg}/1 \times 10^3 \text{ \sout{g}})$$

$$23 \text{ lbs.} = 10 \text{ kg}$$

A conversion problem may include multiple units.

For example, convert the pressure 14 lb/in^2 to g/cm^2.

For problems involving multiple conversions, work with one unit at a time.

Convert the pounds to gram units:

$$14 \text{ lb/in}^2 \times 454 \text{ g/lb}$$

Convert the in^2 unit to cm^2 units.

Set up the conversion without the exponent first, using the conversion factor:

$$1 \text{ in} = 2.54 \text{ cm}$$

Since in^2 and cm^2 are needed, raise each value to the second power:

$$= 14 \text{ lb/in}^2 \times 454 \text{ g/lb} \times 1^2 \text{ in}^2/2.54 \text{ cm}^2$$

$$= 9.9 \times 10^2 \text{ g/cm}^2$$

When the units are squared, the numbers associated with them must be squared.

For 1 in^2, it does not make a difference because of $1^2 = 1$.

Square the quantities (numbers) when the units are squared.

Check the units to verify that the problem has been set up correctly.

Percent mass

Percent mass (% mass) is concentration *comparing the mass of one part of a substance to the mass of the whole.*

The percent mass unit is used for solutions, especially concentrated acid, or base solution, where the concentration is expressed as percent by mass on the bottle.

Percent mass is:

$$\% \text{ mass} = (\text{mass of species of interest} / \text{total mass}) \times 100\%$$

For example, find the % mass of sodium (Na) in sodium bicarbonate ($NaHCO_3$).

Use the periodic table to find the atomic mass of the elements.

Na = 22.99 g/mol

H = 1.01 g/mol

C = 12.01 g/mol

O = 16.00 g/mol

Use the molecular mass of the atoms to calculate the molecular weight of $NaHCO_3$:

1 Na (22.99 g/mol) + 1 H (1.01 g/mol) + 1 C (12.01 g/mol) + 3 O (3 × 16.00 g/mol)

MW $NaHCO_3$ = 84.01 g/mol

Find the percent mass of Na:

% mass Na = (mass of species of interest / total mass) × 100%

% mass Na = (22.99 g/mol / 84.01 g/mol) × 100%

% mass Na = 27.4%

Concentration can be expressed as % volume, calculated like % mass.

Copyright © Sterling Test Prep.

Mole Concept

Stoichiometry

Stoichiometry uses quantitative relationships between products and reactants.

The number of molecules is expressed in the mole unit.

Avogadro's number (N_A) is the number of particles in 1 mole of a substance and equals 6.02×10^{23} particles.

The substance's mass and moles are determined for reactants and products for a reaction.

$$\underline{2\ H_2} \qquad + \qquad \underline{1\ O_2} \qquad \rightarrow \qquad \underline{2\ H_2O}$$

2 moles = 4 g	1 mole = 32 g	2 moles = 36 g
12.04×10^{23} molecules	6.02×10^{23} molecules	12.04×10^{23} molecules

Most stoichiometry problems follow a set strategy using moles:

$$\text{Quantity A} \rightarrow \text{Moles A} \rightarrow \text{Moles B} \rightarrow \text{Quantity B}$$

Many stoichiometry problems are solved using this strategy.

Each step is examined and combined to solve complicated problems.

Moles to moles conversion

For example, how many moles of $CaCO_3$ are in a 25.0 g sample?

Calculate the molar mass of $CaCO_3$ using atomic mass information from the periodic table.

Calculate the molar mass with the same significant figures as the quantity converted:

$$1\ Ca = 40.08\ \text{g/mol} \times 1 = 40.08\ \text{g/mol}$$

$$1\ C = 12.01\ \text{g/mol} \times 1 = 12.01\ \text{g/mol}$$

$$3\ O = 16.00\ \text{g/mol} \times 3 = 48.00\ \text{g/mol}$$

$$CaCO_3 = 40.08\ \text{g/mol} + 12.01\ \text{g/mol} + 48.00\ \text{g/mol}$$

$$CaCO_3 = 100.09 = 100.1\ \text{g/mol}$$

Use the molar mass to convert the 25.0 g mass of $CaCO_3$ to moles $CaCO_3$.

$$CaCO_3 = 25.0\ g \times (1\ \text{mol}\ CaCO_3 / 100.09\ g\ CaCO_3)$$

$$CaCO_3 = 0.250\ \text{mol}$$

Any duplication (copies, uploads, PDFs) is illegal.

Moles to grams conversion

For example, what is the number of grams of $CaCO_3$ in 0.750 mol $CaCO_3$? (Use molecular mass of $CaCO_3$ = 100.09)

$CaCO_3$ = 0.750 mol × (100.09 g $CaCO_3$ / 1 mol $CaCO_3$)

$CaCO_3$ = 75.1 g $CaCO_3$

Moles A to moles B conversion

For example, how many moles of sodium ion (Na^+) does 0.100 mol of sodium carbonate (Na_2CO_3) have?

One mole of sodium carbonate Na_2CO_3 contains 2 moles of Na, 1 mole of C, and 3 moles of O. Na_2CO_3 completely dissociates (i.e., electrolyte) into ions in a solution so the sodium atoms are ions. Compare the moles of Na_2CO_3 and moles of Na^+:

= 0.100 mol Na_2CO_3 × (2 mol Na^+ / 1 mol Na_2CO_3)

= 0.200 mol Na^+

Moles of reactants to moles of products conversion

A reaction may be involved in other examples, and a comparison of two compounds is necessary.

The stoichiometric coefficient is the mole ratio between species for a balanced reaction.

For example, a typical dissociation reaction:

$$2\ KClO_3 \rightarrow 2\ KCl + 3\ O_2$$

In this reaction, 2 moles of potassium chlorate ($KClO_3$) decompose into 2 moles of potassium chloride (KCl) and 3 moles of oxygen (O_2).

How many O_2 molecules are produced during the dissociation reaction above?

If there are 0.400 mol of $KClO_3$ and the number of moles of O_2 is needed, use stoichiometric coefficients to arrange a mole ratio, so moles of $KClO_3$ cancel, and moles of O_2 remain:

= (0.400 mol $KClO_3$) × (3 mol O_2 / 2 mol $KClO_3$)

= 0.600 mol O_2

Avogadro's number converts the number of moles to the number of particles.

= 0.600 mol O_2 × (6.02 × 10^{23} molecules O_2 / 1 mol O_2)

= 3.61 × 10^{23} O_2 molecules

The mole units of O_2 successfully cancel in the answer.

Copyright © Sterling Test Prep.

Density and Concentration

Density and specific gravity

Density (ρ) is the ratio of mass over the volume of a substance.

The SI unit of density is kg/m^3.

Specific gravity is a unitless ratio between the density of a sample and a *reference substance* (often water).

Some units of density:

- Density of water = 1 g/mL = 1 g/cm^3

- Specific gravity of water = 1 g/cm^3 / 1 g/cm^3 = 1

- Density of lead = 11 g/cm^3

- Specific gravity of lead = 11 g/cm^3 / 1 g/cm^3 = 11

The densest known element is osmium at 22.59 g/cm^3.

The least dense known element is hydrogen at $8.99 \times 10^{-5} \, g/cm^3$.

Hydrogen is about 250,000 times less dense than osmium.

Molarity and molality

Molarity (M) is the *moles of solute per liter of solution*.

$$\text{Molarity} = \text{moles of solute} / \text{L of solution}$$

Molarity is the most common concentration unit used for solutions.

The molality (m) is the moles of solute per kilogram of solvent (Phase Equilibria chapter).

$$\text{Molality} = \text{moles of solute} / \text{kg of solvent}$$

Concentration of dilute solutions

Very dilute solutions use the unit of *parts per million* (ppm).

$$\text{ppm} = \frac{\text{grams of solute}}{\text{grams of solution}} \times 10^6$$

The amount of solute relative to the amount of solvent is typically minimal.

Therefore, the density of the solution is, to a first approximation, the same as the density of the solvent.

Any duplication (copies, uploads, PDFs) is illegal.

Parts per million may be expressed as:

$$\text{ppm} = \frac{\text{mg solute}}{\text{kg solution}}$$

If the solvent is water (density of 1.00 kg/L), the expression for ppm:

$$\text{ppm} = \frac{\text{mg solute}}{\text{L solution}}$$

Another measurement unit to express the concentration of more dilute solutions is parts per billion (ppb).

The expression for *parts per billion*:

$$\text{ppb} = \frac{\text{grams of solute}}{\text{grams of solution}} \times 10^9$$

The density of a dilute solution approximates the solvent's density.

Thus, parts per billion may be expressed as:

$$\text{ppb} = \frac{\mu\text{g solute}}{\text{kg solution}}$$

$$\text{ppb} = \frac{\mu\text{g solute}}{\text{L solution}}$$

Mole fraction and mole percent

The *mole fraction* (X) is the ratio between the moles of a specific molecule over the total moles of components in the mixture.

The *mole percent* ($\%X$) is the mole ratio multiplied by 100.

$$X_{\text{solute}} = \frac{\text{mol solute}}{\text{total moles of all components}}$$

$$X_{\text{solute}}\% = \frac{\text{mol solute}}{\text{total moles of all components}} \times 100$$

When converting between units, start by choosing an arbitrary amount of solution in the denominator of the concentration to be converted.

Copyright © Sterling Test Prep.

Percent mass to molarity

For example, if converting percent mass to molarity, assume 100 grams of solution.

If converting molarity to percent mass, assume one liter of solution.

For example, what are the molarity, molality, and mole fraction of HCl for a concentrated HCl solution known to be 37.0% HCl by mass, and its density is 1.19 g/ml?

Begin with the valid assumption that the HCl solution is 100 g.

Since the HCl percent mass is 37%, 37.0 g of the solution is HCl (grams of solute), the remaining 63.0 g is water (grams of solvent).

To find molarity, determine the moles of HCl (solute) per liter of solution.

First, convert the mass of HCl to moles:

mol HCl = 37.0 g HCl × 1 mol HCl / 36.5 g HCl

mol HCl = 1.01 mol HCl

Using the solution's density, convert the known mass of the solution (100 g) to liters of solution.

$$\text{L solution} = 100\,\text{g solution} \times \frac{1\,\text{mL solution}}{1.19\,\text{g solution}} \times \frac{1\,\text{L solution}}{1000\,\text{mL solution}}$$

$$\text{L solution} = 100\,\cancel{\text{g solution}} \times \frac{1\,\cancel{\text{mL solution}}}{1.19\,\cancel{\text{g solution}}} \times \frac{1\,\text{L solution}}{1000\,\cancel{\text{mL solution}}}$$

L solution = 0.0840 L solution

Using the moles of solute (HCl) and volume of solution in liters, calculate molarity (M) of the solution as moles of solute per liter of solution:

$$M = \frac{1.01\,\text{mol HCl}}{0.0849\,\text{L solution}}$$

M = 12.0 M HCl/L solution

Molality to mole fraction

From above, find the molality of the HCl solution.

The moles of solute are known (1.01 mol HCl).

Determine the mass of the solvent (H_2O) in kilograms:

$$63.0 \text{ g } H_2O \times (1 \text{ kg } H_2O \text{ / } 1{,}000 \text{ g } H_2O) = 0.0630 \text{ kg } H_2O$$

Using the moles and the mass of solvent, calculate molality (m or b):

$$m = 1.01 \text{ mol HCl / } 0.0630 \text{ kg } H_2O$$

$$m = 16.0 \text{ mol HCl/kg } H_2O$$

$$m = 16.0 \text{ m } H_2O$$

Finally, determine the mole fraction of HCl.

From before, there are 1.01 moles of HCl. Calculate moles of H_2O.

$$\text{mol } H_2O = 63.0 \text{ g } H_2O \times (1 \text{ mol } H_2O \text{ / } 18.0 \text{ g } H_2O)$$

$$\text{mol } H_2O = 63.0 \text{ g } H_2O \times (1 \text{ mol } H_2O \text{ / } 18.0 \text{ g } H_2O)$$

$$\text{mol } H_2O = 63.0 \times (1 \text{ mol } H_2O \text{ / } 18.0)$$

$$\text{mol } H_2O = 3.50 \text{ mol}$$

The moles of the molecules in the problem are known.

Calculate the mole fraction of HCl:

$$\chi_{solute} = \frac{\text{mol solute}}{\text{total moles of all components}}$$

$$\chi_{HCl} = 1.01 \text{ mol HCl / } [(1.01 \text{ mol HCl} + 3.50 \text{ mol } H_2O)]$$

$$\chi_{HCl} = 0.244$$

Copyright © Sterling Test Prep.

Oxidation States

Oxidation number

The *oxidation number* (or *oxidation state*) is the charge an atom has or appears to have when a set of rules counts the compound's electrons.

Oxidation numbers are typically used in ionic compounds for oxidation-reduction reactions, which involve the transfer of electrons.

For nonionic compounds, oxidation numbers determine if the compound's chemical formula is written correctly.

Assigning oxidation numbers

1. The oxidation state of an element is zero.

2. For main group metals (Groups 1-2, 13-18), the ion's charge equals the valence electrons or group number.

   Transition metals do not obey this rule and may have multiple oxidation states.

   Oxidation states are sometimes not by group number

3. The oxygen ion in a compound is typically -2 except the peroxide ion (O_2^{2-}), which is -1.

4. The sum of oxidation states equals the polyatomic ion's (or neutral molecule's) overall charge for polyatomic ions (or compounds).

Determining oxidation states

For example, determine the oxidation states of the elements in potassium permanganate ($KMnO_4$).

Solution: apply rule two.

Since potassium is a Group IA alkali metal, its oxidation state is $+1$.

Assign x to Mn for now since manganese may exist in several oxidation states.

There are four oxygen atoms in the permanganate ion, with oxidation states of -2 per O atom.

The overall charge of the neutral compound equals zero:

$$K \quad Mn \quad O_4$$

$$+1 \quad x \quad 4(-2)$$

The algebraic expression is:

$$1 + x - 8 = 0$$

Solving for x gives the oxidation state of manganese:

$$x - 7 = 0$$

$$x = +7$$

$$K \quad Mn \quad O_4$$

$$+1 \quad +7 \quad 4(-2)$$

For example, what is the oxidation state of chromium in dichromate ion $Cr_2O_7^{2-}$?

Start by assigning -2 as the oxidation state for oxygen.

Since the oxidation state for chromium is not known, and two chromium atoms are present, assign the algebraic value of $2x$ for chromium:

$$Cr_2 \quad O_7^{2-}$$

$$2x \quad 7(-2)$$

Use the algebraic equation to solve for x.

Since the overall charge of the ion is -2, the expression is set equal to -2, rather than 0:

$$2x + 7(-2) = -2$$

Solve for x:

$$2x - 14 = -2$$

$$2x = 12$$

$$x = +6$$

Each chromium in the ion has an oxidation state of $+6$.

What are the oxidation states of the elements in the polyatomic compound $Fe_2(CO_3)_3$?

Two elements (iron and carbon) have more than one oxidation state.

When considering molecules formed from cations (positive ions) and anions (negative ions), the molecule is split into constituent ions. Determine the charge on each ion.

Copyright © Sterling Test Prep.

For example, iron (Fe) has more than one oxidation state, but carbonate ion has an oxidation state of -2 (CO_3^{2-}).

With this information, Fe's oxidation state can be determined:

$$Fe_2 \quad (CO_3)^3$$

$$2x \quad 3(-2)$$

$$2x - 6 = 0$$

$$2x = 6$$

$$x = 3$$

Each iron ion in the compound has an oxidation state of $+3$.

Next, consider the carbonate ion independent of the iron (III) ion:

$$CO_3^{2-}$$

$$x \quad 3(-2)$$

$$x - 6 = -2$$

$$x = +4$$

The oxidation state of carbon is $+4$, and each oxygen is -2.

Oxidation-reduction reactions

An *oxidation-reduction reaction* occurs when a transfer of electrons occurs between two chemicals.

For an oxidation-reduction (or redox) reaction to proceed, one substance in a reaction is oxidized while another substance is reduced; reduction or oxidation processes cannot occur separately.

Oxidation is a loss of electrons, which is typically associated with metals and may involve the addition of oxygen or the removal of hydrogen.

Reduction is a gain of electrons, which is typically observed in nonmetals and may involve the addition of hydrogen or removal of oxygen.

In biological systems, cells oxidize and reduce metals. Cytochrome c protein plays a vital role in ATP production. The cytochrome c protein contains a Fe^{2+} cation that undergoes oxidation to form Fe^{3+}, then reduced into Fe^{2+}.

In organic reactions, look for the movement of oxygen and hydrogen.

Combustion (i.e., burning) is a reaction with oxygen as an example of a redox reaction.

Any duplication (copies, uploads, PDFs) is illegal.

Oxidizing and reducing agents

An *oxidizing agent* (or *oxidant*) is a substance that oxidizes another substance by removing electrons.

The oxidizing agent gains the removed electrons, thus reducing itself.

Thus, the oxidation number of the oxidizing agent becomes *less positive*.

A *reducing agent* is a substance that reduces another by donating its electrons to the substance being reduced.

The reducing agent loses these electrons and is oxidized.

Thus, the reducing agent's oxidation number becomes *more positive* (or less negative).

Identifying reduced and oxidized species

The number line below helps identify the reduced or oxidized substances and whether they are oxidizing or reducing agents, respectively.

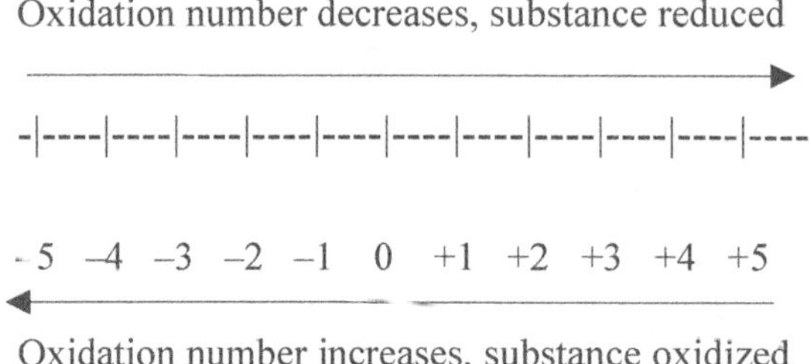

The image below summarizes the relationship between reducing agents and oxidizing agents.

Essentially, atoms that *lose electrons* during a chemical reaction undergo oxidation, and atoms that *gain electrons* during a chemical reaction undergo reduction.

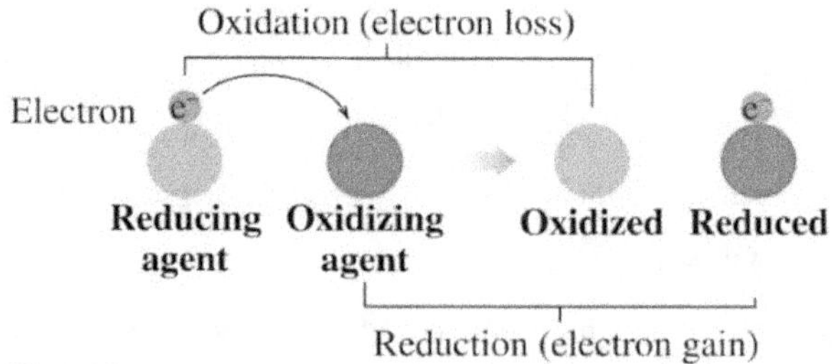

The oxidized *species* is the reducing *agent*, and the reduced *species* is the oxidizing *agent*.

Copyright © Sterling Test Prep.

Loss and gain of electrons

The OIL RIG mnemonic helps remember what is happening to the electrons in a redox reaction.

OIL RIG = **O**xidation **I**s **L**oss; **R**eduction **I**s **G**ain

Remember, this mnemonic device refers to the *loss and gain of electrons*.

For example. identify the substance being oxidized, the substance being reduced, the oxidizing agent, and the reducing agent in the following reaction:

$$Cu\ (s) + 4\ HNO_3\ (aq) \rightarrow Cu(NO_3)_2\ (aq) + 2\ NO_2\ (g) + 2\ H_2O\ (l)$$

Determine the oxidation state for the reactants and products.

Elemental copper, Cu (*s*), has an oxidation state of zero (0).

For nitric acid (HNO_3) set N equal to x:

H   N O_3

+1 x 3(–2)

$1 + x - 6 = 0$

$x = +5$

So,

H = +1, N = +5, O = –2

For copper (II) nitrate, $Cu(NO_3)_2$:

$Cu(NO_3)_2$

Nitrates are –1, while Cu might be +1 or +2.

$Cu(NO_3)_2$

Cu $(NO_3)_2$

$x + 2(-1) = 0$

$x = 2$

Cu's oxidation state is +2, oxygen is –2

For nitrate ion (NO_3^-) set N equal to x:

N O_3^-

x 3(–2)

$x - 6 = -1$

$x = +5$

So, $Cu = +2$, $N = +5$, $O = -2$

For nitrogen dioxide (NO_2) set N equal to x:

      N O_2

      x $2(-2)$

      $x - 4 = 0$

      $x = +4$

So, $N = +4$, $O = -2$

For water (H_2O):

      H_2 O

      $2(+1)$ -2

So, $H = +1$, $O = -2$

The reaction is summarized in the following table.

$$Cu\ (s) + 4\ HNO_3\ (aq) \rightarrow Cu(NO_3)_2\ (aq) + 2\ NO_2\ (g) + 2\ H_2O\ (l)$$

Element	Oxidation State: reactants	Oxidation State: products
Cu	0	+2
H	+1	+1
N	+5	+5 (in NO_3^-), +4 (in NO_2)
O	-2	-2

Oxidation states of reactants and products for the example reaction above

Identify the molecules that changed their oxidation state.

Copper's oxidation number increased from 0 to +2, so copper has been oxidized (loses electrons), and therefore, is the reducing agent.

Copper reduced nitric acid by donating electrons to the nitrogen in nitric acid.

Nitrogen in nitric acid changed from +5 to +4 in nitrogen dioxide, which means that copper has reduced nitric acid. Since nitric acid is reduced, it is the oxidizing agent. Oxidation cannot occur without a reduction.

Copyright © Sterling Test Prep.

Common oxidizing and reducing agents

Common oxidizing agents	Common reducing agents
Oxygen O_2	Hydrogen H_2
Ozone O_3	metals (such as K)
Permanganates MnO_4^-	Zn/HCl
Chromates $CrO4_2^-$	Sn/HCl
Dichromates $Cr_2O_7^{2-}$	LAH (lithium aluminum hydride)
peroxides H_2O_2	$NaBH_4$ (sodium borohydride)
Lewis acids	Lewis bases
compounds with many oxygens	compounds with many hydrogens

Common oxidizing and reducing oxidation agents

Disproportionation reactions

In *disproportionation reactions*, an atom undergoes oxidation and reduction to form two atoms with different oxidation states.

For example, consider:

$$2\ Cu^+ \rightarrow Cu + Cu^{2+}$$

The Cu^+ acts as oxidizing and reducing agents and simultaneously reduces and oxidizes itself.

The oxidized Cu^+ becomes Cu^{2+}

The reduced Cu^+ becomes Cu

For example, which of the following equations is a disproportionation reaction?

$$2\ H_2O \rightarrow 2\ H_2 + O_2$$

$$H_2SO_3 \rightarrow H_2O + SO_2$$

$$HNO_2 \rightarrow NO + HNO_3$$

$$Mg + H_2SO_4 \rightarrow MgSO_4 + H_2$$

Disproportionation is a redox reaction in which a species is simultaneously reduced and oxidized to form two different products.

Any duplication (copies, uploads, PDFs) is illegal.

Consider: $HNO_2 \rightarrow NO + HNO_3$

Unbalanced reaction: $HNO_2 \rightarrow NO + HNO_3$

Balanced reaction: $3\ HNO_2 \rightarrow 2\ NO + HNO_3 + H_2O$

The other reactions are classified as:

$2\ H_2O \rightarrow 2\ H_2 + O_2$: decomposition

$H_2SO_3 \rightarrow H_2O + SO_2$: decomposition

$Mg + H_2SO_4 \rightarrow MgSO_4 + H_2$: single replacement

Redox titration

Titration (or *volumetric analysis*) is a standard laboratory procedure to determine an *unknown concentration.*

The *titrant* is the analytical reagent of known concentration carefully added to the sample from a buret until the *equivalence point* (i.e., the point of neutralization) is reached.

If the titrant's volume and concentration are known, its number of moles can be calculated.

The *endpoint* is marked by an indicator's color change when a solution transitions from acidic to slightly basic.

An *indicator* is added to the reaction to determine the endpoint of a redox titration (i.e., when sample molecules have been depleted).

A *redox indicator* undergoes a color change at the neutralization reaction's equivalence point.

For example, ascorbic acid (vitamin C) titrations use iodine undergoing a redox reaction.

Ascorbic acid is *oxidized* (i.e., loses electrons) to dehydroascorbic acid, while iodine is *reduced* (i.e., gains electrons) to an iodide ion.

Starch is an *indicator* because excess iodine reacts with starch and turns the solution from clear to dark blue.

The ascorbic acid concentration is calculated using the initial volume of ascorbic acid (analyte) and iodine's measured volume (titrant).

Copyright © Sterling Test Prep.

Chemical Equations

Classifying chemical reactions

Predicting the products formed in each reaction is no easy task.

Since many chemical compounds exist, memorizing every reaction is not necessary.

Most chemical reactions are classified into several groups.

Identifying the reaction type is a significant step towards predicting a chemical reaction's products.

The important classes of reactions encountered in general chemistry are shown:

Reaction Type	General Reaction
Synthesis	$A + B \rightarrow AB$
Decomposition	$AB \rightarrow A + B$
Replacement	$AB + C \rightarrow AC + B$ (single) $AB + CD \rightarrow AD + CB$ (double)

1. ***Combination reactions*** (or *synthesis reactions*): two (or more) substances form a single product.

   The general form is $A + B \rightarrow AB$, in which two substances combine to form one compound.

2. ***Decomposition reactions***: one reactant forms two or more products, the reverse of a combination reaction.

   The general form is $AB \rightarrow A + B$, compound decomposes to form constituent elements.

3. ***Replacement reactions*** (*substitution reactions* or *displacement reactions*): one atom (or group) replaces another species in a compound.

   o In a ***single replacement reaction***, an element replaces the corresponding element.

   The general form is $AB + C \rightarrow AC + B$, which replaces element B.

Any duplication (copies, uploads, PDFs) is illegal.

- o In a ***double replacement reaction*** (or *metathesis reaction*), two compounds react and switch partners; cations and anions in one compound exchange with counterparts in the other compound.

  The general form is $AB + CD \rightarrow AD + CB$

<u>Note</u>: In each type of replacement reaction, the number of substances on the reactant side of the equation is the *same* as on the product side.

Specific reaction types

- **Oxidation-reduction reactions** (or *redox*) *transfer electrons* between the two reactants.

  A *loss of electrons* (or increase in oxidation number) is oxidation.

  A *gain of electrons* (or decrease in oxidation number) is reduction.

  - o Many combination and decomposition reactions use oxidation/reduction and single displacement reactions.

- *Combustion* involves a carbon-containing molecule (e.g., a hydrocarbon) reacting with oxygen to produce carbon dioxide (CO_2) and water (H_2O).

  - o Combustion reactions tend to be highly exothermic and are considered irreversible.

  - o Combustion is decomposition and oxidation-reduction reactions.

  - o Combustion uses O_2 to produce energy, dissipating as heat or performing work.

  - o General formula for combustion of a hydrocarbon:

    $$C_xH_y + O_2 \rightarrow CO_2 + H_2O$$

- *Condensation* involves water produced from a double replacement reaction.

  Condensation reactions occur among functional groups that contain ~H and ~OH that break from their compounds and form H_2O.

Copyright © Sterling Test Prep.

Adenosine triphosphate (ATP)

Adenosine diphosphate (ADP) + Energy

Hydrolysis (addition of water) for ATP conversion to ADP + energy (inorganic phosphate)

- **Hydrolysis** (hydro- means "water," lysis means "break") is the reverse process of condensation, where water is a reactant, while the other reactant molecule is split into two smaller product molecules.

The general form for *condensation and hydrolysis* is shown below.

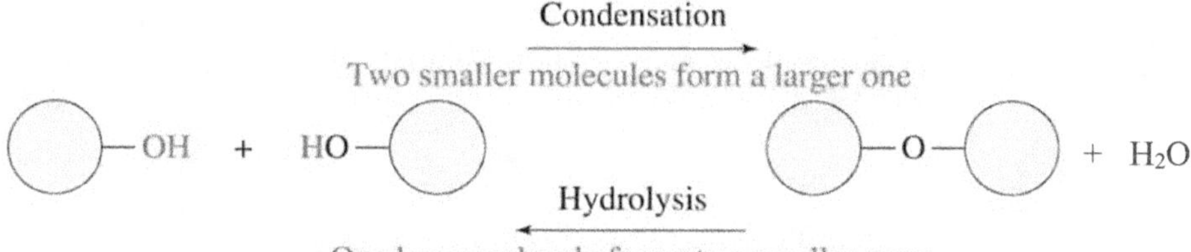

- **Carboxylation reactions** add a carboxyl (C=O) or carboxylic acid (~COOH) group.

  For example, as carbon dioxide (CO_2) moves through the cell, it is added to biomolecules by carboxylase and removed by decarboxylase (~*ase* for enzymes).

Any duplication (copies, uploads, PDFs) is illegal.

Carboxylation is shown below.

bicarbonate   pyruvate   oxaloacetate   water

Conventions for writing chemical equations

Chemical equations have several important parts:

$$H_2SO_4\ (aq) + 2\ NaOH\ (aq) \rightarrow 2\ Na^+\ (aq) + SO_4^{2-}\ (aq) + 2\ H_2O$$

phase   *coefficient*   *direction*   *charge*

The *phase* is indicated with a subscript, solid (*s*), liquid (*l*), gas (*g*), or aqueous (*aq*).

The *coefficient* indicates the relative number of moles of reactants or products.

The *direction* is represented as a single-headed arrow, denoting the forward direction.

Reversible reactions are at a state of *chemical equilibrium* and represented with a double-headed arrow.

At equilibrium, the *rates* of the forward and reverse reactions are equal and constant, and no change in the net number of reactants or products.

A double-sided arrow with one side *larger* denotes a *nonequilibrium condition*, which spontaneously favors the larger arrow's direction.

The *charge* is indicated with a numerical superscript and +/– sign.

It is common not to indicate the charge on a neutral compound or substance.

Copyright © Sterling Test Prep.

Balancing chemical equations

The *law of mass conservation* (refer to the chapter on thermodynamics) states that atoms are neither created nor destroyed in a chemical reaction—they are simply rearranged.

The reactants and products coefficients must balance; there are *equivalent amounts of atoms* on each side of the reaction.

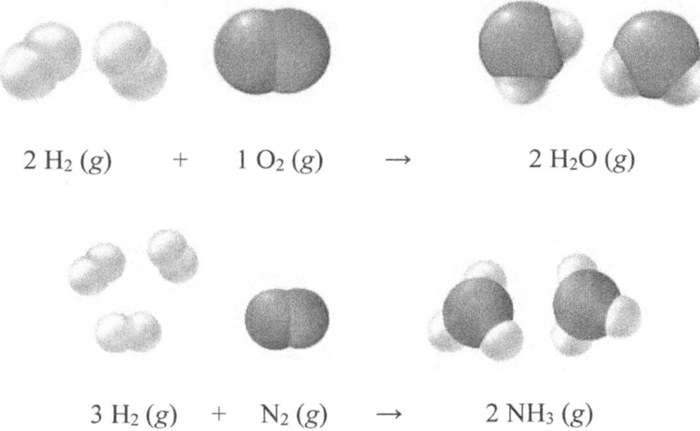

$$2\ H_2\ (g) \quad + \quad 1\ O_2\ (g) \quad \rightarrow \quad 2\ H_2O\ (g)$$

$$3\ H_2\ (g) \quad + \quad N_2\ (g) \quad \rightarrow \quad 2\ NH_3\ (g)$$

Balance a chemical equation by assigning coefficients to each species until atoms balance on both sides.

For example, balance the combustion of propanol (C_3H_8O):

$$C_3H_8O + O_2 \rightarrow CO_2 + H_2O$$

Select the atom (or ion) present in one species on each side of the equation.

For the combustion of propanol, start with carbon.

Add the coefficient 3 to CO_2 on the right side since there are three carbons on the left side, indicated by C_3.

$$C_3H_8O + O_2 \rightarrow \mathbf{3}\ CO_2 + H_2O$$

Hydrogen is the other species present in one molecule on each side so balance the hydrogens.

Add the coefficient 4 to H_2O on the right because $4 \times 2 = 8$ equals the number of H on the left, indicated by H_8.

$$C_3H_8O + O_2 \rightarrow 3\ CO_2 + \mathbf{4}\ H_2O$$

Starting with hydrogen before balancing carbon yields the same result.

First, balance elements present in one species on each side to avoid changing coefficients previously balanced species.

Oxygen is present every term, so it would be more difficult if O were balanced first.

Because of the adjustments to balance other atoms, one oxygen-containing species that has not yet been balanced.

Count the oxygen atoms:

one from C_3H_8O

six from $3\ CO_2$ (because $3 \times 2 = 6$)

four from $4\ H_2O$

Set equation:

$1 + 2x = (3 \times 2) + 4$, where x is the coefficient of the last term, O_2.

Solve for x, which equals $^9/_2$.

$C_3H_8O + x\ O_2 \rightarrow 3\ CO_2 + 4\ H_2O$

$C_3H_8O + {}^9/_2\ O_2 \rightarrow 3\ CO_2 + 4\ H_2O$

Remove fractions, so multiply every term by 2.

$2\ C_3H_8O + 9\ O_2 \rightarrow 6\ CO_2 + 8\ H_2O$

Remove non-integer coefficients, and the equation is balanced.

Copyright © Sterling Test Prep.

Redox Equations

Balancing redox reactions

When solving oxidation-reduction equations, a different approach is required.

Balancing redox reactions is more complicated, splitting the redox reaction into half-reactions.

Two methods for balancing half-reactions:

>The ***ion-electron method*** balances the elements, then the charge by adding electrons.

>The ***oxidation-state method*** treats the species of interest as a single element (i.e., changes oxidation number) and balances it.

Half-reactions balance oxidation-reduction reactions

Use the following procedure when balancing oxidation-reduction reactions:

1. Separate the equation into two *half-reactions*. A half-reaction contains only the species of interest (i.e., those containing the atom that changes oxidation state).

    Each half-reaction corresponds to the oxidation half-reaction or reduction half-reaction.

    Covalently attached moieties are not part of the species of interest.

    A species that does not change its oxidation state is a *spectator ion* (i.e., an ion present in solution but not involved in the reaction of interest).

2. Balance each half-reaction for the *charge* and *number* of atoms.

    >Under *acidic conditions*: add H_2O to the side needing the oxygen atom, then add H^+ to the other.

    >Under *basic conditions*: add 2 OH^- to the side needing oxygen atoms, then add H_2O to the other.

3. After the half-reactions are balanced, recombine the half-reactions:

    Multiply each half-reaction by a factor, so the electrons cancel.

    Like solving a simultaneous equation, the electron term must be eliminated.

4. Lastly, perform these additional steps:

Combine identical species on the same side of the equation.

Cancel identical species on opposite sides of the equation.

Add back in the spectator ions.

For the oxidation-state method, balance the oxygen and hydrogen elements.

Check that each side of the equation has an equal number of atoms and a neutral net charge.

For example, how many electrons are needed to balance the following half-reaction in a basic solution?

$$C_8H_{10} \rightarrow C_8H_4O_4{}^{2-}$$

Rules for balancing a half-reaction in basic conditions:

The first few steps are identical to balancing reactions in acidic conditions.

Step 1: Balance atoms except for H and O

$$C_8H_{10} \rightarrow C_8H_4O_4{}^{2-} \qquad C \text{ is balanced}$$

Step 2: To balance oxygen, add H_2O to the side with fewer oxygen atoms

$$C_8H_{10} + 4\ H_2O \rightarrow C_8H_4O_4{}^{2-}$$

Step 3: To balance hydrogen, add H^+ to the opposing side of H_2O added in the previous step

$$C_8H_{10} + 4\ H_2O \rightarrow C_8H_4O_4{}^{2-} + 14\ H^+$$

This next step is the unique additional step for basic conditions.

Step 4: Add equal amounts of OH^- on both sides. The number of OH^- should match the number of H^+ ions. Combine H^+ and OH^- on the same side to form H_2O. If there are H_2O molecules on each side, subtract accordingly to end up with H_2O on one side only.

There are 14 H^+ ions on the right, so add 14 OH^- ions on both sides:

$$C_8H_{10} + 4\ H_2O + 14\ OH^- \rightarrow C_8H_4O_4{}^{2-} + 14\ H^+ + 14\ OH^-$$

Combine H^+ and OH^- ions to form H_2O:

$$C_8H_{10} + 4\ H_2O + 14\ OH^- \rightarrow C_8H_4O_4{}^{2-} + 14\ H_2O$$

H_2O molecules are on both sides, which cancel, and some H_2O remain on one side:

$$C_8H_{10} + 14\ OH^- \rightarrow C_8H_4O_4{}^{2-} + 10\ H_2O$$

Copyright © Sterling Test Prep.

Step 5: Balance charges by adding electrons to the side with a greater/more positive total charge

Total charge on left side: $14(-1) = -14$

Total charge on the right side: -2

Add 12 electrons to the right side:

$$C_8H_{10} + 14\ OH^- \rightarrow C_8H_4O_4{}^{2-} + 10\ H_2O + 12\ e^-$$

For example, how many electrons are needed to balance the charge for the following half-reaction in an acidic solution?

$$C_2H_6O \rightarrow HC_2H_3O_2$$

Balancing half-reaction in acidic conditions:

Step 1: Balance atoms except for H and O

$$C_2H_6O \rightarrow HC_2H_3O_2 \text{ (C is balanced)}$$

Step 2: To balance oxygen, add H_2O to the side with fewer oxygen atoms

$$C_2H_6O + H_2O \rightarrow HC_2H_3O_2$$

Step 3: To balance hydrogen, add H^+ to the opposing side of H_2O added in the previous step

$$C_2H_6O + H_2O \rightarrow HC_2H_3O_2 + 4\ H^+$$

Step 4: Balance charges by adding electrons to the side with a greater/more positive total charge

Total charge on the left side: 0

Total charge on right side: $4(+1) = +4$

Add 4 electrons to the right side:

$$C_2H_6O + H_2O \rightarrow HC_2H_3O_2 + 4\ H^+ + 4\ e^-$$

The reaction requires adding 4 electrons to the right side.

Ion-electron method for balancing redox reactions

For example, balance the following redox reaction using the ion-electron method:

$$K_2Cr_2O_7\ (aq) + HCl\ (aq) \rightarrow KCl\ (aq) + CrCl_3\ (aq) + H_2O\ (l) + Cl_2\ (g)$$

Step 1 – Separate into half-reactions:

Reduction: $Cr_2O_7{}^{2-} \rightarrow Cr^{3+}$

Oxidation: $Cl^- \rightarrow Cl_2$

Any duplication (copies, uploads, PDFs) is illegal.

The species of interest for the oxidation reaction is Cl^- (not HCl) because the H^+ is not covalently attached to Cl^- and the ions separate in an aqueous solution.

$Cr_2O_7^{2-}$ is used, not $K_2Cr_2O_7$, because K^+ and H^+ are spectator ions.

Step 2 – Balance each of half-reactions:

For the ion-electron method, balance the elements, then balance the charge.

Balance elements for the reduction half-reaction (ion-electron method):

$$Cr_2O_7^{2-} \rightarrow Cr^{3+}$$

$$Cr_2O_7^{2-} \rightarrow 2Cr^{3+}$$

$$Cr_2O_7^{2-} + 14\ H^+ \rightarrow 2\ Cr^{3+} + 7\ H_2O$$

Balance charge for the reduction half-reaction (ion-electron method):

$$Cr_2O_7^{2-} + 14\ H^+ + 6\ e^- \rightarrow 2\ Cr^{3+} + 7\ H_2O$$

Balance charge for the oxidation half-reaction (ion-electron method):

$$Cl^- \rightarrow Cl_2$$

$$2\ Cl^- \rightarrow Cl_2$$

$$2\ Cl^- \rightarrow Cl_2 + 2\ e^-$$

Step 3 – Recombine the half-reactions:

$$Cr_2O_7^{2-} + 14\ H^+ + 6\ e^- \rightarrow 2\ Cr^{3+} + 7\ H_2O$$

$$2\ Cl^- \rightarrow Cl_2 + 2\ e^-$$

Multiply each species in the second equation by 3:

$$6\ Cl^- \rightarrow 3\ Cl_2 + 6\ e^-$$

Add the two equations:

$$Cr_2O_7^{2-} + 14\ H^+ + 6\ e^- + 6\ Cl^- \rightarrow 2\ Cr_3+ + 7\ H_2O + 3\ Cl_2 + 6\ e^-$$

Step 4 – Complete the process:

Except for electrons, there are no identical species to combine or cancel.

$$Cr_2O_7^{2-} + 14\ H^+ + 6\ Cl^- \rightarrow 2\ Cr^{3+} + 7\ H_2O + 3\ Cl_2$$

Copyright © Sterling Test Prep.

Step 5 – For the ion-electron method, the equation is balanced. The spectator ions need to be added to the equation.

When adding the spectator ions, add equal numbers of ions to each side of the reaction.

To the left side: the dichromate ion was paired with K^+, so add 2 K^+ for the dichromate.

To the right side: match the left side by adding 2 K^+.

$$K_2Cr_2O_7 + 14\ H^+ + 6\ Cl^- \rightarrow 2\ Cr^{3+} + 7\ H_2O + 3\ Cl_2 + \mathbf{2\ K^+}$$

Step 6 – There are 14 H^+ on the left side and 14 on the right, so they are balanced. Referring to the original equation, the H and Cl elements on the left originated from the HCl.

Add 8 Cl elements to the product side.

$$K_2Cr_2O_7 + 14\ HCl \rightarrow 2\ Cr^{3+} + 7\ H_2O + 3\ Cl_2 + 2\ K^+ + \mathbf{8\ Cl^-}$$

Step 7 – The right side shows two Cl^- to be combined with the 2 K^+, and the remaining 6 Cl^- goes with the Cr.

The final balanced redox equation is:

$$K_2Cr_2O_7\ (aq) + 14\ HCl\ (aq) \rightarrow \mathbf{2\ CrCl_3}\ (aq) + 7\ H_2O\ (l) + 3\ Cl_2\ (g) + \mathbf{2\ KCl}\ (aq)$$

Oxidation-state method for balancing redox reactions

For example, balance the redox reaction (from above) using the oxidation-state method:

$$K_2Cr_2O_7\ (aq) + HCl\ (aq) \rightarrow KCl\ (aq) + CrCl_3\ (aq) + H_2O\ (l) + Cl_2\ (g)$$

Step 1 - Separate into half-reactions (same as the ion-electron method):

Reduction: $Cr_2O_7^{2-} \rightarrow Cr^{3+}$

Oxidation: $Cl^- \rightarrow Cl_2$

Step 2 - Balance each half-reaction:

Balance the elements of interest first when using the oxidation-state method.

Balance the elements for the reduction half-reaction (oxidation-state method):

1. $Cr_2O_7^{2-} \rightarrow Cr^{3+}$

2. $Cr_2O_7^{2-} \rightarrow 2\ Cr^{3+}$

3. Each oxygen is 2^- so the 2 Cr on the left must be 6^+

4. $2\ Cr^{6+} \rightarrow 2\ Cr^{3+}$

Any duplication (copies, uploads, PDFs) is illegal.

Balance charge for the reduction half-reaction (oxidation-state method):

1. $2\,Cr^{6+} + 6\,e^- \rightarrow 2\,Cr^{3+}$

Balance charge for the oxidation half-reaction (oxidation-state method):

1. $Cl^- \rightarrow Cl_2$

2. $2\,Cl^- \rightarrow Cl_2$

3. $2\,Cl^- \rightarrow 2\,Cl^\circ$

4. $2\,Cl^- \rightarrow 2\,Cl^\circ + 2\,e^-$

Step 3 - Recombine the half-reactions:

$2\,Cr^{6+} + 6\,e^- \rightarrow 2\,Cr^{3+}$

$2\,Cl^- \rightarrow 2\,Cl^\circ + 2\,e^-$

To cancel the electrons, multiply each term in the second equation by 3:

$2\,Cr^{6+} + 6\,e^- \rightarrow 2\,Cr^{3+}$

$6\,Cl^- \rightarrow 6\,Cl^\circ + 6\,e^-$

Add the two equations:

$2\,Cr^{6+} + 6\,e^- + 6\,Cl^- \rightarrow 2\,Cr^{3+} + 6\,Cl^\circ + 6\,e^-$

Step 4 - Except for the electrons, no like terms combine or cancel.

$2\,Cr^{6+} + 6\,Cl^- \rightarrow 2\,Cr^{3+} + 6\,Cl^\circ$

Convert the elements into species by referring to the original equation.

$K_2Cr_2O_7 + 6\,HCl \rightarrow 2\,CrCl_3 + 3\,Cl_2$

Unlike the ion-electron method, where the equation is balanced, and spectator ions are added, the oxidation-state method requires balancing the equation again. After the elements are combined to recreate the molecules, the equation is no longer balanced.

Step 5 - Oxygen: there are seven O atoms on the left, so add 7 H_2O molecules to the right.

(Remember that this method applies to acidic reactions – see explanation for basic reactions).

$K_2Cr_2O_7 + 6\,HCl \rightarrow 2\,CrCl_3 + 3\,Cl_2 + \mathbf{7\ H_2O}$

Copyright © Sterling Test Prep.

Step 6 - Hydrogen: there are 6 H atoms on the left but 14 H atoms on the right.

Eight H atoms should be added to the left for 14 H atoms.

14 H atoms on the left should be HCl (refer to the original equation).

$$K_2Cr_2O_7 + \textbf{14 HCl} \rightarrow 2\ CrCl_3 + 3\ Cl_2 + 7\ H_2O$$

<u>Important</u>: HCl is the species of interest *and* spectator ion.

Some HCl contributes to the $Cl^- \rightarrow Cl_2$ oxidation, but other HCl does not undergo redox; it provides the H^+ ions for water and Cl^- ions for the KCl and $CrCl_3$.

Step 7 - Chlorine: 14 Cl atoms on the left and 12 Cl atoms on the right.

Add 2 Cl atoms to the right. From the original equation, the right-sided Cl atoms come in KCl; do not modify the Cl_2 since it has already been balanced by the oxidation-state method.

When balancing equations at this stage, only manipulate the water and spectator species.

$$K_2Cr_2O_7 + 14\ HCl \rightarrow 2\ CrCl_3 + 3\ Cl_2 + 7\ H_2O + \textbf{2 KCl}$$

The balanced redox equation is:

$$K_2Cr_2O_7\ (aq) + 14\ HCl\ (aq) \rightarrow 2\ CrCl_3\ (aq) + 3\ Cl_2\ (g) + 7\ H_2O\ (l) + 2\ KCl\ (aq)$$

Ten steps for balancing redox half-reactions

1. Split the equation into two half-reactions.

2. Balance atoms other than O or H.

3. Balance O using H_2O. Add water as needed to balance oxygens on the side deficient in O.

4. If in an acidic solution, balance hydrogens using H^+.

   Add H^+ ions as needed to the side deficient in H.

5. In basic solution, add OH^- ions equal to the number of H^+ ions to each side of the chemical equation.

6. The mass is balanced. To balance charge, determine the charge on each side of the reaction and add needed electrons to the positive side, so the charge is the same.

7. Repeat steps 1 through 4 for each half-reaction.

Any duplication (copies, uploads, PDFs) is illegal.

8. If electrons lost do not equal the number gained, multiply each half-reaction by the necessary factor.

9. Sum half-reactions to obtain the balanced net ionic reaction. Inspect H^+ ions and H_2O molecules to cancel.

10. Check the final equation for mass and charge balance.

Important: If expressing H^+ as H_3O^+, change the number of H^+ into the same number of H_3O^+.

Add that same number of H_2O to the opposite side of the equation.

Copyright © Sterling Test Prep.

Limiting Reactant

Identifying limiting reactants

The *limiting reactant* is a component in the lowest stoichiometric quantity, limiting product formed.

The limiting reactant is the reagent depleted first and terminates the reaction.

For example, identify the limiting reactant when a 50.6 g sample of magnesium hydroxide $Mg(OH)_2$ reacts with 45.0 g of hydrogen chloride HCl, according to the reaction below:

$$Mg(OH)_2 + 2\ HCl \rightarrow MgCl_2 + 2\ H_2O$$

Notice that quantities of each reactant are known.

There are three common methods of solving limiting reagent problems.

Moles to determine the limiting reactant

To identify the limiting reagent, *convert grams of reactants to moles* using molar mass:

$$50.6\ g\ Mg(OH)_2 \times (1\ mol\ Mg(OH)_2\ /\ 58.3\ g\ Mg(OH)_2) = 0.868\ mol\ Mg(OH)_2$$

$$45.0\ g\ HCl \times (1\ mol\ HCl\ /\ 36.5\ g\ HCl) = 1.23\ mol\ HCl$$

Select one reactant and calculate the moles of the other reactant needed to deplete the reactant.

For example, begin with magnesium hydroxide ($Mg(OH)_2$):

$$0.868\ mol\ Mg(OH)_2 \times (2\ mol\ HCl\ needed/1\ mol\ Mg(OH)_2) = 1.74\ mol\ HCl\ needed$$

Compare the moles of HCl needed to the actual moles of HCl available.

In this example, 1.74 moles of HCl are needed, and 1.23 moles of HCl are present.

Even though HCl has more moles than $Mg(OH)_2$, HCl is the limiting reagent.

The HCl is consumed before the $Mg(OH)_2$, limiting the product formed.

Theoretical to actual ratio to determine the limiting reactant

The *theoretical yield* is the amount of product expected from a reaction.

Compare the *theoretical ratio* to the *actual ratio* of reactants.

Determine the moles of each reactant using molar mass:

$$50.6\ g\ Mg(OH)_2 \times (1\ mol\ Mg(OH)_2\ /\ 58.3\ g\ Mg(OH)_2) = 0.868\ mol\ Mg(OH)_2\ available$$

$$45.0\ g\ HCl \times (1\ mol\ HCl\ /\ 36.5\ g\ HCl) = 1.23\ mol\ HCl\ available$$

For example, consider the balanced reaction:

$$Mg(OH)_2 + 2\ HCl \rightarrow MgCl_2 + 2\ H_2O$$

From the balanced equation, the theoretical mole ratio is:

2 moles of HCl needed / 1 mol $Mg(OH)_2$

The actual mole ratio, based on the amounts of reactants present:

1.23 mol HCl / 0.868 $Mg(OH)_2$ = 1.42 mol HCl present / 1 mol $Mg(OH)_2$

There is not enough HCl from these ratios, so HCl is the limiting reagent.

2 mol HCl needed / 1 mol $Mg(OH)_2$ *vs.* **1.42 mol HCl present** / 1 mol $Mg(OH)_2$

Comparing theoretical yield to determine the limiting reactant

Calculate each *reactants' theoretical yield* and choose the lesser as the limiting reagent.

$$\textit{theoretical yield} = 50.6\ g\ Mg(OH)_2 \times \frac{1\ mol\ Mg(OH)_2}{58.3\ g\ Mg(OH)_2} \times \frac{1\ mol\ MgCl_2}{1\ mol\ Mg(OH)_2} \times \frac{95.3\ g\ MgCl_2}{1\ mol\ MgCl_2}$$

theoretical yield = 82.7 g $MgCl_2$

$$\textit{theoretical yield} = 45.0\ g\ HCl \times \frac{1\ mol\ HCl}{36.5\ g\ HCl} \times \frac{1\ mol\ MgCl_2}{2\ mol\ HCl} \times \frac{95.3\ g\ MgCl_2}{1\ mol\ MgCl_2}$$

theoretical yield = 58.6 g $MgCl_2$

HCl produced less product (i.e., limiting reagent); 58.6 g $MgCl_2$ as theoretical yield.

Theoretical yield and percent yield

Reactions often yield less product than predicted for several reasons (e.g., equipment inefficiency or human error).

The *percent yield* is the experimental (actual) ratio to theoretical yields.

percent yield = (experimental yield / theoretical yield) × 100%

For example, if the reaction of 30.0 grams of calcium carbonate ($CaCO_3$) produces 15.0 grams of calcium oxide (CaO), what is the percent yield for the following reaction?

$$CaCO_3 \rightarrow CaO + CO_2$$

The *experimental yield* was 15.0 g, so the theoretical yield must be determined.

For every mole of $CaCO_3$ reacted, one mole of CaO was produced.

Copyright © Sterling Test Prep.

From the periodic table, the molar mass of $CaCO_3$ is 100.0896, and the molar mass of CaO is 56.0774.

Use these conversion factors to perform the stoichiometric calculations and determine the theoretical yield.

theoretical yield = 30.0 g CaCo$_3$ × (1 mol CaCO$_3$ / 100.0869 *g*) × (1 mol CaO / 1 mol CaCO$_3$)
× (56.0774 *g* CaO / 1 mol CaO)

theoretical yield = 16.8 *g* CaO

Use the formula above to calculate the *percent yield*.

% yield = (actual yield / theoretical yield) × 100%

% yield = (15.0 *g* CaO / 16.8 *g* CaO) × 100%

% yield = 89.3%

Notes for active learning

Copyright © Sterling Test Prep.

Practice Questions

1. Which substance listed is the strongest reducing agent, given the following spontaneous redox reaction?

$$Mg\ (s) + Sn^{2+}\ (aq) \rightarrow Mg^{2+}\ (aq) + Sn\ (s)$$

A. Sn

B. Mg^{2+}

C. Sn^{2+}

D. Mg

2. Which of the following is a guideline for balancing redox equations by the oxidation number method?

A. Verify that the total number of atoms and the ionic charge is the same for reactants and products

B. In front of the substance reduced, place a coefficient corresponding to the number of electrons lost by the substance oxidized

C. In front of the substance oxidized, place a coefficient corresponding to the number of electrons gained by the substance reduced

D. All of the above

3. Which of the following represents the oxidation of Co^{2+}?

A. $Co \rightarrow Co^{2+} + 2\ e^-$

B. $Co^{3+} + e^- \rightarrow Co^{2+}$

C. $Co^{2+} + 2\ e^- \rightarrow Co$

D. $Co^{2+} \rightarrow Co^{3+} + e^-$

4. What is the molecular formula of a compound with an empirical formula of CHCl with a molar mass of 194 g/mol?

A. $C_4H_4Cl_4$

B. $C_2H_4Cl_3$

C. $C_3H_5Cl_3$

D. CHCl

5. What is the term for the amount of substance that contains 6.02×10^{23} particles?

A. molar mass

B. mole

C. Avogadro's number

D. formula mass

6. How many grams are in 0.7 moles of $CaCO_3$?

A. 25 g

B. 40 g

C. 70 g

D. 48 g

Any duplication (copies, uploads, PDFs) is illegal.

7. What is the coefficient for CO_2 in the balanced reaction?

$$\_\_C_5H_{12} + \_\_O_2 \rightarrow \_\_CO_2 + \_\_H_2O$$

A. 5   **B.** 7   **C.** 8   **D.** 10

8. What are the products for this double-replacement reaction?

$$BaCl_2 \, (aq) + K_2SO_4 \, (aq) \rightarrow$$

A. $BaSO_3$ and $KClO_4$ **C.** BaS and $KClO_4$

B. $BaSO_4$ and $2 \, KCl$ **D.** $BaSO_3$ and KCl

9. Which coefficients balance the following equation: $\_\_P_4 \, (s) + \_\_H_2 \, (g) \rightarrow \_\_PH_3 \, (g)$?

A. 2, 10, 8   **C.** 1, 6, 4

B. 1, 4, 4   **D.** 4, 2, 3

10. After balancing the following redox reaction in acidic solution, what is the coefficient of H^+?

$$Mg \, (s) + NO_3^- \, (aq) \rightarrow Mg^{2+} \, (aq) + NO_2 \, (aq)$$

A. 1   **C.** 4

B. 2   **D.** 6

11. How many atoms are in a sample of phosphorus trifluoride (PF_3) that contains 1.40 moles?

A. 3.37×10^{24} **C.** 3.46

B. 5.38   **D.** 2.218×10^{24}

12. Which of the following is the percent mass composition of acetic acid (CH_3COOH)?

A. 48% carbon, 8% hydrogen, and 44% oxygen

B. 52% carbon, 12% hydrogen, and 36% oxygen

C. 32% carbon, 6% hydrogen, and 62% oxygen

D. 40% carbon, 7% hydrogen, and 53% oxygen

13. If one mole of Ag is produced in the following reaction, how many grams of O_2 gas is produced?

$$2 \, Ag_2O \rightarrow 4 \, Ag + O_2$$

A. 6 g   **C.** 2 g

B. 8 g   **D.** 12 g

         Copyright © Sterling Test Prep.

14. Which metal in the free state has an oxidation number of zero?

    A. Mg   **C.** Al

    B. Na   **D.** All of the above

15. What is the oxidation number of Cr in $K_2Cr_2O_7$?

    A. +6   **C.** +4

    B. +5   **D.** +2

Any duplication (copies, uploads, PDFs) is illegal.

Notes for active learning

Copyright © Sterling Test Prep.

Detailed Explanations

1. D is correct.

Use the mnemonic OIL RIG: <u>O</u>xidation <u>I</u>s <u>L</u>oss, <u>R</u>eduction <u>I</u>s <u>G</u>ain (of electrons).

Oxidation is the loss of electrons, while reduction is the gain of electrons.

An oxidizing agent undergoes reduction, while a reducing agent undergoes oxidation.

Because the reaction is spontaneous, the reducing agent reactant is the strongest reducing agent.

Mg is the reducing agent being oxidized because it went from 0 as a reactant to +2 as a product.

2. D is correct.

All statements are correct for balancing redox equations by the oxidation number method.

3. D is correct.

Use the mnemonic OIL RIG: <u>O</u>xidation <u>I</u>s <u>L</u>oss, <u>R</u>eduction <u>I</u>s <u>G</u>ain (of electrons).

Oxidation is the loss of electrons, while reduction is the gain of electrons.

An *oxidizing agent* undergoes reduction, while a *reducing agent* undergoes oxidation.

Co^{2+} is the starting reactant because it must lose electrons and produce an ion with a higher oxidation number.

4. A is correct.

Formula mass is a synonym for molecular mass/molecular weight (MW).

Start by calculating the mass of the formula unit (CHCl):

$$CHCl = (12.01 \text{ g/mol} + 1.01 \text{ g/mol} + 35.45 \text{ g/mol})$$

$$CHCl = 48.47 \text{ g/mol}$$

Divide the molar mass by the formula unit mass:

$$194 \text{ g/mol} / 48.47 \text{ g/mol} = 4.0025$$

Round it to the closest whole number: 4

Multiply the formula unit by 4:

$$(CHCl)_4 = C_4H_4Cl_4$$

5. B is correct.

A *mole* is a unit of measurement used to express amounts of a chemical substance.

Avogadro's number of molecules in a mole is 6.02×10^{23}.

However, *Avogadro's number* relates to the number of molecules, not the amount of substance.

Molar mass refers to the mass per mole of a substance.

Formula mass may be used for molecular mass or molecular weight; it refers to the mass of a specific molecule.

6. C is correct.

MW of $CaCO_3$:

$$(Ca = 40.08 \text{ g/mol}) + (C = 12 \text{ g/mol}) + (O = 3 \times 16 \text{ g/mol}) = 100 \text{ g/mol}$$

Mass of 0.7 moles of $CaCO_3$:

$$0.7 \text{ mole } CaCO_3 \times 100 \text{ g/mole} = 70 \text{ g } CaCO_3$$

7. A is correct.

The number of oxygens on each side of the reaction equation must be equal.

There are 5 CO_2 molecules; since each CO_2 molecule contains 2 oxygens, multiply $5 \times 2 = 10$.

On the right side are 6 H_2O molecules; $6 \times 1 = 6$.

Add 10 and 6 to get the total oxygens on the right side: $10 + 6 = 16$.

Since there are 16 oxygens on the right side, there should be 16 oxygens on the left side.

Each O_2 molecule contains 2 oxygens; $16 / 2 = 8$.

Therefore, coefficient 8 is needed to balance the equation.

Balanced equation (combustion):

$$C_5H_{12} + 8 \, O_2 \rightarrow 5 \, CO_2 + 6 \, H_2O$$

8. B is correct.

Balanced reaction:

$$BaCl_2 \, (aq) + K_2SO_4 \, (aq) \rightarrow BaSO_4 \text{ and } 2 \text{ KCl}$$

A *double replacement reaction* indicates an exchange of cations and anions between the reactants.

Separate the reactants into ions and then exchange the cation and anion pairings.

     Copyright © Sterling Test Prep.

9. C is correct.

Balanced equation (synthesis):

$$P_4\,(s) + 6\,H_2\,(g) \rightarrow 4\,PH_3\,(g)$$

10. C is correct.

Balancing Redox Equations

From the balanced equation, the coefficient for the proton can be determined.

Balancing a redox equation requires balancing the atoms in the equation, but the charges must be balanced.

The equations must be separated into two different half-reactions. One half-reaction addresses the *oxidizing component*, and the other addresses the *reducing component*.

Magnesium is oxidized; therefore, the unbalanced oxidation half-reaction is:

$$Mg\,(s) \rightarrow Mg^{2+}\,(aq)$$

Furthermore, the nitrogen is reduced; therefore, the unbalanced reduction half-reaction is:

$$NO_3^-\,(aq) \rightarrow NO_2\,(aq)$$

Each half-reaction must be balanced for each atom, and the net electric charge on each side of the equations must be balanced.

Order of operations for balancing half-reactions:

1) Balance atoms except for oxygen and hydrogen.

2) Balance the oxygen atoms by adding water.

3) Balance the hydrogen atoms by adding protons.

   a) If in basic solution, add equal amounts of hydroxide to each side to cancel the protons.

4) Balance the electric charge by adding electrons.

5) If necessary, multiply the coefficients of one half-reaction equation by a factor that cancels the electron count when the equations are combined.

6) Cancel ions or molecules that appear on each side of the overall equation.

After determining the balanced overall redox reaction, the stoichiometry indicates the moles of protons.

For magnesium:

$$Mg\,(s) \rightarrow Mg^{2+}\,(aq)$$

continued...

Any duplication (copies, uploads, PDFs) is illegal.

The magnesium is already balanced with a coefficient of 1.

There are no hydrogen or oxygen atoms present in the equation.

The magnesium cation has a +2 charge, so 2 moles of electrons are added to the right to balance the charge.

The balanced half-reaction for oxidation:

$$Mg\ (s) \rightarrow Mg^{2+}\ (aq) + 2\ e^-$$

For nitrogen:

$$NO_3^-\ (aq) \rightarrow NO_2\ (aq)$$

Equation is balanced for nitrogen because one nitrogen atom appears on each side of the reaction.

The nitrate reactant has three oxygen atoms, while the nitrite product has two oxygen atoms.

To balance the oxygen, one mole of water should be added to the right side:

$$NO_3^-\ (aq) \rightarrow NO_2\ (aq) + H_2O$$

Adding water to the right side of the equation introduces hydrogen atoms to that side.

Therefore, the hydrogen atom count needs to be balanced.

Water possesses two hydrogen atoms; therefore, two protons need to be added to the left side of the reaction:

$$NO_3^-\ (aq) + 2\ H^+ \rightarrow NO_2\ (aq) + H_2O$$

The reaction occurs in *acidic conditions*.

If the reaction were basic, OH^- would need to be added to each side to cancel the protons.

The net charge needs to be balanced.

The left side has a net charge of +1 (+2 from the protons and −1 from the electron), while the right is neutral.

Therefore, one electron should be added to the left side:

$$NO_3^-\ (aq) + 2\ H^+ + e^- \rightarrow NO_2\ (aq) + H_2O$$

When half-reactions are recombined, the electrons in the overall reaction must cancel.

The *reduction half-reaction* contributes *one electron* to the left side of the overall equation.

The *oxidation half-reaction* contributes *two electrons* to the product side.

continued...

Copyright © Sterling Test Prep.

Therefore, the *coefficients* of the reduction half-reaction should be doubled:

$$2 \times [NO_3^- \, (aq) + 2\,H^+ + e^- \rightarrow NO_2 \, (aq) + H_2O]$$

$$= 2\,NO_3^- \, (aq) + \mathbf{4\,H^+} + 2\,e^- \rightarrow 2\,NO_2 \, (aq) + 2\,H_2O$$

The answer is 4 at this step.

Combining half reactions:

$$Mg \, (s) + 2\,NO_3^- \, (aq) + 4\,H^+ + 2\,e^- \rightarrow Mg^{2+} \, (aq) + 2\,e^- + 2\,NO_2 \, (aq) + 2\,H_2O$$

Cancel electrons on each side of the reaction in the following balanced net equation:

$$Mg \, (s) + 2\,NO_3^- \, (aq) + 4\,H^+ \rightarrow Mg^{2+} \, (aq) + 2\,NO_2 \, (aq) + 2\,H_2O$$

This equation is now fully balanced for mass, oxygen, hydrogen and electric charge.

11. A is correct.

Number of molecules = moles × Avogadro's constant

Number of molecules = 1.40 mol × 6.02×10^{23} molecules/mol

Number of molecules = 8.428×10^{23} molecules

4 atoms in each PF_3 molecule:

$$4 \times 8.428 \times 10^{23} = 3.37 \times 10^{24} \text{ atoms}$$

12. D is correct.

CH_3COOH:

mass of O atoms = 2×16 g/mol = 32 g/mol

mass of H atoms = 4×1 g/mol = 4 g/mol

Therefore, the mass % of O is 8 times (32 / 4) greater than the mass % of H.

Calculate the molecular mass (MW) of acetic acid:

MW of CH_3COOH: = (2 × atomic mass of C) + (4 × atomic mass of H) + (2 × atomic mass of O)

MW of CH_3COOH = (2 × 12.01 g/mole) + (4 × 1.01 g/mole) + (2 × 16.00 g/mole)

MW of CH_3COOH = 60.05 g/mole

continued...

Any duplication (copies, uploads, PDFs) is illegal.

To obtain the *percent mass composition* of each element, calculate the total mass of each element, divide it by the molecular mass, and multiply by 100%.

% mass composition of carbon = [(2 × 12.01 g/mole) / 60.05 g/mole] × 100%

% mass composition of carbon = 40.00%

% mass composition of hydrogen = [(4 × 1.01 g/mole) / 60.05 g/mole] × 100%

% mass composition of hydrogen = 6.71%

% mass composition of oxygen = [(2 × 16.00 g/mole) / 60.05 g/mole] ×100%

% mass composition of oxygen = 53%

13. B is correct.

From the balanced equation:

4 moles of Ag are produced for each mole of O_2.

Therefore, if 1 mole of Ag is produced, ¼ mole of O_2 is produced.

The mass of ¼ mole of O_2 is:

(¼ mol)·(2 × 16 g/mol) = 8 g

14. D is correct.

All metals (and elements) have an oxidation number of zero in their elemental state.

15. A is correct.

Assign *oxidation number* to each species:

K_2	Cr_2	O_7
2(+1)	$2x$	7(–2)

The sum of charges in a neutral molecule is zero:

$2 + (2x) + (7 × –2) = 0$

$2 + (2x) + (–14) = 0$

$2x = –2 + 14$

$2x = +12$

Oxidation number of Cr =+6

Copyright © Sterling Test Prep.

Notes for active learning

Notes for active learning

Notes for active learning

Copyright © Sterling Test Prep.

CHAPTER 6

Chemical Kinetics & Equilibria

- Reaction Rates

- Rate Dependence on Concentrations of Reactants

- Rate Order

- Reaction Order

- Rate Mechanisms

- Reaction Rate Depends on Temperature

- Kinetic *vs.* Thermodynamic Control of Reactions

- Catalysts and Enzymes

- Equilibrium in Reversible Reactions

- Mass Action

- Equilibrium Constants

- Equilibrium Variables

- Equilibrium Constant and ΔG° Relationship

- Practice Questions & Detailed Explanations

Copyright © Sterling Test Prep.

Reaction Rates

Measuring reaction rate

The *reaction rate* is the rate of change in the concentration of reactants or products (i.e., how fast a reactant is consumed and how fast a product is formed).

The *reaction rates*:

$$\text{rate of reaction} = \frac{\text{change in concentration of reactant or product}}{\text{change in time}}$$

or

$$\text{rate of reaction} = \frac{\Delta \text{ concentration}}{\Delta \text{ time}}$$

The rate is concentration divided by time; the unit is molarity per second (M/s).

Expressing reaction rates

The *rate of reaction* can be written for the forward and reverse directions of a chemical reaction.

A subscript "fwd" or "rev" specifies the direction of the reaction rate.

Forward rate of reaction:

$$\text{rate}_{\text{fwd}} = \frac{-\Delta \text{reactant}}{\Delta \text{time}} \qquad \textit{reactant disappears}$$

$$\text{rate}_{\text{fwd}} = \frac{\Delta \text{product}}{\Delta \text{time}} \qquad \textit{product forms}$$

Reverse rate of reaction:

$$\text{rate}_{\text{rev}} = \frac{-\Delta \text{product}}{\Delta \text{time}} \qquad \textit{product disappears}$$

$$\text{rate}_{\text{rev}} = \frac{\Delta \text{reactant}}{\Delta \text{time}} \qquad \textit{reactant reforms}$$

Any duplication (copies, uploads, PDFs) is illegal.

Collision theory

A reaction between two molecules occurs *if* the molecules

> 1) have sufficient *kinetic energy,* and

> 2) molecules are *appropriately oriented* to start a reaction in the image below.

The *collision theory* focuses on gas-phase chemical reactions.

Three *collision-related* factors affect the rate of a chemical reaction:

1. *Collision **frequency**:* Increasing the occurrence of reactants colliding increases the reaction rate. The more collisions, the higher the probability that a collision produces the product.

2. *Collision **energy**:* Reactants must collide with enough energy to *form new bonds* for a reaction to occur.

3. *Collision **orientation**:* Reactants need the correct orientation (i.e., properly aligned) for products to form.

A reaction forms a product from two molecules if the molecules have:

1) sufficient kinetic energy and

2) appropriately oriented molecules.

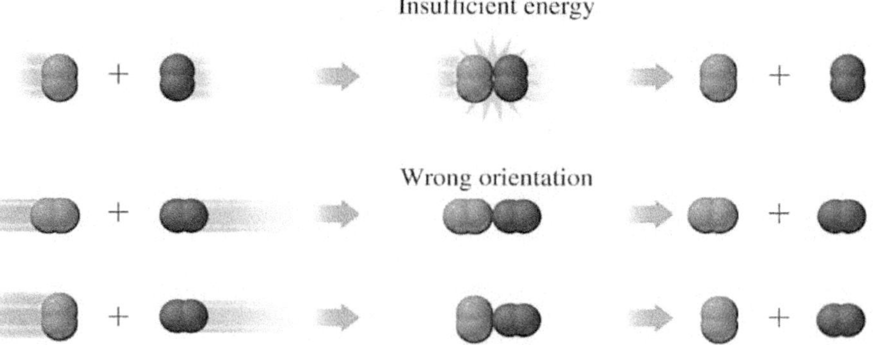

The two requirements of sufficient kinetic energy and appropriately oriented molecules are satisfied when N_2 + O_2 transforms into NO

Collisions that do not form products

Copyright © Sterling Test Prep.

Rate Dependence on Concentration of Reactants

Rate law

The *rate law expression* describes how the reaction rate changes with concentrations of reactants and products.

For example, consider the general expression for a homogeneous reaction:

$$a\,A + b\,B \longrightarrow c\,C + d\,D$$

Its *rate law* expression is:

$$rate = -\Delta[A]\,/\,\Delta t$$

$$-\Delta[A]\,/\,\Delta t = k[A]^{x}\cdot[B]^{y}$$

where $[A]$ and $[B]$ are initial concentrations of the reactants, x and y are partial rate orders of the reaction determined experimentally, and k is the rate constant

Rate constant k

The rate constant (k) is *empirically* derived and quantifies a chemical reaction.

It accounts for factors influencing reaction rate, including *temperature* and *solvent*.

For example, consider the general expression for a homogeneous reaction:

$$a\,A + b\,B \longrightarrow c\,C + d\,D$$

The *rate constant* (k) expression is:

$$rate = k(T)[A]^{x}\cdot[B]^{y}$$

$$k(T) = [A]^{x}\cdot[B]^{y}\,/\,rate$$

where $[A]$ and $[B]$ are initial concentrations of the reactants, x and y are partial rate orders of the reaction determined experimentally, and $k(T)$ is the rate constant dependent on the reaction mechanism

Therefore, the constant is specific to the *experimental conditions* (i.e., *empirically* derived).

There is *no correlation* between the stoichiometric coefficients (a, b) and the rate exponents (x, y).

The *empirically derived exponents* are often small integers, fractions, or zero.

Any duplication (copies, uploads, PDFs) is illegal.

Relative reactant and product concentrations

If the reactants' initial concentration increases, there are more reactant molecules; thus, molecules are closer and collide more frequently.

A *higher collision frequency* results in a *higher rate* of reaction.

As the reaction proceeds, the concentration of *reactant* molecules decreases, and the *forward rate decreases*.

Thus, the concentration of *product* molecules increases, and the *reverse reaction rate increases*.

Copyright © Sterling Test Prep.

Rate Order

Partial rate order

The *partial rate order* is the exponent for each concentration term.

- For a reaction that is ***zero-order*** for a reactant, the rate is constant and does *not* depend on the reactant's concentration.

  The reactant's exponent is zero.

  Thus, the term equals one (since anything to the zeroth power is one).

  Since it equals one, it is excluded from the rate law expression, as the reaction rate does not depend on a zero-order reactant.

- For a reaction that is ***first-order*** for a reactant, the reaction rate is directly proportional to the concentration of that reactant.

  The exponent is one.

- For a reaction that is ***second-order*** for a reactant, the reaction rate is directly proportional to the square of the concentration of that reactant.

  The exponent is two.

Differential rate laws

The *overall order of the reaction* is the sum of the partial orders.

Most differential rate laws are zero-, first-, or second-order overall:

- *Zero-order overall*:

  The differential rate law is generally $r = k$

  The units of the rate constant k are mole $\times$ L^{-1} $\times$ sec^{-1} (or M/s)

- *First-order overall*:

  The differential rate law is generally $r = k[A]$

  The units of the rate constant k are sec^{-1}

- *Second-order overall*:

  The differential rate law is generally $r = k[A]^2$

  The units of the rate constant k are L $\times$ mole^{-1} $\times$ sec^{-1} (or M$^{-1} \cdot$ s^{-1})

Any duplication (copies, uploads, PDFs) is illegal.

Notes for active learning

Copyright © Sterling Test Prep.

Reaction Order

Overall order of reaction

The example below illustrates how to sum the partial orders to obtain an equation's overall order.

For example, consider the reaction:

$$2\ NO\ (g) + O_2\ (g) \leftrightarrow 2\ NO_2\ (g)$$

The *rate law* was experimentally determined to be:

$$rate\ law = k[NO]^2 \cdot [O_2]$$

In the rate law, the partial order for NO is 2, and the partial order for O_2 is 1.

The overall order is three; this reaction is *third order overall*.

Note that the orders match the reaction coefficients, but rate orders and coefficients are *not necessarily* correlated. The coefficients only equal rate orders for elementary-step reactions. It is unknown if the reaction is an elementary or multi-step reaction in most problems unless it is explicitly stated.

Molecularity

In defining the overall order of a reaction, *molecularity* is important.

A reaction mechanism's molecularity is determined by the number of molecules (typically the coefficients) or ions participating in the rate-determining step.

A *unimolecular* reaction has a single reactant in the transition state.

A *bimolecular* reaction is a mechanism involving two reacting species.

A *termolecular* reaction is a mechanism when three independent molecules collide and react. Termolecular reactions are rare as the necessary factors are so specific that they are unlikely.

Reaction Type	Overall Reaction Order	Rate Law(s)
Unimolecular	1	$r = k[A]$
Bimolecular	2	$r = k[A]^2$, $r = k[A] \cdot [B]$
Termolecular	3	$r = k[A]^3$, $r = k[A]^2 \cdot [B]$, $r = k[A] \cdot [B] \cdot [C]$
Zero order	0	$r = k$

Any duplication (copies, uploads, PDFs) is illegal.

Differential *vs.* integrated rate law

The *differential rate law* and *integrated rate law* express specific rate laws.

Differential rate laws (above) express the rate of reaction *vs.* concentration.

Integrated rate laws express the change in concentration of components *vs.* time.

The *integrated rate* laws are:

Integrated rate law for 0th order

$$[A]_t = -kt + [A]_0$$

Integrated rate law for 1st order

$$\ln[A]_t = -kt + \ln[A]_0$$

Integrated rate law for 2nd order

$$1 / [A] = 1 / [A]_0 + kt$$

Initial experimental rate method

The *method of initial rates* determines the rate law using experimental data to determine the rate order for reactants.

For example, the results of an experiment show the reaction rates relative to the initial concentrations of reactants:

Concentrations (M)

$[O_2]$		Rate (M/s)	initial $[NO]$
0.10	Trial 1	1.2×10^{-8}	0.10
0.20	Trial 2	2.4×10^{-8}	0.10
0.10	Trial 3	1.08×10^{-7}	0.30

Based on the data above, determine:

a) reaction orders of the reactants

b) value and units of k

Copyright © Sterling Test Prep.

Rate law by concentrations

To determine the rate law, compare two trials in which the concentration of one reactant changes while the other reactant's concentration is held constant.

It is necessary to determine the reactant's partial order for changed concentrations.

Keep the other reactant concentration constant, so the change in rate is due to one reactant's concentration change.

Consider the results for Trial 1 and Trial 2.

Comparing these two trials shows that NO concentration is constant while O_2 is doubled.

To determine the partial order of O_2, note that doubling the concentration of O_2 doubles the reaction rate.

Expressed as:

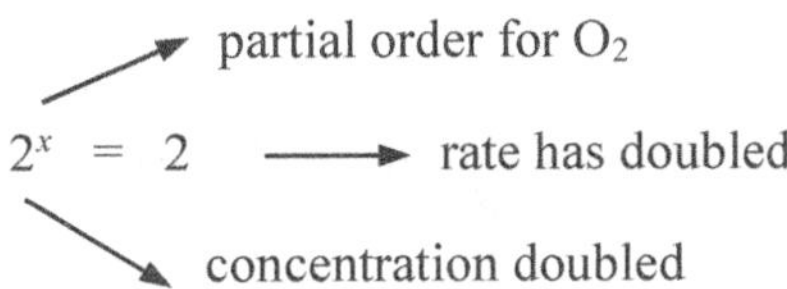

For equality, x equals one.

Therefore, the partial order for O_2 is first-order.

To determine the partial order of NO, select two trials with the concentration of O_2 constant while [NO] varies.

These criteria are satisfied with Trial 1 and Trial 3.

The concentration of O_2 is held constant, while the concentration of NO has tripled.

Under these conditions, the rate of the reaction increased nine-fold.

Expressed as:

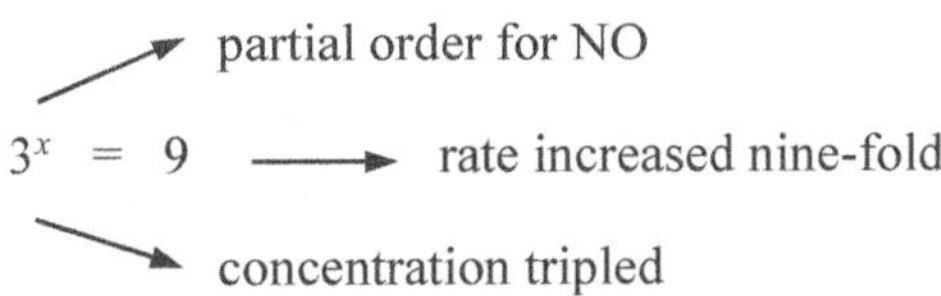

For equality, x equals two.

Therefore, the partial order for NO is second-order.

Any duplication (copies, uploads, PDFs) is illegal.

Calculating rate order experimentally

The calculated rate orders are substituted into the rate law:

$$rate\ law = k[A]^x \cdot [B]^y$$

$$rate\ law = k[NO]^2 \cdot [O_2]$$

Partial first orders are typically omitted, so $[O_2]$ does not have an exponent; the exponent is assumed as one.

With partial rate orders, it is possible to solve the rate constant, k.

For example, choose an experimental trial, and substitute the values of *rate* and *concentrations* of NO and O_2.

Using data from Trial 1:

$$rate = k[NO]^2 \cdot [O_2]$$

$$k = rate\ /\ [NO]^2 \cdot [O_2]$$

$$k = 1.2 \times 10^{-8}\ M^{-2} \cdot s^{-1}\ /\ [(0.10\ M)^2 \cdot (0.10\ M)]$$

$$k = 1.2 \times 10^{-5}\ M^{-2} \cdot s^{-1}$$

Thus, the rate law becomes:

$$rate = 1.2 \times 10^{-5}\ M^{-2} \cdot s^{-1}[NO]^2 \cdot [O_2]$$

The reaction is third order overall, and the units are $M^{-2} \times s^{-1}$.

Notice that the units of k cancel to yield M/s for the reaction rate.

Copyright © Sterling Test Prep.

Rate Mechanisms

Activation energy

The reactants must have enough energy to *break existing* and *create new bonds* for a reaction to proceed.

This minimum energy threshold is the *activation energy* (E_a).

The *activation energy barrier* (E_a as the hill on the reaction profile energy *vs.* reaction progress graph) is the difference between the reactants' energy and the highest point on the reaction profile toward product formation.

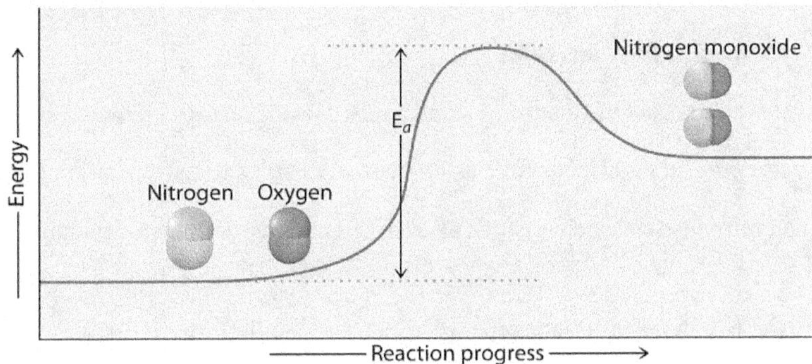

The reacting molecules need sufficient energy to proceed over the energy barrier to transform into products.

For the *forward* reaction:

> the activation energy E_a is the difference between the reactants' energy and the top of the reaction profile (i.e., transition state).

For the *reverse* reaction:

> the activation energy E_a is the difference between the products' energy and the top of the reaction profile (i.e., transition state).

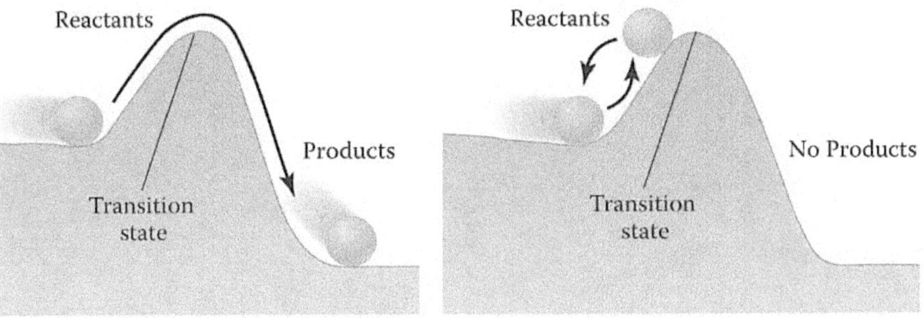

Exergonic reaction forms product   *Exergonic reaction without TS*

Any duplication (copies, uploads, PDFs) is illegal.

Unimolecular mechanisms

Unimolecular mechanisms involve *intermediates* (e.g., carbocation, radicals) during the reaction profile (two peaks on the reaction progress diagram).

The formation of an intermediate of the first step is often the rate-determining (i.e., slow) step.

The intermediate is consumed as the reaction forms products (via the second transition state).

Bimolecular mechanisms

The left image above is a completed bimolecular (E_2 or Sn_2) reaction.

The right image above shows a bimolecular reaction whereby the activation energy (energy barrier) is not overcome: no transition state (TS) forms and no products are generated.

Reaction mechanism influences rates

Most chemical reactions occur in several steps, known as a *reaction mechanism*.

The *reaction mechanism* is the sequence as reactant molecules transform into products.

Each step in a multiple-step mechanism is an *elementary step* with an individual rate constant and rate law.

For example, the following reaction steps were experimentally observed:

$$(CH_3)_3CCl\ (aq) + OH^-\ (aq) \rightarrow (CH_3)_3COH\ (aq) + Cl^-\ (aq)$$

Step 1:

$$(CH_3)_3CCl\ (aq) \rightarrow (CH_3)_3C^+\ (aq) + Cl^-\ (aq) \quad (slow)$$

$$r_1 = k[(CH_3)_3CCl]$$

Step 2:

$$(CH_3)_3C^+\ (aq) + OH^-\ (aq) \rightarrow (CH_3)_3COH\ (aq) \quad (fast)$$

$$r_2 = k[(CH_3)_3C^+] \cdot [OH^-]$$

Copyright © Sterling Test Prep.

Rate-determining step

The *rate-determining step* has the slowest rate and most significant contributor to the reaction's overall rate.

In most cases, the overall reaction rate is almost equal to the rate-determining step's rate because the other steps are fast (almost instantaneous) and do not significantly affect the overall rate.

For example, the first step, the formation of the charged carbonium ion $(CH_3)_3C^+$, is slow compared to the second step, where the carbocation immediately reacts with the OH^- ion to form the neutral product $(CH_3)_3COH$.

$$(CH_3)_3CCl\ (aq) \rightarrow (CH_3)_3C^+\ (aq) + Cl^-\ (aq) \qquad \text{(slow – first step)}$$

$$(CH_3)_3C^+\ (aq) + OH^-\ (aq) \rightarrow (CH_3)_3COH\ (aq) \qquad \text{(fast – second step)}$$

Therefore, the first step is the rate-determining (slow) step.

The rate law of this (slow) step determines the overall rate:

$$r_1 = k[(CH_3)_3CCl]$$

There are reactions when the *rate-determining step* (or *slowest step*) is *not* the first step in the reaction mechanism.

In these examples, an intermediate may appear in the rate-determining step.

The intermediate must be substituted so that it is not in the rate law, requiring a detailed approach for determining the reaction mechanism's rate law.

Any duplication (copies, uploads, PDFs) is illegal.

Notes for active learning

Copyright © Sterling Test Prep.

Reaction Rate Depends on Temperature

Kinetic energy influences rate

Increasing temperature increases the colliding molecules' *average kinetic energy*; the molecules move faster.

These faster molecules are more likely to possess the energy necessary to overcome the activation energy and successfully collide and react.

Faster molecules (i.e., higher kinetic energy) collide at an increased frequency.

Therefore, using collision theory, increasing temperature increases the reaction rate.

However, some reactions might slow or stop as the temperature increases. This is common among reactions in biological systems because enzyme proteins involved in the reaction may be damaged by higher temperatures, reducing the enzyme's capability to participate in reactions.

Transition state as energy of activation

The *transition state* (or *activated complex*) is the energy peak (y-axis) of the energy profile when bonds begin to *break* (reactants) and other bonds begin to *form* (products).

Transition states form if the reactants have enough energy to achieve the needed activation energy (highest energy location on the reaction progress graph).

A molecule in the transition state *either* reverts to reactants or forms products.

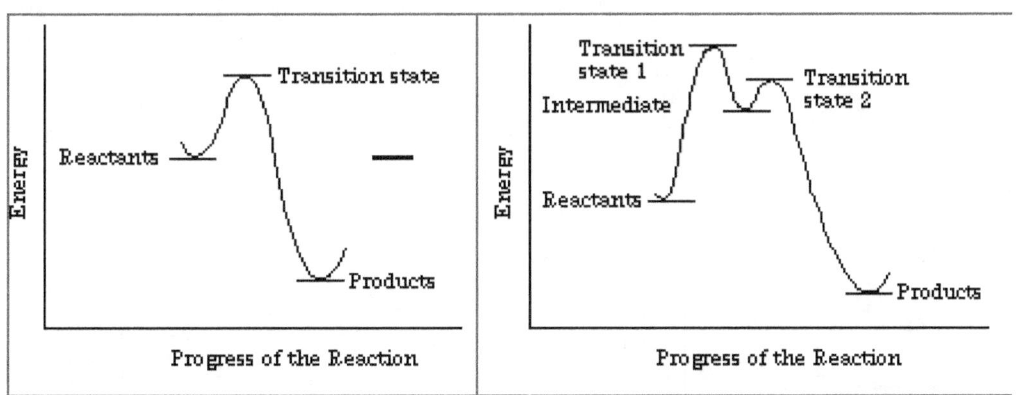

Bimolecular reactions   *Unimolecular reactions*

Molecules in the transition state *cannot be isolated.*

Any duplication (copies, uploads, PDFs) is illegal.

Transition state bond making and breaking events

Transition states signify bond-breaking, and bond-making events, as shown below.

Transition states are denoted by square brackets [ ] and a double dagger ($\ddagger$) around the fleeting (i.e., unable to be isolated) molecule.

Dashed lines symbolize temporary bonds simultaneously *formed and broken*, shown in brackets.

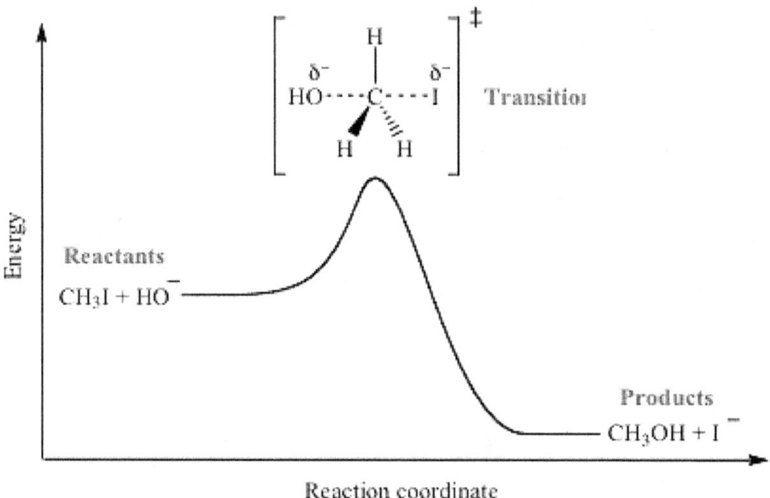

Energy profiles for activation energy E_{ac}

In a reaction, the products and reactants often have different energy levels.

The energy differential is released ($-\Delta H$) or absorbed ($+\Delta H$) from the system as heat (enthalpy or ΔH).

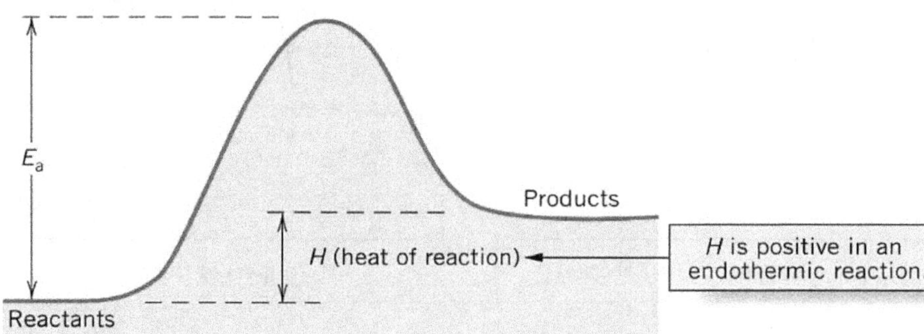

The graph's high point indicates the energy of the activation E_{ac} barrier during the reaction

---

Copyright © Sterling Test Prep.

Reactions are *endothermic* or *exothermic,* based on enthalpy.

In *endothermic* reactions,

the products have higher potential energy than reactants ($+\Delta H$), favoring the *reverse reaction.*

In *exothermic* reactions,

the products are lower potential energy than reactants ($-\Delta H$), favoring the *forward reaction.*

The forward reaction has a $-\Delta H$ (release energy) in the forward direction and a $+\Delta H$ in the reverse direction.

Enthalpy (ΔH)

In *endothermic reactions*, the potential energy of products is *higher* (i.e., less stable) than for reactants ($+\Delta H$).

Heat is required as a reactant to drive the reaction forward.

Heat is *absorbed* by the system from its surroundings to *break bonds.*

This is an *energetically unfavorable* reaction, and the temperature of the surroundings *decreases.*

In *exothermic reactions*, the potential energy of products is *lower* (i.e., more stable) than for reactants ($-\Delta H$).

Excess heat formed by the reaction is released by the system towards its surroundings.

Energy is *released* when by the system to *form bonds.*

This is an *energetically favorable* reaction, and the temperature of the surroundings *increases.*

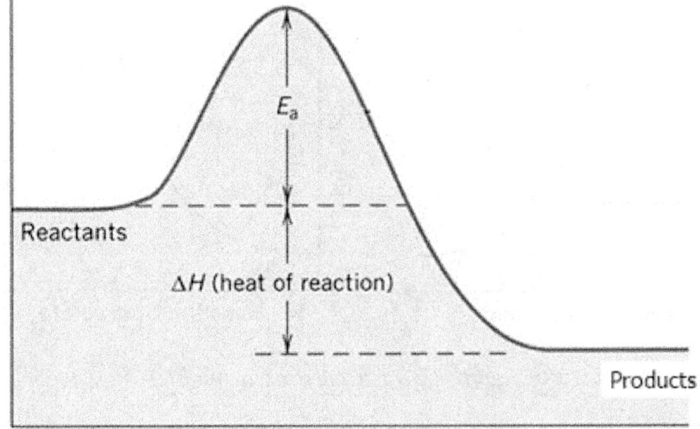

Exothermic reactions form products with lower energy than reactants

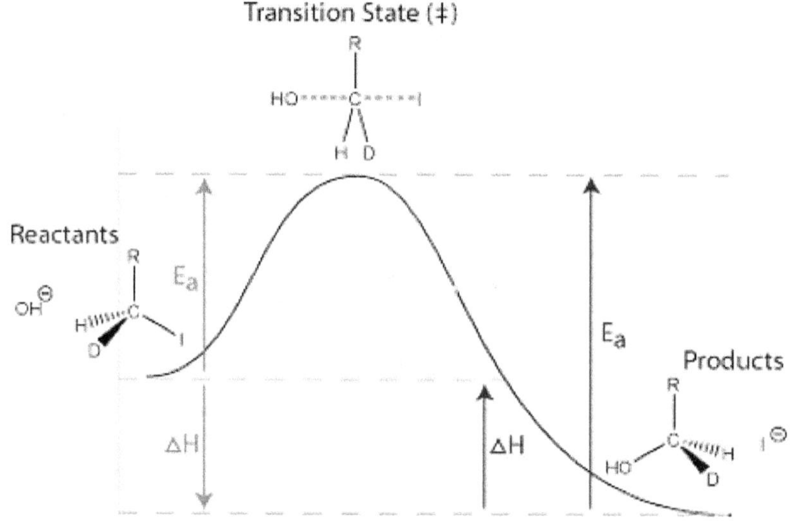

Energy profile diagrams summarize the reaction energies (ΔH and E_{ac})

Hammond postulate

Hammond's postulate (or *axiom*) states that *the transition state resembles either the reactants or the products, to whichever it is closer in energy.*

In an *exothermic* reaction, the transition state (TS) is *closer to the reactants* than to the products in energy.

Therefore, according to Hammond's postulate, the transition state *resembles the reactants.*

In *endothermic* reactions, the transition state is *closer to the products* in energy than to reactants.

Therefore, according to Hammond's postulate, the transition state *resembles the products.*

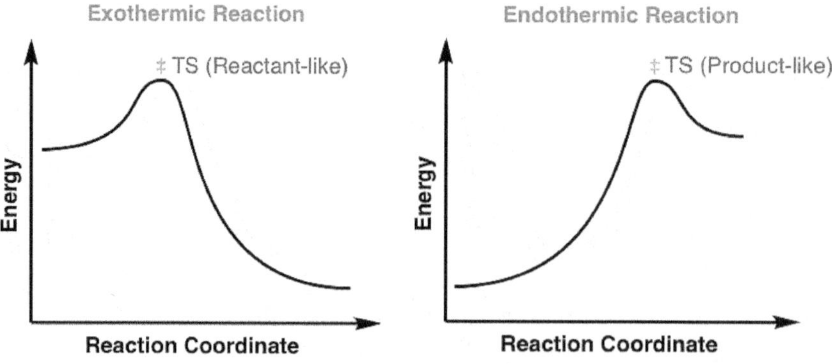

The Hammond postulate states that the transition state resembles the species closer in energy

The E_{ac} is lower (measured from the energy of reactants) to the peak of the energy barrier compared to the E_{ac} in the reverse reaction (measured from the energy of the products).

Copyright © Sterling Test Prep.

Arrhenius equation described temperature dependence of reaction rate

In 1889, Swedish Nobel laureate Svante Arrhenius (1859-1927) described the *temperature dependence* of a reaction rate whereby chemical reactions generally occur rapidly at higher temperatures.

Arrhenius equation uses the relationship between rate constant (k), temperature (T), and activation energy (E_a):

$$k = Ae^{-Eac / RT}$$

where k is the rate constant, R is the ideal gas constant ($R = 8.314$ J mol^{-1} K^{-1}), E_{ac} is the activation energy, T is the temperature in Kelvin, and A is the *Arrhenius frequency factor*

The *Arrhenius frequency factor* is the pre-exponential factor (or collision factor) representing collision frequency.

For the Arrhenius equation, ensure that the quantities' units are proper. T is in Kelvin, E_{ac} is in J/mol, and A has the same units as the rate constant k

Point-slope form for the Arrhenius equation

Taking the natural logarithm of each side of the Arrhenius equation yields the point-slope form for the equation:

$$\ln k = -(E_a / RT) + \ln A$$

A *straight-line slope* is produced if data is graphed, where $\ln k$ is on the y-axis, and $1 / T$ is on the x-axis.

On the $\ln k$ *vs.* $1 / T$ plot, the line's slope is E_a / R, and the y-intercept corresponds to $\ln A$.

Determine the slope of the line mathematically to calculate E_a, whereby the slope of a line is the change in y over the change in x (i.e., rise over run).

By extrapolation, the y-intercept and $\ln A$ collision factor are calculated.

The reaction rate constant at two temperatures is required to calculate the point-slope form.

The rate constants relationship at two temperatures is expressed by:

$$\ln k_2 - \ln k_1 = \left(\ln A - \frac{E_a}{RT_2} \right) \left(\ln A - \frac{E_a}{RT_1} \right) = \frac{E_a}{R} \left(\frac{1}{T_1} - \frac{1}{T_2} \right)$$

The equation calculates the rate constant k of a reaction at two temperatures (in Kelvin).

The y-intercept ($\ln A$) can be ignored because it is a constant and not part of the slope calculation.

Eliminate the $\ln A$ term by subtracting the expressions for the two $\ln$-k.

Any duplication (copies, uploads, PDFs) is illegal.

Simplified Arrhenius equation expresses activation energy

Since the slope embodies the activation energy (E_{ac}), the simplified Arrhenius equation determines the E_{ac}.

Rearrange the expression to isolate activation energy.

$$E_a = \frac{R \ln \frac{k_2}{k_1}}{\frac{1}{T_1} - \frac{1}{T_2}}$$

For example, a reaction rate for a chemical process is investigated at two temperatures. The reaction rate at 25 °C is 1.55×10^{-4} s^{-1}. At 50 °C, the reaction rate is 3.88×10^{-4} s^{-1}. Based on this data, what is the energy of activation for the chemical process expressed in J/mol? (Use R = 8.314 J/mol·K)

Inspect the units. The units for rates, s^{-1}, cancel. Since the activation energy is to be expressed in J/mol, the gas constant R is expressed as 8.314 J/mol·K. This choice of the gas constant R dictates temperature units.

The R constant has temperature units expressed in K; temperatures are converted to Kelvin:

$T_1 = 25$ °C + 273 = 298 K

$T_2 = 50$ °C + 273 = 323 K

$$E_a = \frac{R \ln \frac{k_2}{k_1}}{\frac{1}{T_1} - \frac{1}{T_2}}$$

Substitute the rate constants and temperatures into the Arrhenius equation:

$$E_{ac} = \frac{8.314 \text{ J/mol·K} \times \ln(3.88 \times 10^{-4} \text{ s}^{-1} / 1.55 \times 10^{-4} \text{ s}^{-1})}{(1/298 \text{ K}) - (1/323 \text{ K})}$$

$$E_{ac} = \frac{8.314 \text{ J/mol·K} \times \ln(2.503)}{(2.6 \times 10^{-4} \text{ K})}$$

$$E_{ac} = 2.93 \times 10^4 \text{ J/mol}$$

Copyright © Sterling Test Prep.

Van 't Hoff equation for K_{eq}

The Van 't Hoff equation relates the change in the equilibrium constant (K_{eq}) of a chemical reaction to the change in temperature.

The Van 't Hoff equation calculates the *equilibrium constant K* (uppercase K for equilibrium; lowercase k denotes rate) of a reaction at *two Kelvin temperatures*.

The Van 't Hoff equation modifies the Arrhenius equation by replacing E_{ac} with ΔH.

This modified Arrhenius equation is the Van 't Hoff equation:

$$\ln\left(\frac{K_{T_2}}{K_{T_1}}\right) = \frac{\Delta H^{\,o}}{R}\left(\frac{1}{T_1} - \frac{1}{T_2}\right)$$

The above equation is *not* the Arrhenius equation.

Notes for active learning

Copyright © Sterling Test Prep.

Kinetic *vs.* Thermodynamic Control of Reactions

Reaction selectivity

Several chemical reactions have two possible products, and the *reaction conditions* affect selectivity.

Thermodynamics explains how ΔG and other variables affect if a reaction occurs spontaneously.

Kinetics describes how fast a reaction occurs (based on the activation energy), and kinetics has no impact on the spontaneity or thermodynamics of the reaction.

The *kinetic product* requires lower activation energy and is formed preferentially (according to the transition state being of lower energy) at a lower temperature.

The *thermodynamic product* has a greater negative ΔG (favorable) formed preferentially at a higher temperature.

The energy profile diagram displays a reaction under kinetic *vs.* thermodynamic control.

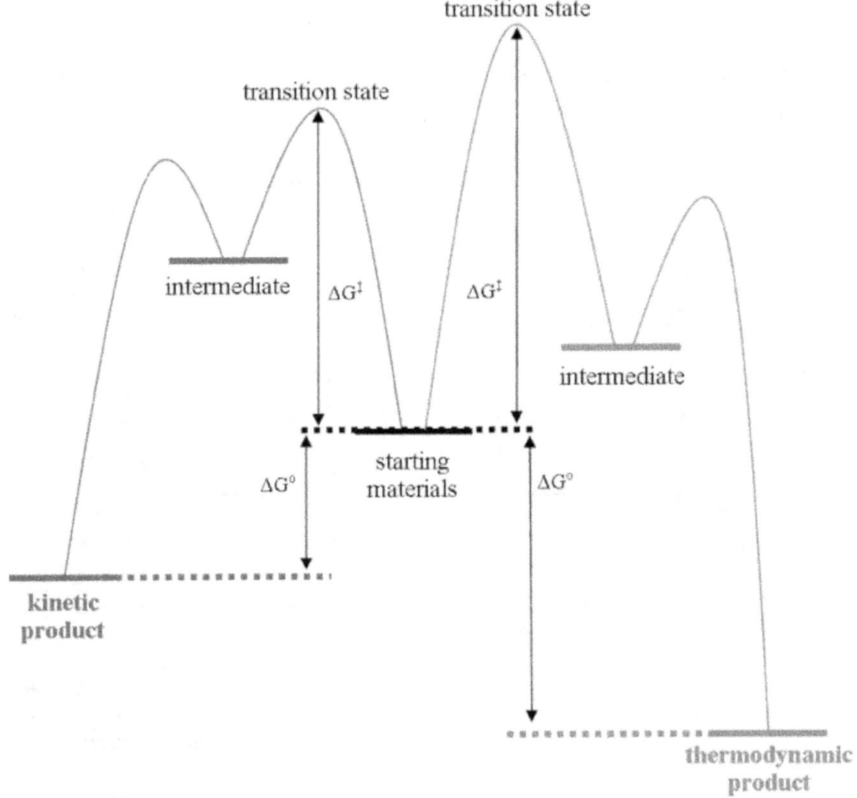

Kinetic vs. thermodynamic control for products depends on energy of transition state

Any duplication (copies, uploads, PDFs) is illegal.

The reactants (i.e., starting material) are in the center of the graph above.

On the left is an intermediate with lower energy than the intermediate on the right.

The intermediate with lower energy forms faster (I.e., kinetic control).

However, the final product from this lower energy intermediate yields a higher energy product than the more stable product on the right (thermodynamic product).

The product on the right on the graph above is the thermodynamic product (overall more stable than the kinetic product). However, it requires a higher (more unstable) transition state before forming a product.

Kinetic products form *fast* because of a *lower energy* transition state.

Thermodynamic products form *slowly* because of a *higher energy* transition state.

Reaction mechanism drives product formation

However, the ΔG, the transition states, and the intermediates and transition states' energy are different.

The product is different, depending on whether the rxn is under kinetic control (top) or thermodynamic (bottom).

When strong acids are added to certain conjugated dienes (e.g., butadiene), two products form the 1,2- product and the 1,4-product.

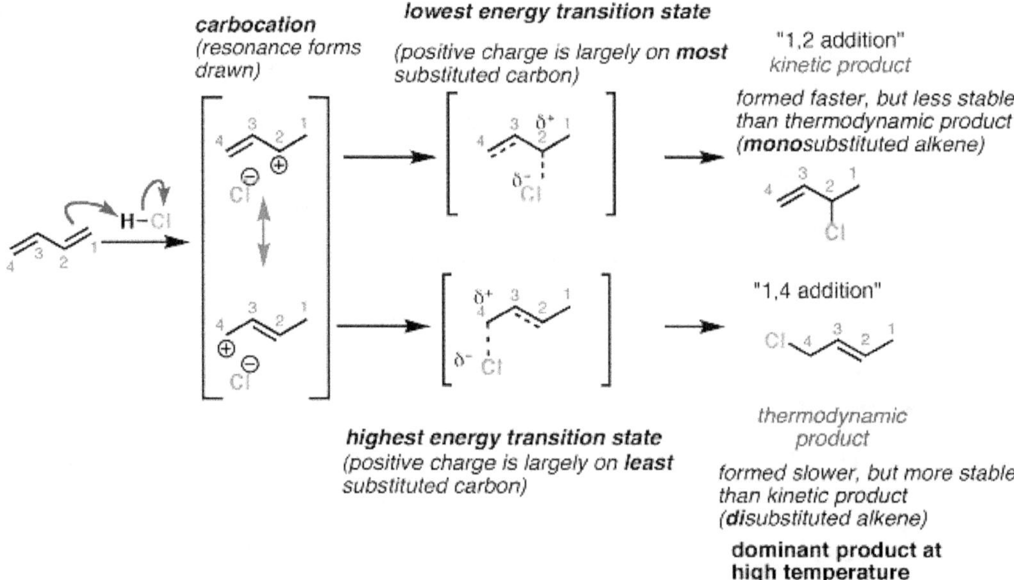

1,2 vs. 1,4 addition of conjugated dienes form kinetic (less stable) and thermodynamic (more stable) products

Copyright © Sterling Test Prep.

In the top transition state, the carbocation's positive charge is localized on the most substituted (2°) carbon, and the transition state has lower energy.

The lower potential energy transition state indicates that the reaction has lower activation energy.

Kinetic (1,2) product

The 1,2 *kinetic product* is preferentially formed at *lower temperatures*.

The bottom transition state is higher in energy because the carbocation's positive charge is localized on the less substituted (1°) carbon.

Therefore, this transition state has higher energy. This intermediate indicates that the bottom reaction has higher activation energy.

Thermodynamic (1,4) product

The 1,4 *thermodynamic product* is preferentially formed at higher temperatures needed to proceed with the less stable (2°) carbon intermediate.

The 1,4-product is more stable than the 1,2-product because it is an internal alkene instead of a terminal alkene.

The 1,4-addition produces the thermodynamically stable product (i.e., lower potential energy) but proceeds slower through the transition state of the higher energy (less substituted carbocation).

The 1,4-addition product is the major product formed at higher temperatures needed to proceed through the less stable transition state.

Any duplication (copies, uploads, PDFs) is illegal.

Terminal *vs.* internal alkene

The energy profile diagram for an E_1 *vs.* E_2 reaction is shown:

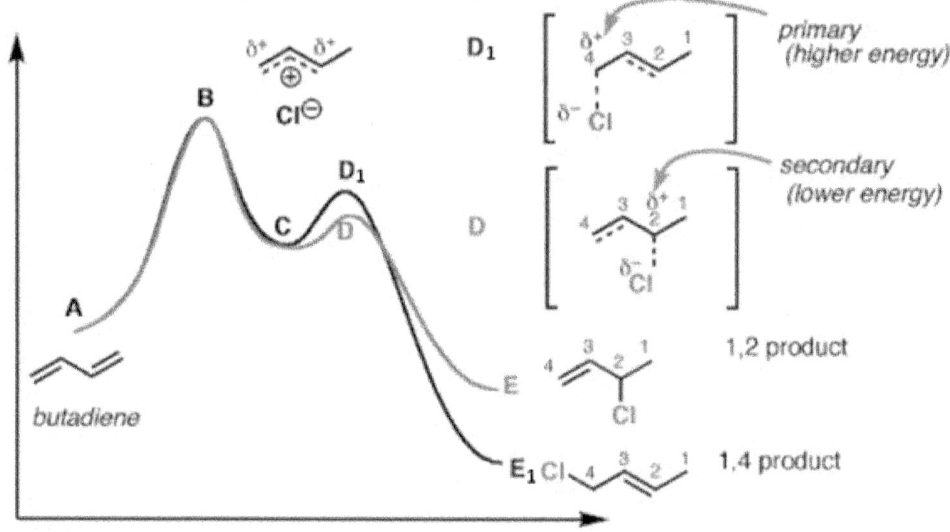

The energy of HCl addition to 1,2-addition vs. 1,4-addition to butadiene

The height of transition states D and D_1 determines the reaction rate once the carbocation C forms.

Lower energy transition states (D *vs.* D_1) proceed faster to products.

Therefore, E (terminal alkene) forms faster from the secondary carbocation intermediate C since the energy of transition is less for D (secondary carbocation) than D_1 (primary carbocation).

The energy of E (terminal alkene) and E_1 (internal alkene) is related to greater stability for the 1,4-addition compared to the 1,2-addition.

E_1 has a more substituted (internal) double bond than E (terminal alkene), which is more stable.

E_1 represents the thermodynamic (more stable) product that is *slower* to form (higher energy transition state) but is *more stable* than the kinetic product E, which formed faster (lower energy transition state) but is less stable than E_1.

Copyright © Sterling Test Prep.

Catalysts and Enzymes

Catalysts lower energy of activation

Catalysts are substances in a chemical reaction that increase the reaction rate by lowering the activation energy (often through an alternative pathway) needed to proceed.

A catalyst participates in a chemical reaction but remains chemically unchanged and used repeatedly.

For a reaction mechanism separated into elementary steps, a catalyst appears at the start and reappears at the end of the reaction. A catalyst is neither a reactant nor a product in the reaction.

Acids, bases, and metal ions often act as catalysts.

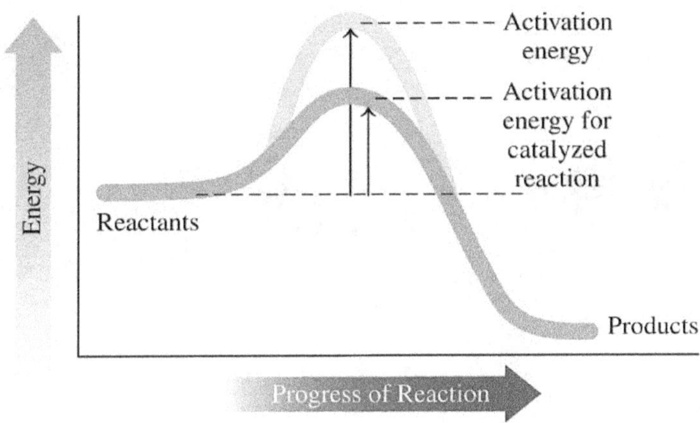

Homogeneous *vs.* heterogeneous catalysts

There are homogeneous and heterogeneous catalysts.

Homogeneous catalysts are in the same phase as the reactants.

Heterogeneous catalysts are in another phase than the reactants. These catalysts often immobilize reactants near each other, increasing the probability of a collision and, therefore, the corresponding reaction rate.

An example of a heterogeneous catalyst is a dissolved acid that catalyzes esters' hydrolysis in an aqueous solution.

Most heterogeneous catalysts are solids, and the reactants are often gases or liquids.

For example, iron catalyzes NH_3 synthesis from gaseous N_2 and H_2.

Modern cars have *catalytic converters* in their exhaust systems as heterogeneous catalysts providing a solid surface for exhaust molecules to bind and react. Catalytic converters reduce pollutants such as carbon monoxide (CO), nitrogen oxide (NO), and hydrocarbons (e.g., octane as C_8H_{18}).

The catalyst usually consists of solid particles, including platinum (Pt) and palladium (Pd).

For example, the reaction:

$$2\ NO\ (g) \rightarrow N_2\ (g) + O_2\ (g)$$

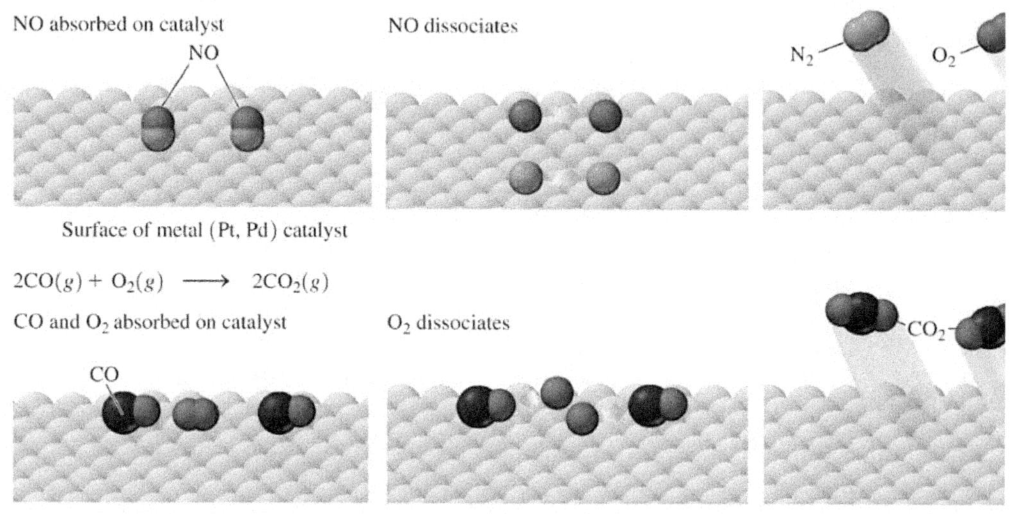

Biological enzymes

Enzymes (e.g., proteins) are *biological catalysts* that increase the reaction rates of biochemical reactions.

Biological enzymes are more efficient than chemical catalysts; enzymes can increase a reaction rate by a factor of more than ten million (1×10^7).

Enzymes are highly specific; they form enzyme-substrate complexes with reactants.

A *substrate* is a molecule on which the enzymes act.

For example, when hydrogen peroxide (H_2O_2) is applied to a wound, oxygen (O_2) gas is produced by decomposing hydrogen peroxide by the catalase enzyme in the blood.

Decomposition of hydrogen peroxide by catalase:

$$2\ H_2O_2\ (l) \xrightarrow{\ catalase\ } 2\ H_2O\ (l) + O_2\ (g)$$

Catalysts are written above or below the arrow in the reaction.

Copyright © Sterling Test Prep.

Catalyzed *vs.* non-catalyzed reactions

Changing variables (i.e., reactant concentration, temperature) could increase the reaction rate.

Catalyst has a similar effect on the reaction rate but uses a different method.

Factor	Result
Increasing reactant concentration	More collisions
Increasing temperature	More collisions exceeding the energy of activation
Adding a catalyst	Lowers energy of activation

Any duplication (copies, uploads, PDFs) is illegal.

Notes for active learning

Copyright © Sterling Test Prep.

Equilibrium in Reversible Reactions

Reversible reactions

Some chemical reactions are reversible; they proceed in either direction, resulting in an equilibrium mixture of reactants and products.

If a chemical bond forms, it can be broken with sufficient energy input.

Reversible chemical reactions are indicated by a double-headed arrow or two arrows facing opposite directions.

$$A + B \leftrightarrow AB$$

$$A + B \leftrightarrows AB$$

The forward reaction occurs when A and B form AB.

When AB (product) accumulates, the reverse reaction occurs from the decomposition of AB, and A and B (reactants) are reformed.

When the forward and the reverse reactions have the *same rate*, the reaction is in chemical *equilibrium*.

Equilibrium

Equilibrium does not mean equal amounts of reactants and products but is the point where the number of reactants and products *remain constant* because the *forward and reverse reactions proceed at equal rates*.

For example, consider the following reaction:

$$2\,SO_2\,(g) + O_2\,(g) \;\rightleftarrows\; 2\,SO_3\,(g)$$

The reaction proceeds forward with SO_2 and O_2, and SO_3 products form until equilibrium (i.e., the equal rate for the forward and reverse reaction).

From the SO_3 products, the reaction proceeds in the reverse direction and forms SO_2 and O_2 until equilibrium.

As shown, the equilibrium concentrations of SO_2, O_2, and SO_3 are the same for the forward and reverse reactions.

Any duplication (copies, uploads, PDFs) is illegal.

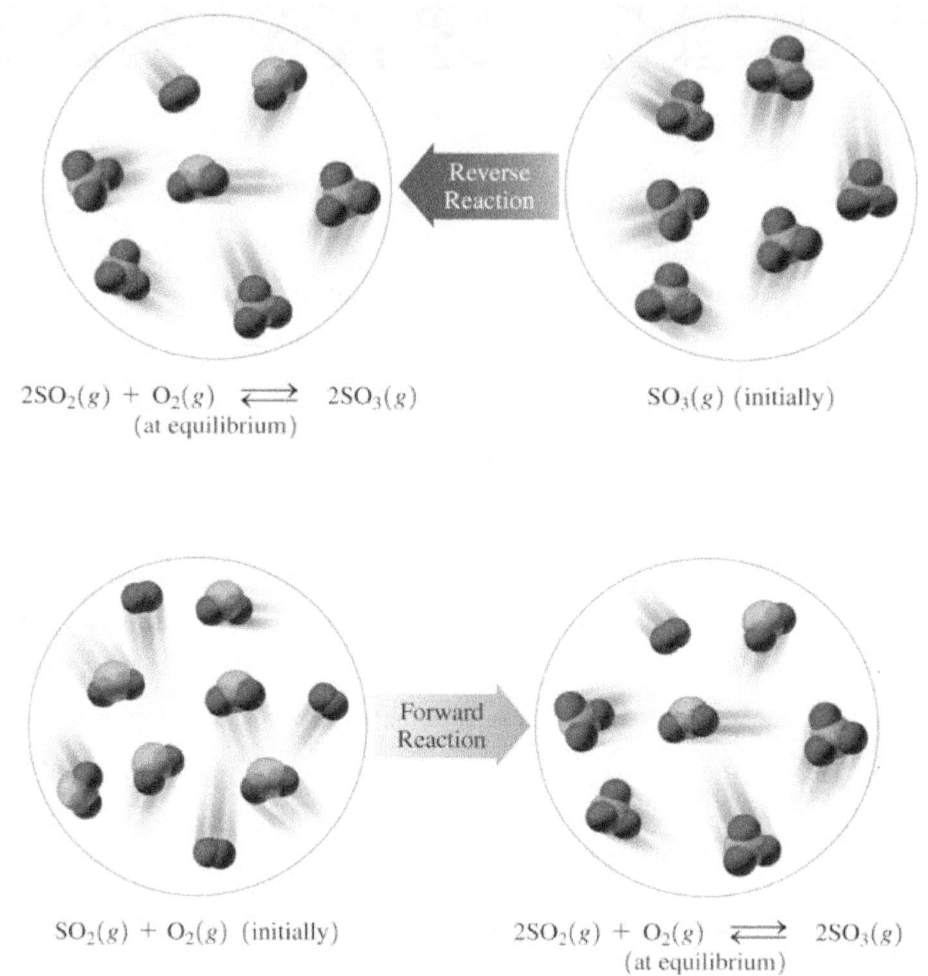

$$2SO_2(g) + O_2(g) \rightleftharpoons 2SO_3(g)$$
(at equilibrium)

$$SO_3(g) \text{ (initially)}$$

$$SO_2(g) + O_2(g) \text{ (initially)}$$

$$2SO_2(g) + O_2(g) \rightleftharpoons 2SO_3(g)$$
(at equilibrium)

Experimental evidence for equilibrium

For example, consider the reversible reaction for the evaporation and condensation of water. Suppose water is placed over a gas-filled chamber, separated by a moving piston as diagramed.

Work *on* or *by* the system is observed by measuring the gas's compression or expansion.

As water evaporates, the water's mass decreases, and the gas expands.

When water vapor condenses, it increases water weight and compresses the gas.

Once the system reaches equilibrium, the liquid-to-vapor ratio remains constant. The gas neither compresses nor expands.

Therefore, no work is done *on* or *by* the system at equilibrium.

Copyright © Sterling Test Prep.

Water molecules constantly transform between liquid and gas phases, but the evaporation rate matches the condensation rate, so there is no noticeable change in composition.

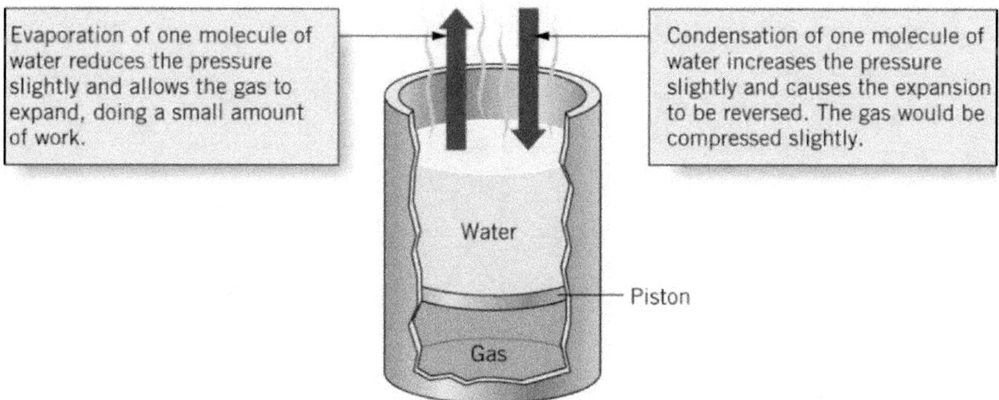

Piston system detects equilibrium for water

Any duplication (copies, uploads, PDFs) is illegal.

Notes for active leaning

Copyright © Sterling Test Prep.

Mass Action

Law of Mass Action

The *Law of Mass Action* states that the *reaction rate depends on the concentration of the participating substances.*

$$aA + bB \rightleftharpoons cC + dD$$

$$\text{rate}_{\text{rxn}} = \left(-\frac{1}{a}\right)\frac{\Delta A}{\Delta t} = \left(-\frac{1}{b}\right)\frac{\Delta B}{\Delta t} = \left(+\frac{1}{c}\right)\frac{\Delta C}{\Delta t} = \left(+\frac{1}{d}\right)\frac{\Delta D}{\Delta t}$$

Deriving the equilibrium constant

Law of Mass Action *derives the equilibrium constant* by setting the forward reaction rate equal to the reverse reaction rate (i.e., equilibrium).

For the reaction:

$$aA + bB \rightleftharpoons cC + dD$$

The following relationships exist:

$$r_{\text{forward}} = r_{\text{reverse}}$$

$$k_{\text{forward}} \cdot [A]^a \cdot [B]^b = k_{\text{reverse}} \cdot [C]^c \cdot [D]^d$$

$$K_c = k_{\text{forward}} / k_{\text{reverse}} = [C]^c \cdot [D]^d / [A]^a \cdot [B]^b$$

Phases for Law of Mass Action

The Law of Mass Action is based on the equilibrium constant K_c, which is the ratio of the concentration of products to the concentration of reactants (i.e., [products] / [reactants]).

The mass action expression of equilibrium *only* includes the equilibrium components in the *same phase.*

Pure solids and liquids do *not* appear in the mass action expression.

The absence of pure solids and liquids from the equation is:

$$CaCO_3 \ (s) \ \rightleftharpoons \ CaO \ (s) + CO_2 \ (g)$$

$$K_c = [CO_2]$$

Any duplication (copies, uploads, PDFs) is illegal.

For example, consider the following equation:

$$2\ HCl\ (aq) + CaCO_3\ (s) \leftrightarrows CaCl_2\ (aq) + CO_2\ (g) + 2\ H_2O\ (l)$$

The *net ionic equation* for the above reaction is:

$$2\ H^+\ (aq) + CaCO_3\ (s) \leftrightarrows Ca^{2+}\ (aq) + CO_2\ (g) + 2\ H_2O\ (l)$$

Cl^- ions were present on each side; therefore, they are spectator ions and *removed* from the net equation.

Mass Action equilibrium constants

There are two forms of *mass action equilibrium constants.*

K_{c1} is expressed as the concentrations of the aqueous species.

$$K_{c1} = [Ca^{2+}] / [H^+]^2$$

K_{c2} is expressed as the concentration of the gaseous species.

$$K_{c2} = [CO_2]$$

The K_{c1} and K_{c2} constants have different values.

The commonly reported K values are based on the most accessible phase to measure experimentally.

Do not mix phases, and do not include solids in either expression. When solving mass action problems, check the molecules' phases.

Copyright © Sterling Test Prep.

Equilibrium Constants

Equilibrium constant K_c

The *equilibrium constant* K_c is the ratio of the molar concentrations of products divided by reactants at equilibrium.

The K_c is abbreviated as K_{eq}, a general term for an equilibrium constant for concentration or partial pressures.

For the reaction $a\text{A} + b\text{B} \rightleftharpoons c\text{C} + d\text{D}$, the equilibrium constant is:

$$K_c = \frac{[\text{C}]^c[\text{D}]^d}{[\text{A}]^a[\text{B}]^b} = \frac{[\text{products}]}{[\text{reactants}]}$$

The equilibrium constant K_c is a temperature-specific constant, and it has the same value, regardless of the molar concentrations, if the temperature remains constant.

The units of K_c depend on the reaction; K_c is without units.

Calculating K_c

Guidelines for calculating the K_c value:

1. State the given and needed qualities.

2. Write the K_c expression for the equilibrium.

3. Substitute equilibrium (molar) concentrations and calculate K_c.

For example, what is the value of K_c at 443 °C for the equilibrium concentrations:

$$\text{H}_2\,(g) + \text{I}_2\,(g) \rightleftharpoons 2\,\text{HI}\,(g)$$

$[\text{H}_2] = 1.2$ mol/L

$[\text{I}_2] = 1.2$ mol/L

$[\text{HI}] = 0.35$ mol/L

Step 1: State the given and needed quantities

Given values		Needed quantities
Reactants	*Products*	
$[\text{H}_2] = 1.2$ mol/L	$[\text{HI}] = 0.35$ mol/L	
$[\text{I}_2] = 1.2$ mol/L		K_c

Step 2: Write the K_c expression for the equilibrium

$$K_c = [HI]^2 / [H_2]{\cdot}[I_2]$$

Step 3: Substitute equilibrium (molar) concentrations and calculate K_c

$$K_c = [0.35]^2 / [1.2]{\cdot}[1.2]$$

$$K_c = 8.5 \times 10^{-2}$$

K_c magnitude indicates reaction direction

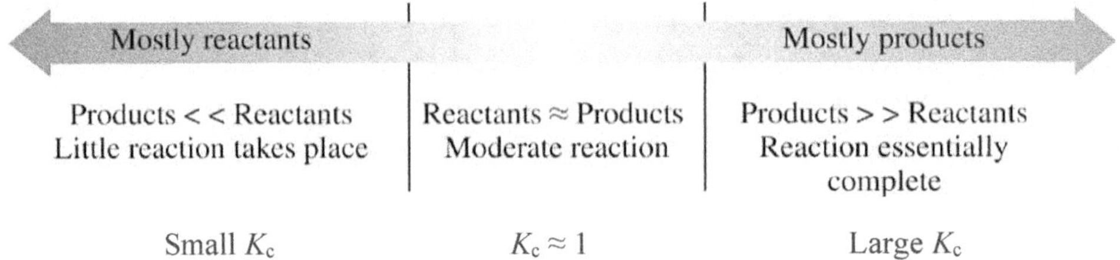

The K_c value depends on whether the equilibrium has a greater quantity of products or reactants.

However, the equilibrium constant's magnitude does not affect how fast equilibrium is reached.

Reactions with a large K_c have more products created from the forward reaction, and products predominate at equilibrium.

Reactions with small K_c have *reactants* predominate at equilibrium.

For example, the equilibrium for the reaction of SO_2 and O_2 has a large K_c.

$$2\,SO_2\,(g) + O_2\,(g) \;\leftrightarrows\; 2\,SO_3\,(g)$$

At equilibrium, the reaction contains mostly *products*.

$$K_c = [SO_3]^2 / [SO_2]^2{\cdot}[O_2]$$

$$K_c = [\text{many products}] / [\text{few reactants}]$$

$$K_c = 3.4 \times 10^2$$

Copyright © Sterling Test Prep.

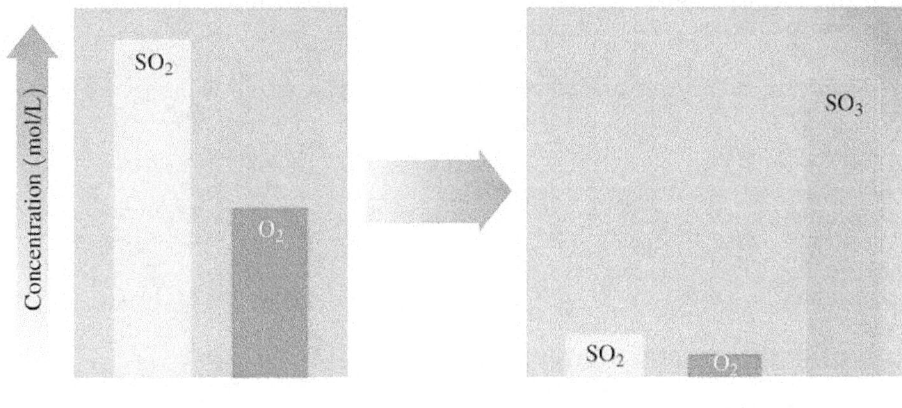

Initially   *At equilibrium*

For example,

$$2\ SO_2\ (g) + O_2\ (g) \rightleftharpoons 2\ SO_3\ (g)$$

Reactions with a small K_c have an equilibrium mixture with a *low concentration of products* and a *high concentration of reactants*.

The equilibrium constant for the reaction of N_2 and O_2 has a small K_c.

$$N_2\ (g) + O_2\ (g) \rightleftharpoons 2\ NO\ (g)$$

At equilibrium, the reaction mixture contains mostly *reactants*.

$$K_c = [NO] / [N_2] \cdot [O_2]$$

$$K_c = [\text{few products}] / [\text{mostly reactants}]$$

$$K_c = 2 \times 10^{-9}$$

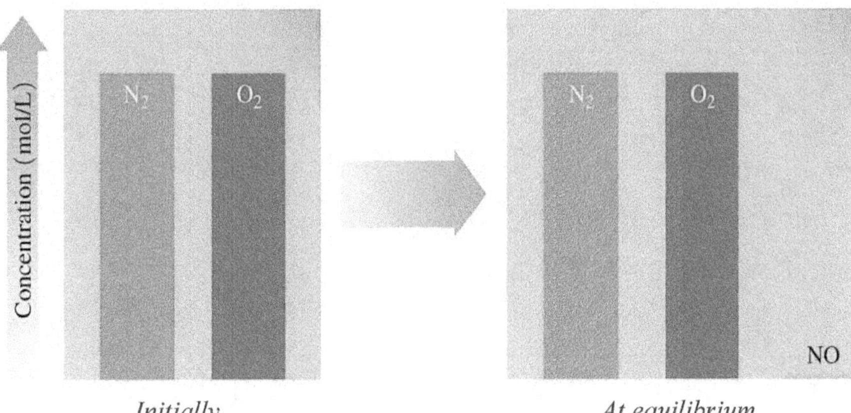

Initially   *At equilibrium*

Homogeneous *vs.* heterogeneous equilibrium

A *homogeneous equilibrium* is a reaction with products and reactants in the same physical state, such as:

$$2 \, SO_2 \, (g) + O_2 \, (g) \rightleftarrows 2 \, SO_3 \, (g)$$

$$K_c = [SO_3]^2 / [SO_2]^2 \cdot [O_2]$$

A *heterogeneous equilibrium* is a reaction with substances in different physical states:

$$C \, (s) + H_2O \, (g) \rightleftarrows CO \, (g) + H_2 \, (g)$$

$$K_c = [CO] \cdot [H_2] / [H_2O]$$

The concentrations of liquids and solids do not change and are omitted from the equilibrium constant K_c expression.

This omission of liquids and solids is like mass-action equilibrium.

Reaction quotient Q_c

The reaction quotient Q_c is calculated using the method for K_c, but Q_c calculation can be performed during the reaction.

Q_c is a "snapshot" of the system, representing the ratio between products and reactants at a given time.

Comparing Q_c and K_c determines the direction of the reaction:

If $Q_c > K_c$, the reaction creates more reactants

If $Q_c < K_c$, the reaction creates more products

If $Q_c = K_c$, the reaction is at equilibrium, $\Delta G = 0$

Copyright © Sterling Test Prep.

Equilibrium Variables

Le Châtelier's Principle for equilibrium

Le Châtelier's Principle states that when a reversible reaction at equilibrium is stressed by a change in a variable (e.g., concentration, pressure, volume, or temperature), equilibrium shifts to relieve the effect of the change.

For the equilibrium between colorless N_2O_4 and brown NO_2:

$$N_2O_4\ (g) \rightleftarrows 2\ NO_2\ (g)$$

Effect of concentration on equilibrium

If the N_2O_4 *reactant increases*, the reaction shifts to the right to produce more NO_2.

If the NO_2 *product increases*, the reaction shifts to the left to produce more N_2O_4.

Effect of pressure on equilibrium

Increasing pressure shifts the reaction to the side with fewer gas molecules in a gaseous equilibrium.

For the reaction:

$$N_2O_4\ (g) \rightleftarrows 2\ NO_2\ (g)$$

Increasing pressure shifts the reaction to the left, producing N_2O_4.

Changing pressure does *not* shift equilibrium for equal moles of gases as reactants and products.

Effect of decreasing volume on equilibrium

A change in the volume of a gas mixture at equilibrium changes the concentration of gases in the mixture.

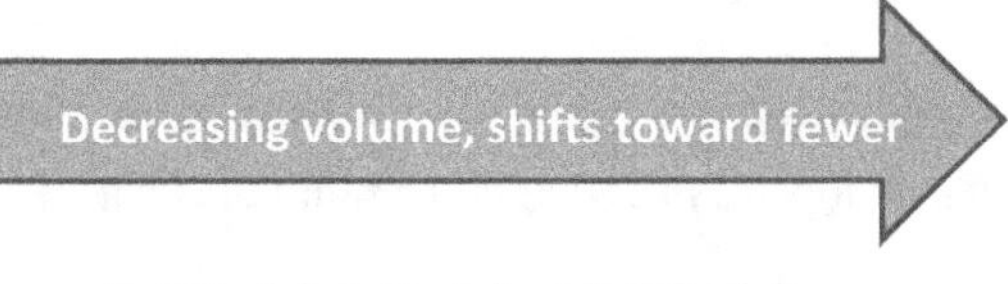

$$2\ CO_2\ (g) + O_2\ (g) \rightleftarrows 2\ CO_2\ (g)$$

Decreasing volume increases the concentration of the gases as the system shifts toward the smaller number of moles from the decrease in volume.

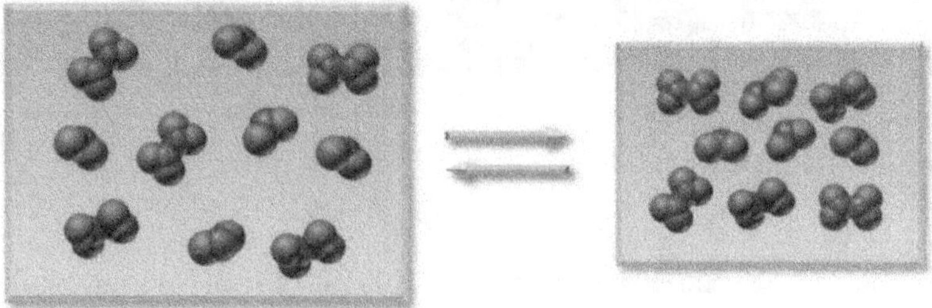

1.00 L at equilibrium   0.750 L at equilibrium

Effect of increasing volume on equilibrium

A change in the volume of a gas mixture changes its concentration.

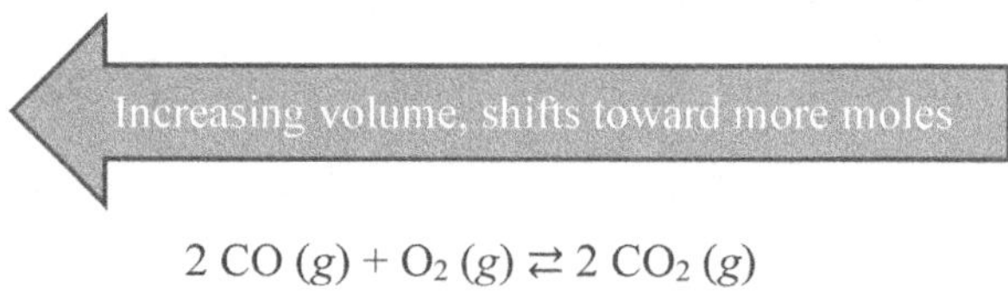

$$2\ CO\ (g) + O_2\ (g) \rightleftharpoons 2\ CO_2\ (g)$$

Increasing the volume decreases gas concentration as the system shifts toward the larger number of moles to compensate.

Temperature effect equilibrium depends on the ΔH (i.e., enthalpy) of a reaction (i.e., endothermic or exothermic).

Endothermic reaction equilibrium and temperature

Decreasing temperature of an endothermic reaction ($+\Delta H$ with heat as a reactant) causes the system to shift the reaction toward heat; toward *reactants*, increasing heat in the system.

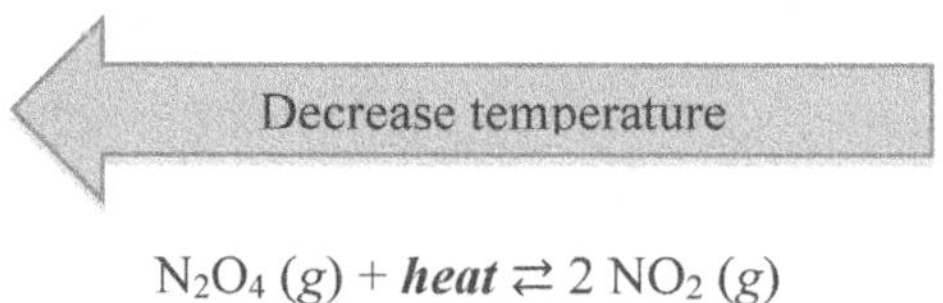

$$N_2O_4\ (g) + \textbf{\textit{heat}} \rightleftharpoons 2\ NO_2\ (g)$$

Increasing temperature of an endothermic reaction ($+\Delta H$ with heat as a reactant) causes the system to shift the reaction to remove heat; toward *products* consuming the heat.

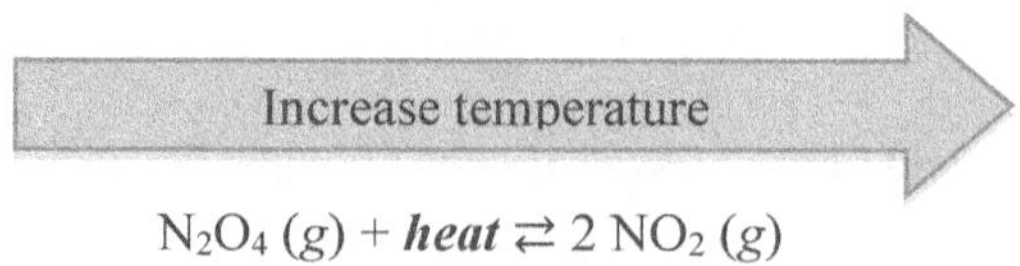

$$N_2O_4\ (g) + \textbf{\textit{heat}} \rightleftharpoons 2\ NO_2\ (g)$$

                    Copyright © Sterling Test Prep.

Exothermic reaction equilibrium and temperature

Decreasing temperature of an exothermic reaction ($-\Delta H$ with heat as a product) causes the system to shift the reaction toward heat; it shifts the reaction toward *products*, increasing heat in the system.

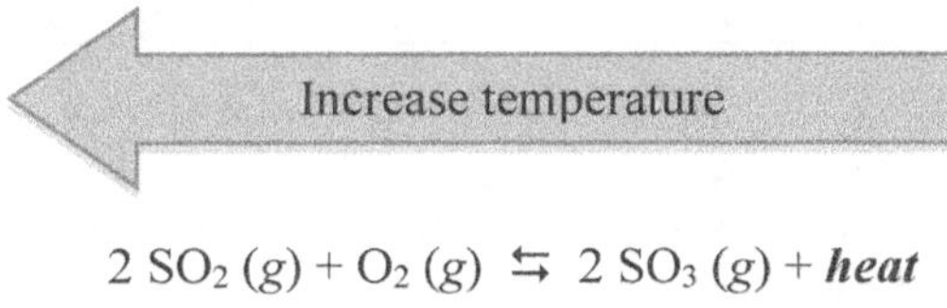

$$2\ SO_2\ (g) + O_2\ (g) \rightleftharpoons 2\ SO_3\ (g) + \boldsymbol{heat}$$

Increasing temperature of an exothermic reaction ($-\Delta H$ with heat as a product) causes the system to shift the reaction toward removing heat; it shifts the reaction toward *reactants*, decreasing heat in the system.

$$2\ SO_2\ (g) + O_2\ (g) \rightleftharpoons 2\ SO_3\ (g) + \boldsymbol{heat}$$

Variables affecting equilibrium

Variable	Stress	Equilibrium shifts toward
Concentration	Add a reactant	Products (forward reaction)
	Remove a reactant	Reactants (reverse reaction)
	Add a product	Reactants (reverse reaction)
	Remove a product	Products (forward reaction)
Volume (container)	Decrease volume	Side with lower total coefficients (*aq* and *g*)*
	Increase volume	Side with higher total coefficients (*aq* and *g*)
Pressure	Decrease pressure	Side with higher total coefficients (*aq* and *g*)
	Increase pressure	Side with lower total coefficients (*aq* and *g*)
Temperature	**Endothermic Rxn**	
	Raise Temperature	Products (forward reaction to *remove heat*)
	Lower Temperature	Reactants (reverse reaction to *add heat*)
	Exothermic Rxn	
	Raise Temperature	Reactants (reverse reaction to *add heat*)
	Lower Temperature	Products (forward reaction to *remove heat*)
Catalyst	Increases rates equally	No effect

* Pure solids and pure liquids are not included in the equilibrium expression.

Copyright © Sterling Test Prep.

Biological equilibrium

Equilibrium is an important concept in many systems, including biological ones.

For example, O_2 transport uses an equilibrium between hemoglobin (Hb), oxygen, and oxyhemoglobin (HbO_2).

$$Hb \ (aq) + O_2 \ (g) \rightleftarrows HbO_2 \ (aq)$$

$$K_c = [\text{products}] \ / \ [\text{reactants}]$$

$$K_c = [HbO_2] \ / \ [Hb] \cdot [O_2]$$

If there is a high concentration of O_2 (reactant) in the lungs' alveoli, the reaction shifts to the right to make more oxyhemoglobin (product).

When the concentration of O_2 is low in tissues, the reverse reaction releases O_2 from oxyhemoglobin.

At normal atmospheric pressure, oxygen diffuses into the blood because the partial pressure of oxygen in the alveoli is higher than that in the blood.

At altitudes above 8,000 ft, a decrease in atmospheric pressure results in the lower partial pressure of O_2.

Hypoxia may occur at high altitudes where the $[O_2]$ is low. At an altitude of 18,000 ft., a person obtains 29% less oxygen and may experience hypoxia.

According to Le Châtelier's Principle, a decrease in O_2 shifts the equilibrium in the direction of the reactants and depletes the concentration of HbO_2, which causes hypoxia.

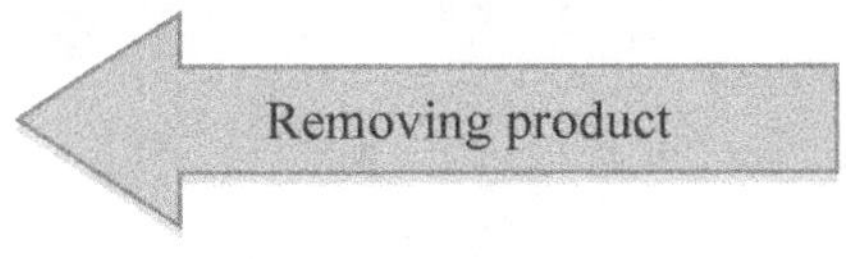

$$Hb \ (aq) + O_2 \ (g) \rightleftarrows HbO_2 \ (aq)$$

Another reaction that applies Le Châtelier's Principle follows.

For example, indicate the shift in equilibrium caused by each change:

$$2\ NO_2 \ (g) + \text{heat} \rightleftarrows 2\ NO \ (g) + O_2 \ (g)$$

The following changes cause the equilibrium to shift toward reactants or products (*answers below*)?

A. adding NO

B. lowering the temperature

C. removing O_2

D. increasing the volume

E. removing NO

Answers:

A: reactants	C: products	
B: reactants	D: products	E: products

Copyright © Sterling Test Prep.

Equilibrium Constant and $\Delta G°$ Relationship

Direction and extent of reaction

Gibbs free energy ($\Delta G°$) is an expression for a system's free energy.

The equations relate Gibbs free energy to the equilibrium constant:

$$\Delta G° = -RT \ln K$$

Taking the antilog (e^x) of each side gives:

$$K = e^{-\Delta G° / RT}$$

The ΔG and K imply the direction and extent of a reaction, *not* the rate.

ΔG at equilibrium

If $\Delta G° = 0$, the reaction is at equilibrium; no net change to the system or surroundings.

$$\Delta G° = 0$$

The reaction appears spontaneous in either direction if some small change is made to the system.

The system remains at equilibrium if no heat is added or removed.

Phases exist indefinitely.

ΔG affects equilibrium

ΔG affects the position of equilibrium in the following ways:

$$\Delta G° > 0 \text{ (positive)}$$

If $\Delta G° > 0$, equilibrium is closer to reactants; reverse reaction (towards reactants) is spontaneous.

$$\Delta G° < 0 \text{ (negative)}$$

If $\Delta G° < 0$, equilibrium is closer to products; forward reaction (towards products) is spontaneous.

For example, consider the freezing of water at 0 °C:

$$H_2O \ (l) \rightarrow H_2O \ (s)$$

Below 0 °C:

$$\Delta G < 0 \text{ and freezing is spontaneous}$$

Above 0 °C:

$$\Delta G > 0 \text{ and freezing is nonspontaneous}$$

Any duplication (copies, uploads, PDFs) is illegal.

Energy profile for spontaneous reactions

The energy profile displays a reaction with a $+\Delta G$ (i.e., nonspontaneous).

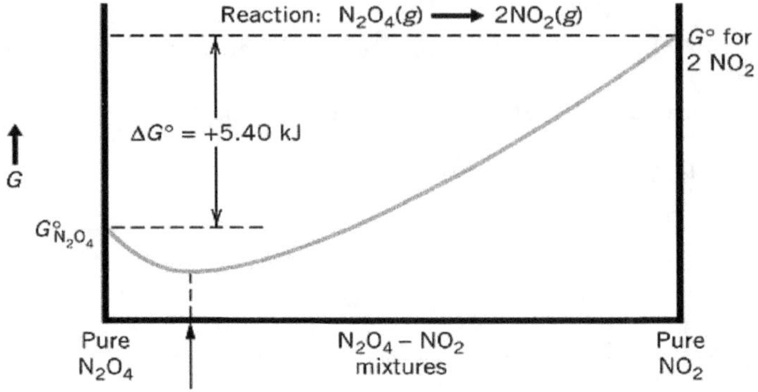

Energy profile for nonspontaneous reactions

The following energy profile represents a reaction with $-\Delta G$ (i.e., spontaneous).

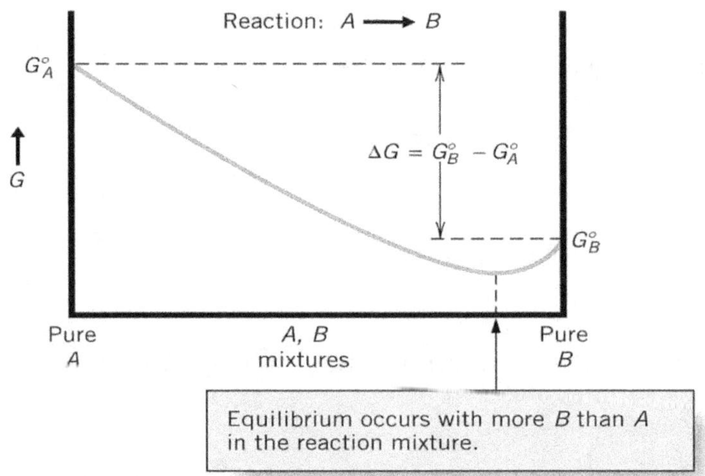

Copyright © Sterling Test Prep.

$\Delta G°$ and equilibrium constant K_c relationship

The equilibrium constant Kc is the ratio of the equilibrium concentrations of products over the equilibrium concentrations of reactants, each raised to the power of their stoichiometric coefficients.

$$Kc = [products]^x / [reactants]^y$$

where [ ] represents concentration and exponents x and y are stoichiometric coefficients

For example, calculate the equilibrium constant K_c at 25 °C for the decarboxylation ($\sim CO_2$) of liquid pyruvic acid ($CH_3COCOOH$) to form gaseous acetaldehyde (CH_3COH) and carbon dioxide (CO_2).

$$CH_3COCOOH \ (l) \rightleftharpoons CH_3COH \ (g) + CO_2 \ (g)$$

Compound	$\Delta G°_f$ (kJ/mol)
CH_3COH	-33.30
$CH_3COCOOH$	-463.38
CO_2	-394.36

Equilibrium constant K_{eq} from ΔG

K_{eq} is a measure of the ratio of the concentrations of products to the concentrations of reactants.

$\Delta G°$ is the standard change in free energy or the change in free energy *under standard conditions*.

$$\Delta G° = -RT \ln K_{eq}$$

where $\Delta G°$ is the standard change in free energy, K_{eq} is the ratio of the concentrations of products to the concentrations of reactants, $R = 8.314$ J mol^{-1} K^{-1} and T is temperature in Kelvin

Sum $\Delta G°_f$ for each reaction to obtain total $\Delta G°$.

$$\Delta G° = \Sigma \Delta G°_f$$

$$\Delta G° = \Delta G°_f \ (CH_3COH) + \Delta G°_f \ (CO_2) - \Delta G°_f \ (CH_3COCOOH)$$

$$\Delta G° = (-133.30 \ (kJ/mol) + (-394.36 \ (kJ/mol) - (-463.38 (kJ/mol)$$

$$\Delta G° = -64.28 \ kJ/mol$$

Any duplication (copies, uploads, PDFs) is illegal.

Use the $\Delta G°$ value to calculate K_{eq}

$$K_{eq} = e^{-\Delta G° / RT}$$

where K is the equilibrium constant, R is the gas constant, and T is the temperature in Kelvin

Solve for the exponent

$$\Delta G° / RT = (-64.28 \text{ kJ/mol}) / [(8.314 \text{ J/K}) \times (298 \text{ K}) \times (1,000 \text{ J/kJ})]$$

$$\Delta G° / RT = -25.94$$

Substitute exponent value

$$K_{eq} = e^{-(-25.945)}$$

$$K_{eq} = e^{25.945}$$

$$K_{eq} = 1.85 \times 10^{11}$$

The value for K_{eq} means that the ratio for concentration of products to reactants is 1.85×10^{11}

Therefore, product formation is favored.

Copyright © Sterling Test Prep.

Practice Questions

1. Which statement is NOT a correct characterization for a catalyst?

    A. Catalysts are not consumed in a reaction

    B. Catalysts do not actively participate in a reaction

    C. Catalysts lower the activation energy for a reaction

    D. Catalysts can be either solids, liquids, or gases

2. What is the rate law for the following reaction that was found to be first-order in each of the two reactants and second-order overall?

$$2\,NO\,(g) + O_2\,(g) \rightarrow 2\,NO_2\,(g)$$

   A. rate $= k[NO]^2 \cdot [O_2]^2$ **C.** rate $= k[NO] \cdot [O_2]$

   B. rate $= k[NO_2]^2 \cdot [NO]^{-2} \cdot [O_2]^{-\frac{1}{2}}$ **D.** rate $= k[NO]^2$

3. Which equilibrium constant applies to a reversible reaction involving a gaseous mixture at equilibrium?

    A. Ionization equilibrium constant, K_w

    B. General equilibrium constant, K_p

    C. Solubility product equilibrium constant, K_{sp}

    D. Ionization equilibrium constant, K_i

4. What is the equilibrium constant (K_{eq}) expression for the following reaction:

$$CaCO_3\,(s) \leftrightarrow CaO\,(s) + CO_2\,(g)$$

   A. $K_{eq} = [CO_2]$ **C.** $K_{eq} = [CaO] \cdot [CO_2] / [CaCO_3]$

   B. $K_{eq} = 1 / [CO_2]$ **D.** $K_{eq} = [CaO] \cdot [CO_2]$

5. Which of the following statements about catalysts is NOT true?

    A. A catalyst does not change the energy of the reactants or the products

    B. A catalyst increases the rate of slow reactions

    C. A catalyst is consumed in a reaction

    D. A catalyst alters the rate of a chemical reaction

Any duplication (copies, uploads, PDFs) is illegal.

6. Increasing the temperature of a chemical reaction increases the rate of reaction because:

    A. both the collision frequency and collision energies of reactant molecules increase

    B. the collision frequency of reactant molecules increases

    C. the activation energy increases

    D. the activation energy decreases

7. If $K_{eq} = 6.1 \times 10^{-11}$, which statement is true?

    A. Slightly more products are present   **C.** Mostly products are present

    B. The number of reactants equals products   **D.** Mostly reactants are present

8. Which is true before a reaction reaches chemical equilibrium?

    A. The amounts of reactants and products are equal

    B. The amounts of reactants and products are constant

    C. The amount of products is decreasing

    D. The amount of products is increasing

9. Why might increasing temperature alter the rate of a chemical reaction?

    A. The molecules combine with other atoms at high temperature to save space

    B. The density decreases as a function of temperature that increases volume and decreases the reaction rate

    C. The molecules have higher kinetic energy and have more force when colliding

    D. The molecules are less reactive at higher temperatures

10. In the following reaction, how does increasing the pressure affect the equilibrium?

$$2\ SO_2\ (g) + O_2\ (g) \leftrightarrow 2\ SO_3\ (g) + heat$$

    A. Remains unchanged, but the reaction mixture gets warmer

    B. Remains unchanged, but the reaction mixture gets cooler

    C. Shifts to the right towards products

    D. Shifts to the left towards reactants

11. According to Le Châtelier's Principle, which shifts the equilibrium to the left for the following reactions?

$$N_2\ (g) + 3\ H_2\ (g) \leftrightarrow 2\ NH_3\ (g) + heat$$

    A. Decreasing the temperature   **C.** Increasing $[N_2]$ the pressure

    B. Increasing $[H_2]$ **D.** Decreasing the pressure on the system

Copyright © Sterling Test Prep.

12. What is the term for a dynamic state of a reversible reaction in which the rates of the forward and reverse reactions are equal?

    A. chemical equilibrium   **C.** reversible equilibrium

    B. rate equilibrium   **D.** concentration equilibrium

13. When a system is at equilibrium, the:

    A. reaction rate of the forward reaction is low compared to the reverse

    B. number of products and reactants are equal

    C. reaction rate of the forward reaction equals the rate of the reverse

    D. reaction rate of the reverse reaction is low compared to the forward

Any duplication (copies, uploads, PDFs) is illegal.

Notes for active learning

Notes for active learning

Copyright © Sterling Test Prep.

Detailed Explanations

1. B is correct.

Enzymes (i.e., *biological catalysts*) bind to substrates to form an enzyme-substrate complex. While in this complex, enzymes often align reactive chemical groups and hold them closer.

Enzymes can induce structural changes that strain substrate bonds. Therefore, catalysts can actively participate in reactions.

Catalysts *lower the reaction's activation energy* (energy barrier), thus increasing its rate (k).

Although catalysts decrease the amount of energy required to reach the rate-limiting transition state, they do *not* decrease the relative energy of the products and reactants.

Therefore, a catalyst does *not* affect ΔG.

A catalyst provides an *alternative pathway* for the reaction to proceed to product formation. It lowers activation energy (i.e., relative energy between reactants and transition state) and, therefore, *speeds* the reaction rate.

Catalysts do not affect the Gibbs free energy (ΔG: stability of products *vs.* reactants) or the enthalpy (ΔH: bond breaking in reactants or bond making in products).

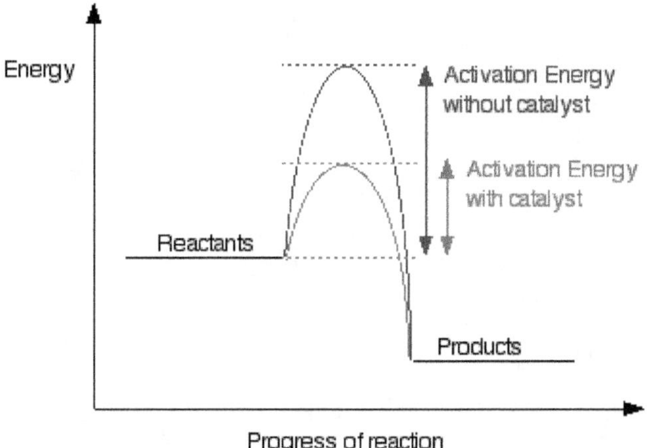

2. C is correct.

The *rate law* is calculated by comparing trials and determining how changes in the initial concentrations of the reactants affect the reaction rate.

$$\text{rate} = k[\text{A}]^{x} \cdot [\text{B}]^{y}$$

where k is the rate constant, and the exponents x and y are the partial reaction orders (i.e., determined experimentally). They are *not* equal to the stoichiometric coefficients

To write a complete rate law, each reactant's order is required. *continued...*

The problem provided each reactant's order (first order in each).

The total order is a sum of the individual orders: $1 + 1 = 2$.

However, the total order is usually not expressed in the rate law.

3. B is correct.

To determine the amount of each compound at equilibrium, consider the chemical reaction written in the form:

$$aA + bB \leftrightarrow cC + dD$$

The *equilibrium constant* is:

$$K_{eq} = ([C]^c \times [D]^d / ([A]^a \times [B]^b)$$

K_w is the self-ionization (i.e., autoionization) constant for water.

K_{sp} is used for "solubility" and is only applicable for equilibriums for solids dissolving in liquids.

K_i is used for "ionization" and only applies to equilibrium systems involving liquids.

For equilibria in the gas phase, the equilibrium equation (K_p) is a function of the partial pressures (P) of the reactants and products.

Therefore:

$$K_p = ([P_C]^c \times [P_D]^d) / ([P_A]^a \times [P_B]^b)$$

where P represents partial pressure, usually in atm

4. A is correct.

General formula for the *equilibrium constant* of a reaction:

$$aA + bB \leftrightarrow cC + dD$$

$$K_{eq} = ([C]^c \times [D]^d) / ([A]^a \times [B]^b)$$

For equilibrium constant calculation, only include species in *aqueous* or *gas phases*:

$$K_{eq} = [CO_2]$$

Copyright © Sterling Test Prep.

5. C is correct.

The primary function of catalysts is lowering the reaction's activation energy (energy barrier), thus increasing its rate (k).

Although catalysts decrease the amount of energy required to reach the rate-limiting transition state, they do *not* decrease the relative energy of the products and reactants.

Therefore, a catalyst does *not* affect ΔG.

A catalyst is never consumed in a reaction; reagents are consumed.

A catalyst lowers the energy of the high-energy transition state (i.e., the activation energy), but it does not change the energy of the reactants or the products.

Catalysts *increase the rate* of chemical reactions.

6. A is correct.

Temperature measures the molecules' average kinetic energy (KE).

Increasing the temperature increases the collision frequency (i.e., due to an increased probability of molecules striking each other) *and* collision energies (kinetic energy = $\frac{1}{2}mv^2$) of reactant molecules.

7. D is correct.

For the general equation:

$$a\text{A} + b\text{B} \leftrightarrow c\text{C} + d\text{D}$$

The equilibrium constant is:

$$K_{eq} = ([\text{C}]^c \times [\text{D}]^d) / ([\text{A}]^a \times [\text{B}]^b)$$

or

$$K_{eq} = [\text{products}] / [\text{reactants}]$$

If K_{eq} is less than 1 (e.g., 6.1×10^{-11}), then the numerator (i.e., products) is smaller than the denominator (i.e., reactants), and fewer products have formed relative to the reactants.

If the reaction favors reactants compared to products, the equilibrium lies to the left.

8. D is correct.

A reaction proceeds with the formation of products.

Therefore, the forward reaction rate is greater than the reverse reaction rate until the reaction achieves equilibrium.

At equilibrium, the rate (not relative concentrations of products and reactants) decreases for the forward reaction while the reverse reaction rate increases.

At equilibrium, the forward reaction rate equals the reverse reaction rate.

An *endergonic reaction* (i.e., $+\Delta G$) has fewer products forming than reactants remaining (i.e., the reaction is nonspontaneous) with higher energy than the reactants.

An *exergonic reaction* (i.e., $-\Delta G$) has more products formed than reactants remaining (i.e., the reaction is spontaneous), with the products lower in energy than the reactants.

Catalysts lower the reaction's activation energy (energy barrier), thus increasing its rate (k).

Although catalysts decrease the amount of energy required to reach the rate-limiting transition state, they do *not* decrease the relative energy of the products and reactants.

Therefore, a catalyst does not affect ΔG.

A catalyst provides an alternative pathway for the reaction to proceed to product formation.

Catalysts lower the energy of activation (i.e., relative energy between reactants and transition state) and, therefore, speed the reaction rate.

Catalysts do *not* affect the Gibbs free energy (ΔG: stability of products *vs.* reactants) or the enthalpy (ΔH: bond breaking in reactants or bond making in products).

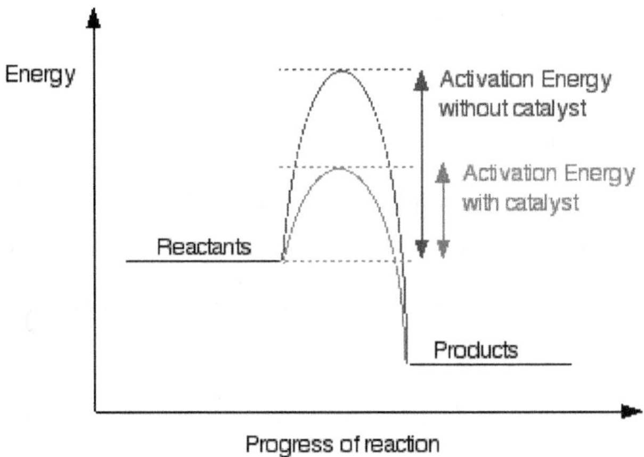

Copyright © Sterling Test Prep.

9. C is correct.

As the average kinetic energy (i.e., $KE = \frac{1}{2}mv^2$) *increases*, the particles move faster, collide more frequently per unit time, and possess greater energy when they collide.

This *increases* the *reaction rate*. Hence the *reaction rate* of most *reactions increases* with *increasing temperature.*

10. C is correct.

When changing the reaction conditions, Le Châtelier's Principle states that the position of equilibrium shifts to counteract the change.

If the reaction temperature, pressure, or volume change, the position of equilibrium changes.

In general, increasing the pressure tends to favor the side of the reaction with a lower molar coefficient sum (i.e., toward the products in this example).

11. D is correct.

When changing the reaction conditions, Le Châtelier's Principle states that the position of equilibrium shifts to counteract the change.

If the reaction temperature, pressure, or volume changes, the equilibrium changes.

Removing reactants or adding products shifts the equilibrium to the left.

In this reaction, reducing the pressure of the system favors the reactants because of their respective molar concentrations of reactants (i.e., $1 + 3$) and products (i.e., 2).

All other modifications listed would shift the equilibrium toward products (i.e., to the right).

12. A is correct.

Equilibrium refers to the state when the forward reaction rate equals the reverse reaction rate.

Equilibrium does not describe the state when the relative energy of the reactants and products is the same, nor does it describe the state when the relative amounts of reactants and products are the same.

13. C is correct.

Chemical equilibrium refers to a dynamic process whereby the rate of reactant molecules transforming into products equals the rate for a product molecule to be transformed into a reactant.

Therefore, the reaction rate of the forward reaction equals the reverse rate.

Notes for active learning

Notes for active learning

Copyright © Sterling Test Prep.

Notes for active learning

Notes for active learning

Copyright © Sterling Test Prep.

CHAPTER 7

Solution Chemistry

- Solutions
- Ions in Solution
- Solubility Concepts
- Solution Equilibrium
- Common and Complex Ions
- Practice Questions & Detailed Explanations

Copyright © Sterling Test Prep.

Solutions

Solutions as homogenous mixtures

A *solution* is a homogeneous mixture of one or more solutes and a solvent with indistinguishable components.

The formation of a solution involves disrupting the crystalline lattice structure of the solute and mixing solute particles with solvent molecules.

The *solvent* is the substance present in the most substantial quantity.

Solute particles diffuse into the solution and become uniformly dispersed.

Aqueous solutions have water as the solvent.

Enthalpy considerations for solutions

There must be sufficiently strong interactions between solute particles and solvent molecules for solute particles to diffuse into the solution.

The interactions between solute and solvent molecules must overcome the attractive intermolecular forces between solute particles in the liquid or solid.

Solution processes may be divided into three main stages:

1. pure solvent $\rightarrow$ separated solvent molecules (endothermic);
   $$\Delta H_1 > 0$$

2. pure solute $\rightarrow$ separated solute particles (endothermic);
   $$\Delta H_2 > 0$$

3. separated solvent and solute molecules $\rightarrow$ solution (exothermic);
   $$\Delta H_3 < 0$$

For example:

$$\text{Solute(s)} + \text{Solvent} \rightarrow \text{Solution}$$

$$\Delta H_{\text{soln}} = \Delta H_1 + \Delta H_2 + \Delta H_3$$

Depending on the magnitude of ΔH_1, ΔH_2 and ΔH_3, the solution process is:

Exothermic: if $| \Delta H_3| > | \Delta H_1 + \Delta H_2|$

Endothermic: if $| \Delta H_3| < | \Delta H_1 + \Delta H_2|$

For solutions, remember the phrase "like dissolves like."

Polar solutes dissolve in polar solvents, while nonpolar solutes dissolve in nonpolar solvents.

Any duplication (copies, uploads, PDFs) is illegal.

Notes for active learning

Copyright © Sterling Test Prep.

Ions in Solution

Hydration energy

Many ionic compounds dissolve in polar water with negative enthalpies (release energy) as *hydration energy*.

Solvation is the attraction of a solvent with solute molecules or ions.

Hydration is the process of attraction of water with solute molecules or ions.

The *hydration energy* produced from ion-dipole interactions provides the energy to overcome the lattice energy and disrupt ions in the crystalline solids (e.g., NaCl).

Water molecules are polar and interact strongly with ionic cations and anions through ion-dipole interactions.

The overall solute–solution formation process may be exothermic or endothermic, depending on if the solute–solute or solute–solution enthalpy is larger.

Ionic compounds dissolve in water and dissociate (i.e., ionize) to produce free ions (cations and anions).

For example, NaCl dissolves and produces Na^+ and Cl^- ions, which become hydrated with water molecules.

Sodium chloride and many other ionic compounds dissolve in water because of the strong ion-dipole interactions between the charged solute particles and the polar solvent (e.g., water) molecules.

For example, for NaCl, the *lattice energy* (i.e., ionic solid to gaseous ions) is slightly higher than the sum of *hydration energy* for Na^+ and Cl^- ions. The solution process for NaCl is slightly endothermic.

Electrolytes

Strong electrolytes (solutions of ionic compounds) completely dissociate, and the dissolved ions are good conductors of electric current.

Weak electrolytes (solutions of polar covalent compounds) partially dissociate in solution.

Nonelectrolytes do not dissociate into ions (e.g., sugar as a nonelectrolyte) and dissolve in the solution as the original molecules rather than separate into cations and anions.

Any duplication (copies, uploads, PDFs) is illegal.

Anions and cations

Anions are *negatively charged* species, with a single negative sign indicating a charge of -1.

Cations are *positively charged* species, with a single positive sign indicating a charge of $+1$.

The charge is written as a superscript.

Example ions include:

ammonium (NH^{4+})

phosphate (PO_4^{3-})

sulfate (SO_4^{2-})

Common anions and cations	Formula
Anions	
Hydroxide	OH^-
Chloride	Cl^-
Hypochlorite	ClO^-
Chlorite	ClO_2^-
Chlorate	ClO_3^-
Perchlorate	ClO_4^-
Halide, hypohalite, etc.	X^-, XO^-, etc.
Carbonate	CO_3^{2-}
Hydrogen Carbonate (Bicarbonate)	HCO_3^-
Sulfate	SO_4^{2-}
Hydrogen Sulfate (Bisulfate)	HSO_4^-
Sulfite	SO_3^{2-}
Thiosulfate	$S_2O_3^{2-}$
Nitrate	NO_3^-
Nitrite	NO_2^-
Phosphate	PO_4^{3-}

Copyright © Sterling Test Prep.

Hydrogen Phosphate	HPO_4^{2-}
Dihydrogen Phosphate	$H_2PO_4^-$
Phosphite	PO_3^{3-}
Cyanide	CN^-
Thiocyanate	SCN^-
Peroxide	O_2^{2-}
Oxalate	$C_2O_4^{2-}$
Acetate	$C_2H_3O_2^-$
Chromate	CrO_4^{2-}
Dichromate	$Cr_2O_7^{2-}$
Permanganate	MnO_4^-
Cations	
Hydronium	H_3O^+
Ammonium	NH_4^+
Metal	M^{n+}

Common ions, associated formulas, and charges shown

Identifying ions experimentally

Qualitative analysis is a laboratory method for separating and identifying ions in a mixture.

The technique uses solubility differences for ionic compounds in an aqueous solution and a particular cation's ability to form *complex ions* with ligands.

The approach in qualitative analysis of cations separates them sequentially into *ion groups*.

The complex ions of many transition metals are often colored and used in visual identity.

Ion group 1: Insoluble chlorides

Treating the mixture with 6 M HCl precipitates Ag^+, Hg_2^{2+}, and Pb^{2+} as chlorides, with other cations in solution.

The formation of a *white precipitate* indicates at least one of these cations in the mixture.

Any duplication (copies, uploads, PDFs) is illegal.

Ion group 2: Acid-insoluble sulfides

The supernatant from the above treatment with HCl is adjusted to pH ≈ 0.5 and treated with aqueous H_2S.

The high $[H_3O^+]$ in solution keeps $[HS^-]$ low, which precipitates cation groups: Cu^{2+}, Cd^{2+}, Hg^{2+}, Sn^{2+}, and Bi^{3+}.

Centrifuging and decanting give the next solution.

Ion group 3: Base-insoluble sulfides

The supernatant from acidic sulfide treatment is treated with an NH_3/NH_4^+ buffer to make the solution slightly basic (pH ≈ 8).

The excess OH^- in solution increases $[HS^-]$, which causes precipitation of more soluble sulfides and hydroxides.

The cations that precipitate under this condition of base-insoluble sulfides are Zn^{2+}, Mn^{2+}, Ni^{2+}, Fe^{2+}, and Co^{2+}, as sulfides, and Al^{3+}, Cr^{3+}, and Fe^{3+} as hydroxides.

The precipitate is centrifuged, and the supernatant decanted to give the next solution.

Ion group 4: Insoluble phosphates

The slightly basic supernatant separated from the group 3 ions are treated with $(NH_4)_2HPO_4$, which precipitates $Mg_3(PO_4)_2$, $Ca_3(PO_4)_2$ and $Ba_3(PO_4)_2$.

Ion group 5: Alkali metal and ammonium ions

The final solution contains any of the following ions: Na^+, K^+, and NH_4^+.

Solvation and hydration

Solvation is the attraction of a solvent with solute molecules or ions.

Hydration is the attraction of water with solute molecules or ions, and water forms a shell around the ions in a solution.

For example, the oxygen atom on the water is partially negative, so it surrounds cations.

Conversely, the hydrogen atoms on the water are partially positive, surrounding anions.

Copyright © Sterling Test Prep.

Hydronium ions

H^+ does not exist as a bare proton in water; it exists as the *hydronium ion* (H_3O^+).

The high charge density of a proton attracts it to a nearby molecule with a partial or full negative charge.

Water has two lone pairs of electrons on the oxygen atom; the positive proton interacts with one of the lone pairs, forming a hydronium ion.

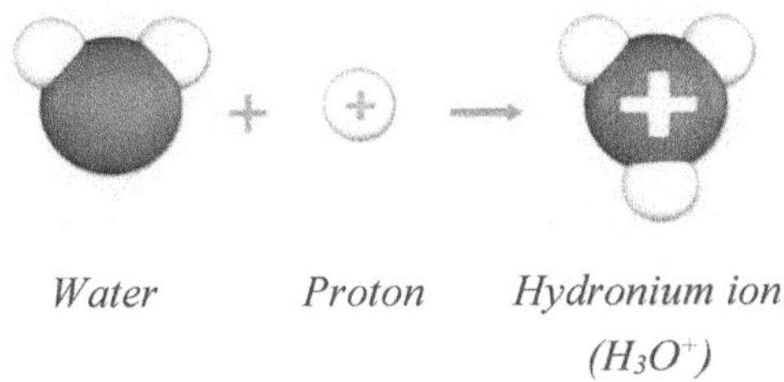

Water   *Proton*   *Hydronium ion*
(H_3O^+)

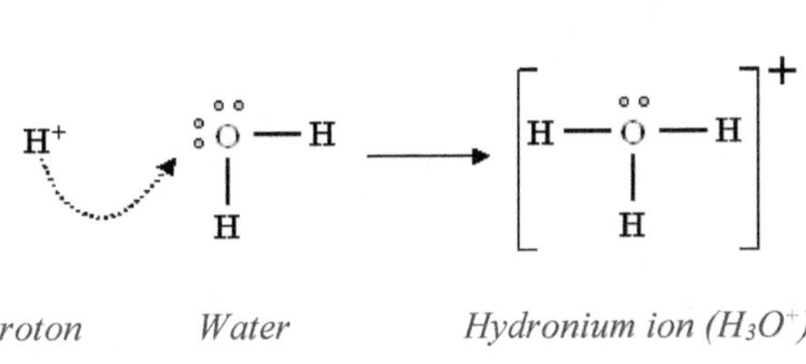

Proton   *Water*   *Hydronium ion (H_3O^+)*

Any duplication (copies, uploads, PDFs) is illegal.

Notes for active learning

Copyright © Sterling Test Prep.

Solubility Concepts

Solubility

A substance's solubility is the amount of solute (e.g., grams) that dissolves in a solvent to give a saturated solution at a specific temperature.

Solubility is *temperature-dependent*.

A *saturated* solution contains the maximum dissolved solute possible at a given temperature, with a state of dynamic equilibrium existing between dissolution and crystallization.

For example, the solubility of KNO_3 is ~30 g per 100 g water at 20 °C and 63 g at 40 °C.

Solubility rules for ionic compounds

Use the following rules to *predict the solubility of ionic compounds* in water. These rules generally relate to the compound's lattice and hydration energies of the individual ions.

One of the factors determining the *lattice energy* is the magnitudes of charges on the ions (i.e., the greater the magnitude of the charges, the greater the lattice energy). Simplified, the greater the lattice energy, the lower the solubility.

The rules focus on the magnitudes of the charges and ignore other influences on the lattice energies, such as hydration energy. In addition, the rules disregard the effect of temperature.

The solubility rules are listed in order of *decreasing importance*. For example, if the first rule applies, disregard the subsequent rules. Similarly, if the second rule is applicable, disregard the third rule. Thus, only use the third rule when rules one and two are not applicable.

The following ions are usually soluble: $C_2H_3O_2^-$, ClO_3^-, ClO_4^-. K^+, Na^+, NH_4^+, and NO_3^-.

General principle: Hydrides (H^-) decompose in water to yield H_2 and ^-OH. Many metal oxides react with water to produce hydroxides (^-OH).

Rule 1: compounds containing a +1 or –1 ion are typically soluble

Examples include: $AlCl_3$, $FeCl_2$, K_2SO_4, $NaBr$, Rb_3PO_4

Exceptions (Insoluble): OH^- (other than with cations that produce strong bases)

         Hg_2^{2+} (other than with NO_3^-, $C_2H_3O_2^-$, ClO_3^-, and ClO_4^-)

         Pb^{2+}, Hg^{2+} with most –1 ions

         IB metals (column 11)

Any duplication (copies, uploads, PDFs) is illegal.

Rule 2: compounds containing a +3 or –3 or higher ion are typically NOT soluble

Examples include: $AlPO_4$, $Ca_3(PO_4)_2$, TiO2

Exceptions (Soluble):   cations combined with sulfate or dichromate

Rule 3: compounds containing –2 ion are typically NOT soluble

Examples include: $CaCO_3$, FeS, ZnC_2O_4

Exceptions (Soluble):

Dichromates and sulfates (insoluble if combined with Ba, Ca, Hg, Pb, and Sr)

There are exceptions to the solubility rules besides those listed. However, the ions commonly encountered follow these rules.

Solubility examples for ionic compounds

AgBr	Insoluble	Rule 1	Ag^+
AgCl	Insoluble	Rule 1	$Ag^+ + Cl^-$
$AgIO_3$	Insoluble	Rule 1	$Ag^+ + IO_3^-$
AgOH	Insoluble	Rule 1	weak base
Ag_2S	Insoluble	Rule 1	Ag^+
Hg_2Cl_2	Insoluble	Rule 1	Hg_2^{2+}
$PbCl_2$	Insoluble	Rule 1	Pb^{2+}
$AgNO_3$	Soluble	Rule 1	$NO_3^- + Ag^+$
$Ba(NO_3)_2$	Soluble	Rule 1	NO_3^-
$Ba(OH)_2$	Soluble	Rule 1	strong base
NaCl	Soluble	Rule 1	$Na^+ + Cl^-$
$NaClO_3$	Soluble	Rule 1	$Na^+ + ClO_3^-$
NaOH	Soluble	Rule 1	strong base
Na_3PO_4	Soluble	Rule 1	Na^+
$(NH_4)_2CO_3$	Soluble	Rule 1	NH_4^+
NH_4IO_3	Soluble	Rule 1	$NH_4^+ + IO_3^-$
$FeAsO_4$	Insoluble	Rule 2	$Fe^{3+} + AsO_4^{3-}$
$Fe_2(SO_4)_3$	Soluble	Rule 2	SO_4^{2-}
$BaCO_3$	Insoluble	Rule 3	CO_3^{2-}
$BaSO_4$	Insoluble	Rule 3	$Ba^{2-} + SO_4^{2-}$

Copyright © Sterling Test Prep.

$MgSO_4$	Insoluble	Rule 3	SO_4^{2-}
$PbCrO_4$	Insoluble	Rule 3	CrO_4^{2-}
ZnS	Insoluble	Rule 3	S^{2-}
$Ba(IO_3)_2$	Insoluble	exception	

Solution concentration

When a solution that is almost saturated at a higher temperature is cooled slowly to a lower temperature where the solubility is less, the excess solute typically precipitates to give a saturated solution at the *lower* temperature.

However, if the solution is cooled rapidly, precipitates do not form, and the resulting solution contains more dissolved solute than it would in a typical saturated solution. This process produces a *supersaturated* solution.

> The *supersaturated solution* is unstable, and crystallization occurs by seeding (i.e., introducing particles providing nuclei for precipitation).

> A *concentrated solution* contains a relatively large amount of dissolved solute; contains 40 g or more of dissolved solute in 100 mL of water.

> A *dilute solution* contains few dissolved solutes in a relatively large amount of solvent; contains less than 10 g of dissolved solute per 100 mL of water.

A *concentrated* solution does *not* necessarily imply a *saturated* solution.

A *saturated* solution does *not* necessarily imply a *concentrated* solution.

Temperature and pressure affect solubility

The solubility of most liquids and solids in water *increases* with temperature.

Gas solubility *decreases* with temperature.

The pressure of a gas strongly affects its solubility, as described by *Henry's Law*, which states that solubility increases as pressure increases (i.e., directly proportional).

Factors affecting solubility include the solute's surface area (e.g., granulated sugar dissolves faster than a sugar cube) and heating or agitating a solution (e.g., stirring accelerates dissolving).

The terms *miscible* and *immiscible* describe liquids that dissolve or do not dissolve in another liquid, respectively.

For example, ethanol and water are *miscible* because they dissolve in each other freely when mixed.

Oil and water are *immiscible* since they remain separate.

Any duplication (copies, uploads, PDFs) is illegal.

Units of concentration

The solution's concentration is expressed for the solute amount dissolved and expressed as:

$$\text{Mass Percent, \% (w/w)} = \frac{\text{mass of solute}}{\text{mass of solution}} \times 100\%$$

$$\text{Volume Percent \% (v/v)} = \frac{\text{volume of solute}}{\text{volume of solution}} \times 100\%$$

$$\text{Molarity (M)} = \frac{\text{number of moles of solute}}{\text{liters of solution}}$$

$$\text{Molality (m)} = \frac{\text{number of moles of solute}}{\text{kg of solvent}}$$

$$\text{Mole fraction, } (X_i) = \frac{\text{moles of a component}}{\text{total moles of components in the solution}}$$

$$\text{Normality } (N) = \frac{\text{number of equivalents of solute}}{\text{liters of solution}}$$

where

$$\text{Number of equivalents of solute} = \frac{\text{grams of solute}}{\text{equivalent weight of solute}}$$

Normality and equivalent weight

Normality is based on the chemical mass unit of equivalent weight.

Equivalent weight is the amount of solute needed to equal one mole of hydrogen ions.

The equivalent weight is thus dependent on the *valence of the solute*.

For solutes with a valence of 1 (e.g., HCl), the molecular and equivalent weights are the same.

For solutes with a valence greater than 1 (e.g., H_3PO_4, valence equals 3), the equivalent weight equals the *molecular weight divided by the valence*.

Therefore, molarity is related to normality, as in the examples:

$$1 \text{ M HCl} = 1 \text{ N HCl}$$

$$1 \text{ M } H_2SO_4 = 2 \text{ N } H_2SO_4$$

$$1 \text{ M } H_3PO_4 = 3 \text{ N } H_3PO_4$$

Copyright © Sterling Test Prep.

Since ionic compounds dissociate into ions, the total number of particles in a solution of an ionic compound is greater than in a solution of a nonionic (molecular) compound.

The *total concentration* of ions in the solution is the sum of the concentrations of cations and anions, depending on the compound's formula.

For example, a 1.0 M aluminum nitrate solution, $Al(NO_3)_3$, contains 1.0 M of Al^{3+} and 3.0 M NO_3^- ions, with the total concentration of ions in the solution of 4.0 M.

Before dissolving   *After dissolving*

$Al(NO_3)_3\,(aq) \quad \rightarrow \quad Al^{3+}\,(aq)\ +\ 3\,NO_3^-\,(aq)$

$(1.0\ \text{M}) \quad \rightarrow \quad (1.0\ \text{M}) \quad (3 \times 1.0\ \text{M})$

For example, a solution of 1.0 M $MgCl_2$ contains 1.0 M of Mg^{2+} and 2.0 M of Cl^- ions, and the total ion concentration of 3.0 M.

Any duplication (copies, uploads, PDFs) is illegal.

Notes for active learning

Notes for active learning

Copyright © Sterling Test Prep.

Solution Equilibrium

Solubility product constant (K_{sp})

Solubility product constants (K_{sp}) describe saturated solutions of relatively *low solubility* ionic compounds.

For example, when a slightly soluble salt such as silver chloride (AgCl) dissolves in water, a saturated solution is obtained rapidly.

A minimal number of solid dissolves, while most AgCl salt remains undissolved.

The equilibrium between AgCl (s) and Ag$^+$ (aq) + Cl$^-$ (aq) ions occurs in solution:

$$AgCl\ (s) \rightleftharpoons Ag^+\ (aq) + Cl^-\ (aq)$$

$$K_{sp} = [Ag^+]{\cdot}[Cl^-]$$

A *general expression* of the solubility product constant (K_{sp}) for solubility equilibrium is:

$$M_aX_b\ (s) \rightleftharpoons aM^{b+}\ (aq) + bX^{a-}\ (aq)$$

$$K_{sp} = [M^{b+}]^a{\cdot}[X^{a-}]^b$$

In some problems, the K_{sp} is given, and the solubility needs to be calculated, or vice versa.

How to proceed depends on the type of equilibria. Sample solubility problems are solved below.

Ionic equilibrium

A. Ionic equilibria of the type:

$$MX\ (s) \rightleftharpoons M^{n+}\ (aq) + X^{n-}\ (aq)$$

$$K_{sp} = [M^{n+}]{\cdot}[X^{n-}]$$

If the solubility (S) of compounds is S mol/L:

$$K_{sp} = S^{\,2}$$

$$S = \sqrt{(K_{sp})}$$

For example, the solubility equilibrium (K_{sp}) for BaSO$_4$ is:

$$BaSO_4\ (s) \rightleftharpoons Ba^{2+}\ (aq) + SO_4^{2-}\ (aq)$$

$$K_{sp} = [Ba^{2+}]{\cdot}[SO_4^{2-}]$$

$$K_{sp} = 1.5 \times 10^{-9}$$

Any duplication (copies, uploads, PDFs) is illegal.

If the solubility (S) of $BaSO_4$ is S mol/L, a saturated solution of $BaSO_4$ has:

$$[Ba^{2+}] = [SO_4^{2-}] = S \text{ mol/L}$$

$$K_{sp} = S^2$$

$$S = \sqrt{(K_{sp})}$$

$$S = \sqrt{(1.5 \times 10^{-9})}$$

$$S = 3.9 \times 10^{-5} \text{ mol/L}$$

B. Ionic equilibria of the type:

$$MX_2\,(s) \rightleftarrows M^{2+}\,(aq) + 2\,X^-\,(aq)$$

$$K_{sp} = [M^{2+}] \cdot [X^-]^2$$

For the type:

$$M_2X\,(s) \rightleftarrows 2M^+\,(aq) + X^{2-}\,(aq)$$

$$K_{sp} = [M^+]^2 \cdot [X^{2-}]$$

For both types if the solubility is S mol/L:

$$K_{sp} = 4S^3$$

$$S = (K_{sp}/4)^{1/3}$$

For example:

$$CaF_2\,(s) \rightleftarrows Ca^{2+}\,(aq) + 2\,F^-\,(aq)$$

$$K_{sp} = [Ca^{2+}] \cdot [F^-]^2$$

$$K_{sp} = 4.0 \times 10^{-11}$$

The solubility (S) of calcium fluoride (CaF_2) is:

$$S = (K_{sp}/4)^{1/3}$$

$$S = (4.0 \times 10^{-11}/4)^{1/3}$$

$$S = 2.2 \times 10^{-4} \text{ mol/L}$$

C. Solubility equilibria of the type:

$$MX_3\,(s) \rightleftarrows M^{3+}\,(aq) + 3\,X^-\,(aq)$$

$$K_{sp} = [M^{3+}] \cdot [X^-]^3$$

Copyright © Sterling Test Prep.

Or of the type:

$$M_3X\,(s) \rightleftarrows 3\,M^+\,(aq) + X^{3-}\,(aq)$$

$$K_{sp} = [M^+]^3 \cdot [X^{3-}]$$

If the solubility (S) of the compound (MX_3 or M_3X) is S mol/L:

$$K_{sp} = 27S^4$$

$$S = \sqrt[4]{(K_{sp}/27)}$$

For example:

$$Ag_3PO_4\,(s) \rightleftarrows 3Ag^+\,(aq) + PO_4{}^{3-}\,(aq)$$

$$K_{sp} = [Ag^+]^3 \cdot [PO_4{}^{3-}]$$

$$K_{sp} = 1.8 \times 10^{-18}$$

The solubility (S) of silver phosphate (Ag_3PO_4) is:

$$S = \sqrt[4]{(K_{sp}/27)}$$

$$S = \sqrt[4]{(1.8 \times 10^{-18}/27)}$$

$$S = 1.6 \times 10^{-5}\ \text{mol/L}$$

D. Solubility equilibria of the type:

$$M_2X_3\,(s) \rightleftarrows 2\,M^{3+}\,(aq) + 3\,X^{2-}\,(aq)$$

$$K_{sp} = [M^{3+}]^2 \cdot [X^{2-}]^3$$

Or of the type:

$$M_3X_2\,(s) \rightleftarrows 3\,M^{2+}\,(aq) + 2\,X^{3-}\,(aq)$$

$$K_{sp} = [M^{2+}]^3 \cdot [X^{3-}]^2$$

If the solubility (S) of the compound M_3X_2 is S mol/L:

$$[M^{2+}] = 3S \text{ and } [X^{3-}] = 2S$$

$$K_{sp} = (3S)^3 \cdot (2S)^2$$

$$K_{sp} = 108S^5$$

$$S = \sqrt[5]{(K_{sp}/108)}$$

Any duplication (copies, uploads, PDFs) is illegal.

For example:

$$Ca_3(PO_4)_2 \ (s) \rightleftharpoons 3 \ Ca^{2+} \ (aq) + 2 \ PO_4^{3-} \ (aq)$$

$$K_{sp} = [Ca^{2+}]^3 \cdot [PO_4^{3-}]^2$$

$$K_{sp} = 1.3 \times 10^{-32}$$

The solubility (S) of calcium phosphate ($Ca_3(PO_4)_2$) is:

$$S = 5\sqrt{(1.3 \times 10^{-32}) / 108}$$

$$S = 1.6 \times 10^{-7} \ mol/L$$

Calculating K_{sp} from solubility

For example, the solubility (S) of $PbSO_4$ in water is 4.3×10^{-3} g/100 mL solution at 25 °C. What is the K_{sp} of $PbSO_4$ at 25 °C?

$$\text{Solubility } (S) \text{ of } PbSO_4 \text{ in mol/L} = \frac{4.3 \times 10^{-3} \ g}{100 \ mL} \times \frac{1000 \ mL/L}{303.26 \ g/mol}$$

$$S = 1.40 \times 10^{-4} \ mol/L$$

A saturated solution of $PbSO_4$, contains:

$$[Pb^{2+}] = [SO_4^{2-}] = 1.4 \times 10^{-4} \ mol/L$$

For the equilibrium:

$$PbSO_4 \ (s) \rightleftharpoons Pb^{2+} \ (aq) + SO_4^{2-} \ (aq)$$

$$K_{sp} = [Pb^{2+}] \cdot [SO_4^{2-}]$$

$$S^2 = (1.4 \times 10^{-4} \ mol/L)^2$$

$$S = 2.0 \times 10^{-8}$$

Calculating solubility from K_{sp}

For example, if the K_{sp} of $Mg(OH)_2$ is 6.3×10^{-10} at 25 °C, what is its solubility in mol/L at 25 °C?

Determine the solubility equilibrium for magnesium hydroxide ($Mg(OH)_2$):

$$Mg(OH)_2 \ (s) \rightleftharpoons Mg^{2+} \ (aq) + 2 \ OH^- \ (aq)$$

$$K_{sp} = [Mg^{2+}] \cdot [OH^-]^2$$

$$4S^3 = 6.3 \times 10^{-10}$$

Copyright © Sterling Test Prep.

Solubility of $Mg(OH)_2$:

$$S = \sqrt[3]{(K_{sp}/4)}$$

$$S = \sqrt[3]{(6.3 \times 10^{-10}/4)}$$

$$S = 5.4 \times 10^{-4} \text{ mol/L}$$

If the K_{sp} of $Mg(OH)_2$ is 6.3×10^{-10} at 25 °C, what is its solubility in a g/100 mL solution at 25 °C?

Solubility in g/100 mL solution:

$$S = (5.4 \times 10^{-4} \text{ mol/L}) \cdot (58.32 \text{ g/mol}) \cdot (0.1 \text{ L/100 mL})$$

$$S = 3.1 \times 10^{-3} \text{ g/100 mL solution}$$

Any duplication (copies, uploads, PDFs) is illegal.

Notes for active learning

Notes for active learning

Copyright © Sterling Test Prep.

Common and Complex Ions

Common ions

Common ions are produced by more than one solute in the solution.

According to *Le Châtelier's Principle*, the following equilibrium exists for acetic acid:

$$HC_2H_3O_2 \, (aq) + H_2O \, (l) \leftrightharpoons H_3O^+ \, (aq) + C_2H_3O_2^- \, (aq)$$

The reaction shifts to the left if $C_2H_3O_2^-$ is introduced from another source. The acetate ion $(C_2H_3O_2^-)$ is the common ion in sodium acetate and acetic acid solution.

This effect reduces the degree of dissociation of the acid, decreases $[H_3O^+]$, and increases the pH of the solution.

For example, the addition of ammonium chloride, NH_4Cl, causes the reaction to shift left in the equilibrium of ammonia in an aqueous solution:

$$NH_3 \, (aq) + H_2O \, (l) \leftrightharpoons NH_4^+ \, (aq) + OH^- \, (aq)$$

NH_4Cl dissociates into NH_4^+ and Cl^-, where NH_4^+ is a common ion in ammonia equilibrium.

The equilibrium shift caused by NH_4^+ ion reduces the extent of ionization of ammonia, which decreases $[OH^-]$ in the system and lowers the pH of the solution.

Common-ion effect

The *common-ion effect* is Le Châtelier's Principle applied to K_{sp} reactions. It states that a common ion *decreases* a slightly soluble ionic compound's solubility.

A common ion is an ion in a solution common to the ionic compound.

For example, in the following equilibrium:

$$PbCl_2 \, (s) \rightleftharpoons Pb^{2+} \, (aq) + 2 \, Cl^- \, (aq)$$

If NaCl is added to a saturated solution of $PbCl_2$, the $[Cl^-]$ increases, and according to Le Châtelier's Principle, the equilibrium shifts in the direction that reduces $[Cl^-]$.

In this example, the equilibrium shifts left to form more $PbCl_2$ solid, decreasing $PbCl_2$ that dissolves into solution. More $PbCl_2$ dissolves in pure water than in water with Cl^- ions.

The common-ion effect precipitates one component in a mixture selectively in laboratory separations.

For example, add NaCl to separate AgCl from a mixture of AgCl and Ag_2SO_4. The addition of NaCl selectively displaces AgCl by the common-ion effect (Cl^- being the common ion).

Complex ion formation

A *complex ion* consists of a central metal ion covalently bonded to two or more *ligands*, which can be anions such as ^-OH, Cl^-, F^- and ^-CN, or neutral molecules such as H_2O, CO, and NH_3.

Metal ions form complex ions with water molecules as ligands in aqueous solutions.

Ligand exchanges occur when another ligand is introduced into the solution, establishing equilibrium.

For example, in the complex ion $[Cu(NH_3)_4]^{2+}$, Cu^{2+} is the central metal ion, with four NH_3 molecules covalently bonded to it.

Complex ions are Lewis adducts (i.e., the addition of a Lewis acid and a Lewis base).

The Lewis base can be charged or uncharged.

Lewis acids and bases

The metal ions act as *Lewis acids* (electron-pair acceptors).

The ligands are *Lewis bases* (electron-pair donors).

$$\text{Metal}^+ + \text{Lewis base} \rightarrow \text{Complex ion}$$

$$M^+ + L \rightarrow M - L_n^+$$

The K_{eq} for this reaction is K_f, or the *formation constant*.

Formation constant K_f

For example, when NH_3 is added to an aqueous solution containing Cu^{2+} ion, the following equilibrium occurs:

$$Cu(H_2O)_6^{2+}\,(aq)\ +\ 4\,NH_3\,(aq) \rightleftarrows\ [Cu(NH_3)_4]^{2+}\,(aq) + 6\,H_2O$$

$$K_f = [Cu(NH_3)_4^{2+}] \,/\, [Cu(H_2O)_6^{2+}]\cdot[NH_3]^4$$

The ligand exchange process occurs stepwise; each water molecule is sequentially replaced with an NH_3 molecule to yield a series of intermediate species, each with its formation constant (K_f).

For convenience, the water molecule is omitted.

1. $Cu^{2+}\,(aq) + NH_3\,(aq) \rightleftarrows Cu(NH_3)^{2+}\,(aq)$

$$K_{f1} = [Cu(NH_3)^{2+}] \,/\, [Cu^{2+}]\cdot[NH_3]$$

2. $Cu(NH_3)^{2+}\,(aq) + NH_3\,(aq) \rightleftarrows Cu(NH_3)_2^{2+}\,(aq)$

$$K_{f2} = [Cu(NH_3)_2^{2+}] \,/\, [Cu(NH_3)^{2+}]\cdot[NH_3]$$

Copyright © Sterling Test Prep.

3. $Cu(NH_3)_2^{2+} (aq) + NH_3 (aq) \rightleftarrows Cu(NH_3)_3^{2+} (aq)$

$$K_{f3} = [Cu(NH_3)_3^{2+}] / [Cu(NH_3)_2^{2+}] \cdot [NH_3]$$

4. $Cu(NH_3)_3^{2+} (aq) + NH_3 (aq) \rightleftarrows Cu(NH_3)_4^{2+} (aq)$

$$K_{f4} = [Cu(NH_3)_4^{2+}] / [Cu(NH_3)_3^{2+}] \cdot [NH_3]$$

The *formation constant* (K_f) is the product of the intermediate formation constants:

$$K_f = K_{f1} \times K_{f2} \times K_{f3} \times K_{f4}$$

$$K_f = [Cu(NH_3)_4^{2+}] / [Cu^{2+}] \cdot [NH_3]^4$$

Solubility of complex ions

The *complex-ion effect* is the opposite of the *common* ion effect.

A ligand *increases* the solubility of slightly soluble ionic compounds if complex ions form with the metal ions.

For example, silver chloride (AgCl) is *more soluble* in ammonia (NH_3) solution because silver ions form complex ions with NH_3:

$$AgCl (s) \rightleftarrows Ag^+ (aq) + Cl^- (aq)$$

$$K_{sp} = 1.6 \times 10^{-10}$$

$$Ag^+ (aq) + 2\,NH_3 (aq) \rightleftarrows Ag(NH_3)_2^+ (aq)$$

$$K_f = 1.7 \times 10^7$$

$$AgCl (s) + 2\,NH_3 (aq) \rightleftarrows Ag(NH_3)_2^+ (aq) + Cl^- (aq)$$

$$K_{net} = K_{sp} \times K_f$$

$$K_{net} = (1.6 \times 10^{-10}) \cdot (1.7 \times 10^7)$$

$$K_{net} = 2.7 \times 10^{-3}$$

The Cl^- ion is reduced when a complex ion forms, so more AgCl dissolves.

Alternatively:

$$AgCl (s) \leftrightarrow Ag^+ (aq) + Cl^- (aq)$$

$$NH_3 + Ag^+ \leftrightarrow Ag\text{-}(NH_3)_n \text{ complex ion}$$

The complex ion formation reduces Ag^+, more AgCl dissolves.

Any duplication (copies, uploads, PDFs) is illegal.

pH and solubility

The pH affects the solubility of slightly soluble compounds containing anions as *conjugate bases of weak acids* (e.g., F^-, NO_2^-, OH^-, SO_3^{2-} and PO_4^{3-}).

The pH *does not* affect the solubility of slightly soluble compounds containing anions as *conjugate bases of strong acids* (SO_4^{2-}, Cl^- and Br^-).

In a saturated solution of calcium fluoride (CaF_2) equilibrium exists:

$$CaF_2\ (s) \rightleftarrows Ca^{2+}\ (aq) + 2\ F^-\ (aq)$$

If a strong acid is added to the saturated solution, the following reaction occurs:

$$H^+\ (aq) + F^-\ (aq) \rightarrow HF\ (aq)$$

The reaction has the net effect of *reducing* F^- ion concentration, causing the equilibrium to shift to the *right*, and more CaF_2 dissolves.

Acids are more soluble in bases.

$$HA \rightarrow H^+ + A^-$$

Placing the above reactions in a base reduces the H^+.

Thus, more HA dissolves, according to Le Châtelier's Principle.

Bases are more soluble in acids.

$$B + H^+ \rightarrow BH^+$$

Putting the above reactions in an acid adds H^+, and thus, more B dissolves, according to Le Châtelier's Principle.

Copyright © Sterling Test Prep.

Practice Questions

1. Which of the following statements describing solutions is NOT true?

    A. Solutions are colorless
    B. The particles in a solution are atomic or molecular
    C. Making a solution involves a physical change in size
    D. Solutions are homogeneous

2. Which of the following describes a saturated solution?

    A. When the ratio of solute to solvent is small
    B. When it contains less solute than it can hold at 25 °C
    C. When it contains as much solute as it can hold at a given temperature
    D. When it contains 1 g of solute in 100 mL of water

3. An ionic compound that strongly attracts atmospheric water is said to be:

    A. immiscible
    B. miscible

    C. diluted
    D. hygroscopic

4. What happens when molecule-to-molecule attractions in solutes are less than in the solvent?

    A. The material has only limited solubility in the solvent
    B. The solution will become saturated
    C. The solute can have infinite solubility in the solvent
    D. The solute does not dissolve in the solvent

5. What does a negative heat of solution indicate about solute-solvent bonds compared to solute-solute bonds and solvent-solvent bonds?

    A. Solute-solute and solute-solvent bond strengths are greater than solvent-solvent bond strength
    B. Solute-solute and solvent-solvent bonds are weaker than solute-solvent bonds
    C. Solute-solute and solvent-solvent bonds are stronger than solute-solvent bonds
    D. Solute-solute and solvent-solvent bond strengths are equal to solute-solvent bond strength

Any duplication (copies, uploads, PDFs) is illegal.

6. Which statement supports that calcium fluoride is much less soluble in water than sodium fluoride?

        I. calcium fluoride is not used in toothpaste

        II. calcium fluoride is not used to fluoridate city water supplies

        III. sodium fluoride is not used to fluoridate city water supplies

A. I only

B. II only

C. I and II only

D. I and III only

7. The principle *like dissolves like* is NOT applicable for predicting solubility when the solute is a/an:

A. nonpolar liquid

B. polar gas

C. nonpolar gas

D. ionic compound

8. Which of the following is the most soluble in benzene (C_6H_6)?

A. glucose ($C_6H_{12}O_6$)

B. sodium benzoate

C. octane (C_8H_{18})

D. hydrobromic acid

9. Apply the *like dissolves like* rule to predict which liquids are immiscible in water.

A. ethanol, CH_3CH_2OH

B. acetone, C_3H_6O

C. acetic acid, CH_3COOH

D. none of the above

10. What is the equilibrium constant expression (K_{sp}) for slightly soluble silver sulfate in an aqueous solution: $Ag_2SO_4\,(s) \leftrightarrow 2\,Ag^+\,(aq) + SO_4^{2-}\,(aq)$?

A. $K_{sp} = [Ag^+]\cdot[SO_4^{2-}]^2$

B. $K_{sp} = [Ag^+]^2\,[SO_4^{2-}]$

C. $K_{sp} = [Ag^+]\cdot[SO_4^{2-}]$

D. $K_{sp} = [Ag^+]\cdot[SO_4^{2-}]^2 / [Ag_2SO_4]$

11. In the reaction $KHS\,(aq) + HCl\,(aq) \rightarrow KCl\,(aq) + H_2S\,(g)$, which ions are spectator ions?

A. H^+ and HS^-

B. K^+ and HS^-

C. K^+ and Cl^-

D. K^+ and H^+

12. What is the mass of a 10.0% blood plasma sample containing 2.50 g of dissolved solute?

A. 21.5 g

B. 25.0 g

C. 0.215 g

D. 0.430 g

E. 12.5 g

Copyright © Sterling Test Prep.

13. What is the molarity of KCl in seawater if KCl is 12.5% (m/m) and the density of seawater is 1.06 g/mL?

A. 1.78 M	**C.** 2.78 M
B. 17.8 M	**D.** 0.845 M

14. If 25.0 mL of urine has a mass of 25.5 g and contains 1.8 g of solute, what is the mass/mass percent concentration of solute in the urine sample?

A. 12.48%	**C.** 7.06%
B. 17.42%	**D.** 3.82%

15. With increasing temperature, many solvents expand to occupy greater volumes. What happens to the concentration of a solution made with such a solvent as temperature increases?

A. Decreases because the solution has a greater ability to dissolve more solute at a higher temperature

B. Increases because the solution has a greater ability to dissolve more solute at a higher temperature

C. Decreases as the volume increases since concentration depends on the mass dissolved in a given volume

D. Increases as the solute fits into the new spaces between the molecules

Notes for active learning

Copyright © Sterling Test Prep.

Detailed Explanations

1. A is correct.

Some solutions are colored (e.g., Kool-Aid powder dissolved in water).

The color of chemicals is a physical property of chemicals from (most common) the excitation of electrons due to the chemical's absorption of energy.

The observer sees not the absorbed color but the wavelength that is reflected.

Most simple inorganic (e.g., sodium chloride) and organic compounds (e.g., ethanol) are colorless.

Transition metal compounds are often colored due to transitions of electrons of d-orbitals of different energy.

Organic compounds tend to be colored when there is extensive conjugation (i.e., alternating double and single bonds), causing the energy gap between the *HOMO* (i.e., highest occupied molecular orbital) and *LUMO* (i.e., lowest occupied molecular orbital) to decrease.

This moves the absorption band from the UV to the visible region.

Color is due to the energy absorbed by compounds when electrons transition from the *HOMO* to the *LUMO*.

2. C is correct.

A *saturated solution* contains the maximum dissolved material in the solvent under normal conditions. Increased heat allows for a solution to become supersaturated.

Supersaturated refers to the vapor of a compound with a higher partial pressure than the vapor pressure of that compound.

A saturated solution forms a precipitate as more solute is added to the solution.

3. D is correct.

A *hygroscopic substance* (e.g., honey, glycerin) readily attracts water from its surroundings through adsorption or absorption.

Adsorption is the process in which atoms, ions, or molecules from a substance (e.g., gas, liquid, dissolved solid) adhere to a surface of the *adsorbent.*

4. C is correct.

Suppose the attraction between solute molecules is less than between solvent molecules. In that case, a solute has (a virtually) infinite solubility because solvent-solute attraction is stronger than solute-solute attraction.

5. B is correct.

Bond formation releases energy: if the heat of the solution is negative, then energy is released.

The *heat of the solution* is the net of *enthalpy* changes for making and breaking bonds.

Solvent-solvent and solute-solute bonds break during solution formation while solute-solvent bonds form.

The *breaking of bonds* absorbs energy, while the formation of bonds releases energy.

6. C is correct.

Fluorides (e.g., BaF_2, MgF_2 PbF_2) are frequently *insoluble*.

7. D is correct.

Like dissolves like means polar substances dissolve in polar solvents, and nonpolar substances in nonpolar solvents.

Molecules that can form hydrogen bonds with water are soluble.

Ionic compounds form anions and cations that bond with the polar water molecule and are soluble.

8. C is correct.

Like dissolves like means polar substances dissolve in polar solvents, and nonpolar substances in nonpolar solvents.

Benzene is a nonpolar molecule, and octane is also nonpolar.

Thus, nonpolar octane is soluble in nonpolar benzene.

9. D is correct.

The "like dissolves like" rule applies when a solvent is miscible (soluble) with a solute of similar properties.

A polar solute is miscible with a polar solvent.

All the molecules are polar and therefore are miscible in water.

Copyright © Sterling Test Prep.

10. B is correct.

The *solubility constant* (K_{sp}) calculation is similar to the equilibrium constant.

For K_{sp}, only aqueous species are included in the calculation.

The concentration of each species is raised to their coefficients' power and multiplied with each other.

Therefore,

$$K_{sp} = [Ag^+]^2 \cdot [SO_4^{2-}]$$

11. C is correct.

Spectator ions do not participate in the chemical reaction and solution before and after the reaction.

Ionic equation of the reaction:

$$K^+ (aq) + HS^- (aq) + H^+ (aq) + Cl^- (aq) \rightarrow K^+ (aq) + Cl^- (aq) + H_2S (g)$$

The *net ionic equation* only shows elements, compounds, and ions directly involved in the chemical reaction and is written without the spectator ions.

Net ionic equation of the reaction:

$$HS^- (aq) + H^+ (aq) \rightarrow H_2S (g)$$

12. B is correct.

Mass % = mass of solute / mass of solution

Rearrange that equation to solve for mass of solution:

mass of solution = mass of solute / mass %

mass of solution = 2.50 g / 10.0%

mass of solution = 2.50 g / 0.1

mass of solution = 25.0 g

Any duplication (copies, uploads, PDFs) is illegal.

13. A is correct.

Assume 1 L of seawater.

Molarity is the number of moles in 1 L of solution, so starting with 1 L makes the calculation easier.

Mass of seawater = volume × density

Mass of seawater = (1 L × 1,000 L/mL) × 1.06 g/mL

Mass of seawater = 1,060 g

Use the mass of the seawater to determine the mass of KCl:

Mass of KCl = mass % of KCl × mass of seawater

Mass of KCl = 12.5% × 1,060 g

Mass of KCl = 132.5 g

Calculate the number of moles:

Moles of KCl = mass of KCl / molar mass of KCl

Moles of KCl = 132.5 g / (39.1 g/mol + 35.45 g/mol)

Moles of KCl = 1.78 moles

Divide moles by volume to calculate molarity:

Molarity of KCl = moles of KCl / volume of KCl

Molarity of KCl = 1.78 moles / 1 L

Molarity of KCl = 1.78 M

14. C is correct.

Mass % of solute = (mass of solute / total urine mass) × 100%

Mass % of solute = (1.8 g / 25.5 g) × 100%

Mass % of solute = 7.06%

15. C is correct.

As the volume of a solution increases, the concentration of a solution decreases

because concentration depends on how much mass is dissolved in a given volume.

Copyright © Sterling Test Prep.

Notes for active learning

Notes for active learning

Copyright © Sterling Test Prep.

CHAPTER 8

Acids & Bases

- Acid–Base Nomenclature

- Acidic and Basic Solutions

- Acid–Base Equilibria

- Acid–Base Ionization

- Periodic Trends of Acids and Bases

- Percent Dissociation

- Salts of Acid–Base Reactions

- pH calculations for Acid–Base Solutions

- Buffered Solutions

- Ionic Equations and Neutralization

- Titration

- Indicators

- Titration Curves

- Practice Questions & Detailed Explanations

Copyright © Sterling Test Prep.

Acid–Base Nomenclature

Characteristics of acids and bases

Before the twentieth century, acids, bases, and salts were characterized by properties such as taste and the ability to change litmus's color (a water-soluble mixture of organic dyes).

Acids taste *sour* (e.g., lemon juice), bases taste *bitter* (e.g., mustard), and salts, as their name suggests, taste *salty* (e.g., sodium chloride as table salt).

Acids cause blue litmus paper to turn *red*, bases turn red litmus paper *blue*, while neutral compounds do not affect litmus paper's color.

Bases are recognized by their slippery feel (e.g., soap).

The position on the pH scale categorically classifies acids and bases.

Some complex metabolic processes in the human body are controlled by physiological pH ($\approx$ 7.4); a small change may lead to severe illness and death.

The chemistry of acids and bases has a significant role in nature and industry processes. For example, soil acidity is essential for plant growth.

Acids and bases are particularly essential in manufacturing industries. Sulfuric acid (H_2SO_4) is the most widely produced chemical, and it is needed to produce fertilizers, polymers, steel, and many other materials.

The extensive use of sulfuric acid has led to environmental problems, such as the phenomenon of acid rain.

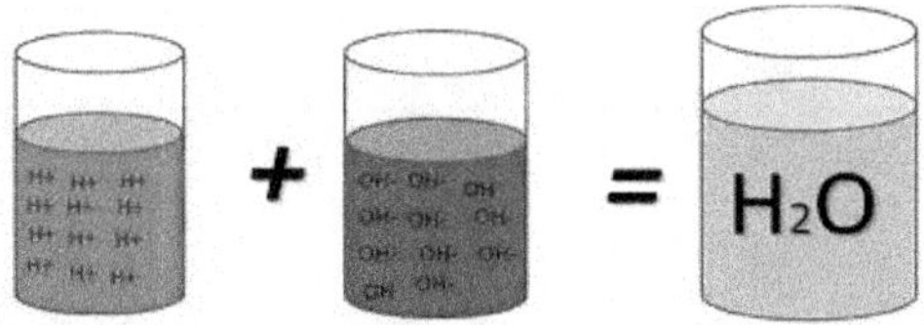

Acids containing   *Bases containing*   *Acids and bases combine*
protons (H⁺)   *hydroxides (⁻OH)*   *to produce water (H₂O)*

Binary acids and oxoacids

There are *two types of acids*:

    binary acids (acids that do not contain oxygen atoms)

    oxoacids (acids containing oxygen atoms).

Acids have slightly different names from the ionic naming rules discussed, while bases follow the ionic naming rules discussed.

Naming binary acids

Binary acids start with *hydro-*, then the first syllable of the anion's name, and end with *~ic*:

HF : *hydro*fluor*ic* acid

HCl : *hydro*chlor*ic* acid

HI : *hydro*iod*ic* acid

HBr : *hydro*brom*ic* acid

H_2S : *hydro*sulfur*ic* acid

HCN : *hydro*cyan*ic* acid

Oxoacid's names derive from the name of the oxyanion of the acid.

Naming associated anions

For anions ending with *~ate*, the acid's name starts with the first syllable of the anion name and ends with *~ic*.

For anions ending with *~ite*, the name of acid starts with the first syllable of the anion's name and ends with *-ous*.

Anion	Name of Anion	Acid	Name of Acid
NO_3^-	nitrate ion	HNO_3	nitric acid
NO_2^-	nitrite ion	HNO_2	nitrous acid
SO_4^{2-}	sulfate ion	H_2SO_4	sulfuric acid
SO_3^{2-}	sulfite ion	H_2SO_3	sulfurous acid
PO_4^{3-}	phosphate ion	H_3PO_4	phosphoric acid
$C_2H_3O_2^-$	acetate ion	$HC_2H_3O_2$	acetic acid
ClO^-	hypochlorite	$HClO$	hypochlorous acid
ClO_2^-	chlorite	$HClO_2$	chlorous acid
ClO_3^-	chlorate	$HClO_3$	chloric acid
ClO_4^-	perchlorate	$HClO_4$	perchloric acid

Copyright © Sterling Test Prep.

Arrhenius acids and bases

In 1884, Swedish chemist Svante Arrhenius defined acids and bases:

Acids dissociate in water to produce *hydrogen ions* (H^+)

Bases dissociate to in water produce *hydroxide ions* (^-OH)

For example, HCl is an *acid*:

$$HCl\ (aq) \rightarrow H^+\ (aq) + Cl^-\ (aq)$$

For example, NaOH is a *base*:

$$NaOH\ (aq) \rightarrow Na^+\ (aq) + {}^-OH\ (aq)$$

Arrhenius acids increase the hydronium ion concentration $[H_3O^+]$ in an aqueous solution.

Arrhenius bases increase the hydroxide ion concentration $[^-OH]$ in an aqueous solution.

Acids, according to the Arrhenius concept:

1. $HCl\ (aq) + H_2O \rightarrow H_3O^+\ (aq) + Cl^-\ (aq)$

2. $HNO_3\ (aq) + H_2O \rightarrow H_3O^+\ (aq) + NO_3{}^-\ (aq)$

3. $CH_3COOH\ (aq) + H_2O \rightleftarrows H_3O^+\ (aq) + CH_3COO^-\ (aq)$

Bases, according to the Arrhenius concept:

1. $\quad NaOH\ (aq) \rightarrow Na^+\ (aq) + OH^-\ (aq)$

2. $\quad Ba(OH)_2\ (aq) \rightarrow Ba^{2+}\ (aq) + 2\ OH^-\ (aq)$

3. $\quad NH_3\ (aq) + H_2O \rightleftarrows NH_4{}^+\ (aq) + OH^-\ (aq)$

Any duplication (copies, uploads, PDFs) is illegal.

Lewis acids and bases

A *Lewis base* provides *a lone pair of electrons* to another reactant to form a coordinate covalent bond.

A *Lewis acid* has an empty orbital to accept *a pair of electrons* from another reactant to form a coordinate covalent bond.

In 1923, Gilbert Newton Lewis (1875-1946) proposed a definition that Lewis acids (e.g., H^+) accept a pair of nonbonding electrons.

Lewis acids are electron-pair *acceptors*.

Bases are electron-pair *donors*.

H+ is the acid while water and ammonia are the Lewis bases in the following reactions:

$$H^+ \; + \; H_2O \; \rightarrow \; H_3O^+ \qquad\qquad H^+ \; + \; NH_3 \; \rightarrow \; NH_4^+$$

Lewis	Lewis	Lewis	Lewis
acid	base	acid	base

In reactions forming new covalent bonds, the species with an incomplete octet (i.e., an electron-deficient molecule) may act as Lewis acids; a lone pair of electrons may act as Lewis bases.

In the following reactions,

BF_3, $AlCl_3$, and $FeBr_3$ are Lewis acids

while

NH_3, Cl^- and Br^- are Lewis bases

$$BF_3 + NH_3 \rightarrow F_3B{:}NH_3 \qquad AlCl_3 + Cl^- \rightarrow AlCl_4^- \qquad FeBr_3 + Br^- \rightarrow FeBr_4^-$$

In the formation of complex ions, the positive ions act as Lewis acids, and the ligands (anions or small molecules) act as Lewis bases:

$$Cu^{2+}\,(aq) \; + \; 4\,NH_3\,(aq) \; \rightleftarrows \; Cu(NH_3)_4{}^{2+}\,(aq)$$

Lewis acid Lewis base

$$Al^{3+}\,(aq) \; + \; 6\,H_2O \; \rightleftarrows \; [Al(H_2O)_6]^{3+}\,(aq)$$

Lewis acid Lewis base

The *ionizable hydrogen* in oxoacids is bonded to the oxygen.

Lewis base *Lewis acid*

Copyright © Sterling Test Prep.

Brønsted–Lowry acids and bases

The *Brønsted-Lowry theory* was proposed, in 1923, independently by Danish chemist Johannes Nicolaus Brønsted (1879-1947) and British chemist Martin Lowry (1874-1936).

The Brønsted-Lowry theory states that:

an acid is a substance that acts as a proton donor, and

a base is a substance that acts as a proton acceptor.

Brønsted–Lowry conjugate acids and conjugate bases

From the exchange of protons,

acids form its *conjugate base*

bases form its *conjugate acid*

Brønsted-Lowry acid-base reactions can be represented as follows:

$$HA + B \rightleftharpoons BH^+ + A^-$$
acid   base   conjugate   conjugate
                 acid   base

Brønsted-Lowry acids, bases, conjugate acids, and conjugate bases

$$HCl + H_2O \rightarrow H_3O^+\,(aq) + Cl^-\,(aq)$$
acid   base   conjugate   conjugate
                       acid   base

$$HC_2H_3O_2 + H_2O \rightleftharpoons H_3O^+\,(aq) + C_2H_3O_2^-\,(aq)$$
acid   base   conjugate   conjugate
                       acid   base

$$NH_3 + H_2O \rightleftharpoons NH_4^+\,(aq) + OH^-\,(aq)$$
base   acid   conjugate   conjugate
                 acid   base

The transfer of protons in a Brønsted-Lowery acid-base reaction

$$HF\,(aq) + NH_3 \rightleftharpoons F^- + {}^+NH_4\,(aq)$$

$HF\,(aq)$	NH_3	F^-	${}^+NH_4\,(aq)$
Acid donates H^+ to NH_3	Base accepts H^+ from HF	Conjugate base accepts H^+ / ${}^+NH_4$	Conjugate acid donates H^+ to F^-

Any duplication (copies, uploads, PDFs) is illegal.

Amphoteric substances

Water is an *amphoteric* substance because it can *act as an acid or a base*. It acts as a base by accepting a hydrogen ion; another water molecule acts as an acid.

Therefore, the ions $H_3O^+ (aq) + OH^- (aq)$ form in pure water.

As ions form, they react to produce water again with the following equilibrium:

$$H_2O \; + \; H_2O \; \rightleftarrows \; H_3O^+ (aq) \; + OH^- (aq)$$

$$\text{acid} \qquad \text{base} \qquad \text{conjugate} \qquad \text{conjugate}$$
$$\text{acid} \qquad\quad \text{base}$$

Autoionization of water

In the autoionization of water, a proton is transferred from one water molecule to another, producing a hydronium ion (H_3O^+) and a hydroxide ion (OH^-).

Ions formed in the auto-ionization of water are:

$$H_2O \; (l) + H_2O \; (l) \; \rightleftarrows \; H_3O^+ \; (aq) + OH^- \; (aq)$$

This process of exchange is the *autoionization of water*.

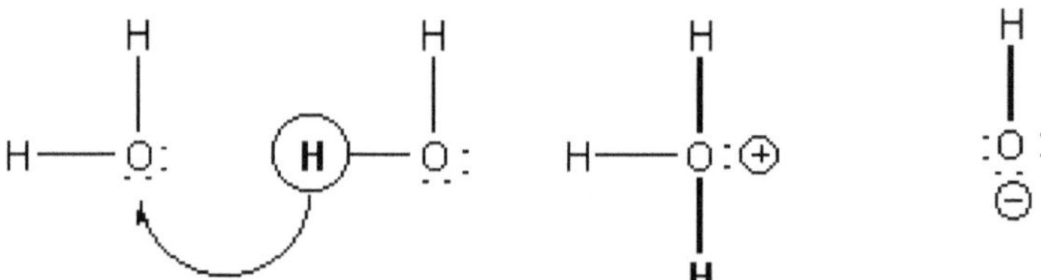

H^+ *acceptor base* H^+ *donor acid*   *Conjugate acid*   *Conjugate base*

Copyright © Sterling Test Prep.

Equilibrium constant K_w for autoionization

The equilibrium constant K_w:

$$K_w = [H_3O^+] \cdot [^-OH]$$

where K_w is the autoionization constant for water

At standard temperature and pressure (25 °C, 1 atm), the equilibrium constant K_w (*ion-product constant*) for water:

$$K_w = [H_3O^+] \cdot [^-OH]$$

$$K_w = 1.0 \times 10^{-14}$$

For example,

If $[H_3O^+]$ increases ($> 1.0 \times 10^{-7} M$)

If $[^-OH]$ decreases ($< 1.0 \times 10^{-7}$ M)

If $[H_3O^+] = [^-OH] = 1.0 \times 10^{-7}$ M, neutral solution (e.g., pure water)

If $[H_3O^+] > 1.0 \times 10^{-7}$ M, $[^-OH] < 1.0 \times 10^{-7}$ M, solution is acid ($[H^+] > [^-OH]$)

If $[H_3O^+] < 1.0 \times 10^{-7}$ M, $[^-OH] > 1.0 \times 10^{-7}$ M, solution is basic ($[H^+] < [^-OH]$)

Any duplication (copies, uploads, PDFs) is illegal.

Notes active learning

Copyright © Sterling Test Prep.

Acidic and Basic Solutions

pH and pOH

The pH scale measures acidity or basicity, especially when the hydrogen ion (H^+) concentration is extremely low.

$$pH = -\log[H^+]$$

$$pOH = -\log[^-OH]$$

Neutral solutions

$$[H^+] = 1.0 \times 10^{-7} \text{ M}$$

$$pH = -\log(1.0 \times 10^{-7})$$

$$pH = 7.00$$

Neutral solutions contain

$$[^-OH] = 1.0 \times 10^{-7} \text{ M}$$

$$pOH = -\log(1.0 \times 10^{-7})$$

$$pOH = 7.00$$

Acidic solutions

$$[H^+] > 1.0 \times 10^{-7} \text{ M}$$

$$pH < 7.00$$

Basic solutions

$$[H^+] < 1.0 \times 10^{-7} \text{ M}$$

$$pH > 7.00$$

pH standards

$$pH = 7 \rightarrow \text{a neutral solution}$$

$$pH < 7 \rightarrow \text{an acidic solution}$$

$$pH > 7 \rightarrow \text{a basic solution}$$

Any duplication (copies, uploads, PDFs) is illegal.

Note

$$K_w = [H^+]\cdot[^-OH]$$

$$K_w = 1.0 \times 10^{-14}$$

$$pK_w = -\log(K_w)$$

$$pK_w = -\log(1.0 \times 10^{-14})$$

$$pK_w = 14.00$$

but

$$pK_w = pH + pOH = 14.00$$

$$pOH = 14.00 - pH$$

$$pH = 14.00 - pOH$$

if

$$pH = 7$$

$$pOH = 7$$

if

$$pH < 7$$

$$pOH > 7$$

if

$$pH > 7$$

$$pOH < 7$$

thus

$$pH < 7$$

$$[H^+] > [^-OH]$$

and

$$pH > 7$$

$$[^-OH] > [H^+]$$

For example, if

$$[H^+] = 1.0 \times 10^{-4} \text{ M}$$

$$pH = -\log(1.0 \times 10^{-4})$$

$$pH = 4.00$$

 Copyright © Sterling Test Prep.

when

$$[^-OH] = 1.0 \times 10^{-4} \text{ M}$$

$$[H^+] = 1.0 \times 10^{-14} / 1.0 \times 10^{-4} \text{ M}$$

$$[H^+] = 1.0 \times 10^{-10} \text{ M}$$

$$pH = -\log(1.0 \times 10^{-10})$$

$$pH = 10.00$$

Alternatively,

$$[^-OH] = 1.0 \times 10^{-4} \text{ M}$$

$$pOH = -\log[^-OH]$$

$$pOH = -\log(1.0 \times 10^{-4} \text{ M})$$

$$pOH = 4.00$$

$$pH = 14.00 - 4.00$$

$$pH = 10.00$$

Reference values for [H$^+$] and pH

[H$^+$], M	pH	[H$^+$], M	pH
1.0×10^{-1}	1.00	1.0×10^{-8}	8.00
1.0×10^{-2}	2.00	1.0×10^{-9}	9.00
1.0×10^{-3}	3.00	1.0×10^{-10}	10.00
1.0×10^{-4}	4.00	1.0×10^{-11}	11.00
1.0×10^{-5}	5.00	1.0×10^{-12}	12.00
1.0×10^{-6}	6.00	1.0×10^{-13}	13.00
1.0×10^{-7}	7.00	1.0×10^{-14}	14.00

Any duplication (copies, uploads, PDFs) is illegal.

pH of an acidic solution

To calculate the pH of an acidic solution, use the expression:

$$pH = -\log[H_3O^+]$$

For example, if

$$[H_3O^+] = 1.0 \times 10^{-2} \text{ M}$$

$$pH = -\log(1.0 \times 10^{-2})$$

$$pH = -(-2.00)$$

$$pH = 2.00 \; (\rightarrow \text{acidic})$$

pH of a basic solution

To calculate the pH of a basic solution, use the expression:

$$pOH = -\log[^-OH]$$

If a solution has

$$[OH^-] = 1.0 \times 10^{-2} \text{ M}$$

$$pOH = -\log(1.0 \times 10^{-2})$$

$$pOH = -(2.00)$$

$$pOH = 2.00 \; (\rightarrow \text{basic})$$

Since at 25 °C

$$K_w = [H_3O^+]\cdot[OH^-]$$

$$K_w = 1.0 \times 10^{-14}$$

$$pK_w = -\log(K_w)$$

$$pK_w = -\log[H_3O^+] + (-\log[^-OH])$$

$$pK_w = -\log(1.0 \times 10^{-14})$$

$$pK_w = -(-14.00)$$

$$pK_w = 14.00$$

$$pK_w = pH + pOH$$

$$pK_w = 14.00$$

$$pOH = 14.00 - pH$$

Copyright © Sterling Test Prep.

Thus, in aqueous solutions

$$pH = 2$$

$$pOH = 12$$

and

$$pOH = 2$$

$$pH = 12$$

Any duplication (copies, uploads, PDFs) is illegal.

Notes for active learning

Copyright © Sterling Test Prep.

Acid–Base Equilibria

Equilibrium constants K_a and K_b

Equilibrium constants K_a and K_b measure the extent to which an acid or base dissociates (dissociation constants).

The strength of an acid is defined by its dissociation (ionization) in an *aqueous* (*aq*) solution.

$$HX\,(aq) + H_2O \rightleftharpoons H_3O^+\,(aq) + X^-\,(aq)$$

The equilibrium constant, K_a, for the acid ionization:

$$K_a = [H_3O^+] \cdot [X^-] \,/\, [HX]$$

For K_a, the products (conjugate acid H_3O^+ and conjugate base X^-) of the dissociation are the numerator, while the parent acid (HX) is the denominator.

The strength of a base is defined by its dissociation (ionization) in an *aqueous* (*aq*) solution.

$$X^-\,(aq) + H_2O \rightleftharpoons OH^-\,(aq) + HX\,(aq)$$

$$K_b = [HX^+] \cdot [^-OH] \,/\, [X]$$

For K_b, the products (conjugate acid HX^+ and conjugate base OH^-) of the dissociation are the numerator, while the parent base (X) is the denominator.

Acid dissociation

The value of K_a measures the extent of *acid dissociation*, hence the acid's relative strength.

For strong acids (e.g., $HClO_4$, HCl, H_2SO_4, HNO_3), K_a is exceptionally large (not in the reported K_a values).

For weak acids, $K_a \ll 10^{-1}$.

Strong bases have larger K_b values.

Weak bases have exceedingly small K_b values.

If the K_a value for a conjugate acid-base pair is known, the K_b can be calculated (and vice versa) using the following relationship:

$$K_a K_b = K_w$$

Sodium acetate dissolves in water dissociating into sodium (Na^+) and acetate ions ($C_2H_3O_2^-$):

$$NaC_2H_3O_2\,(aq) \rightarrow Na^+\,(aq) + C_2H_3O_2^-\,(aq)$$

If the K_a is 1.8×10^{-5} (provided in a reference table), what is the K_b?

Any duplication (copies, uploads, PDFs) is illegal.

The acetate ion reacts with water with the following equilibrium:

$$C_2H_3O_2^- \, (aq) + H_2O \rightleftharpoons HC_2H_3O_2 \, (aq) + OH^- \, (aq)$$

$$K_b = [HC_2H_3O_2] \cdot [OH^-] \, / \, [C_2H_3O_2^-]$$

For the dissociation of acetic acid:

$$HC_2H_3O_2 \, (aq) + H_2O \rightleftharpoons H_3O^+ \, (aq) + C_2H_3O_2^- \, (aq)$$

$$K_a = [H_3O^+] \cdot [C_2H_3O_2^-] \, / \, [CH_3COOH]$$

$$K_a \times K_b = [H_3O^+] \cdot [C_2H_3O_2^-] \, / \, [CH_3COOH] \times [HC_2H_3O_2] \cdot [\, OH^-] \, / \, [C_2H_3O_2^-]$$

$$K_w = [H_3O^+] \cdot [OH^-]$$

$$K_w = 1.0 \times 10^{-14}$$

Thus, for acetate ion $C_2H_3O_2^-$ in solution

$$K_b = K_w \, / \, K_a \, (\text{for } HC_2H_3O_2)$$

$$K_b = (1.0 \times 10^{-14}) \, / \, (1.8 \times 10^{-5})$$

$$K_b = 5.6 \times 10^{-10} \, (> K_w)$$

An aqueous solution of 0.10 M $NaC_2H_3O_2$ has $[OH^-] \sim 7.5 \times 10^{-6}$ M and pH ~ 8.9.

pK_a and pK_b measure acidity and basicity

The operator "p" means "take the negative logarithm of."

Therefore

$$pK_a = -\log K_a$$

As K_a gets larger, pK_a gets smaller.

The *smaller* the pK_a, the *stronger* the acid (e.g., pK_a for $H_2SO_4 = 1.92$).

$$pK_b = -\log K_b$$

As K_b gets larger, pK_b gets smaller.

The *smaller* the value of pK_b, the *stronger* the base.

Copyright © Sterling Test Prep.

Conjugate acids and bases

A *conjugate base* is the species remaining from the acid after it *loses* a proton.

A *conjugate acid* is the species remaining from the base after *gaining* a proton.

The *conjugate acid-base pairs* ($acid_1$–conjugate $base_1$ and $acid_2$–conjugate $base_2$) are related substances by the loss or gain of a single proton (H^+).

H_2O and H_3O^+, and H_2O and OH^- are conjugate acid-base pairs, but H_3O^+ and OH^- are not.

Strong acids have *weak conjugate bases.*

Weak acids have *strong conjugate bases.*

The *weaker the acid*, the stronger its conjugate base.

Strong bases have *weak conjugate acids.*

Weak bases have *strong conjugate acids.*

The *weaker the base*, the stronger the conjugate acid it produces.

HCl is a strong acid, and Cl^- is a weak base.

HF is a weak acid, and F^- is a strong conjugate.

A Brønsted-Lowry acid-base reaction involves a competition between two bases for a proton, in which the stronger base is the most protonated at equilibrium.

$$HCl + H_2O \rightarrow H_3O^+ (aq) + Cl^- (aq)$$

H_2O is a stronger base than the Cl^-.

At equilibrium, the HCl solution contains mainly H_3O^+ and Cl^- ions.

$$HC_2H_3O_2 (aq) + H_2O \rightleftarrows H_3O^+ (aq) + C_2H_3O_2^- (aq)$$

$C_2H_3O_2^-$ is the stronger base.

At equilibrium, acetic acid contains mainly $HC_2H_3O_2$ and few H_3O^+ and $C_2H_3O_2^-$ ions.

$$NH_3 (aq) + H_2O \rightleftarrows NH_4^+ (aq) + {}^-OH (aq)$$

H_2O is an acid. Competition for protons occurs between NH_3 and ^-OH, where ^-OH is the stronger base, and the equilibrium favors the reactants. An aqueous ammonia solution contains *mostly NH_3 molecules* and smaller amounts of NH_4OH, NH_4^+, and ^-OH.

According to Brønsted-Lowry, the net acid-base reactions favor *strong acid-strong base* combinations to *weak acid-weak base* combinations.

Any duplication (copies, uploads, PDFs) is illegal.

Acid-base reaction equilibrium

The following acid-base reactions proceed in the forward direction:

$$HCl\ (aq) + NH_3\ (aq) \rightarrow NH_4^+\ (aq) + Cl^-\ (aq)$$

$$HSO_4^-\ (aq) + CN^-\ (aq) \rightarrow HCN\ (aq) + SO_4^{2-}\ (aq)$$

HSO_4^- ($pK_a = 6.91$) is a stronger acid than HCN ($pK_a = 9.21$)

Many acid-base reactions reach a state of equilibrium.

For the following acid-base reactions, the equilibrium may favor the products or the reactants, depending on the relative strength of the acid:

$$H_2PO_4^-\ (aq) + C_2H_3O_2^-\ (aq) \rightleftarrows HC_2H_3O_2\ (aq) + HPO_4^{2-}\ (aq)$$

Equilibrium shifts to the *left* (reactants):

$HC_2H_3O_2$ ($pK_a = 4.75$) is the stronger acid

HPO_4^{2-} ($pK_a = 7.21$) is the stronger base

$$HNO_2\ (aq) + C_2H_3O_2^-\ (aq) \leftrightarrows HC_2H_3O_2\ (aq) + NO_2^-\ (aq)$$

Equilibrium shifts to the *right* (products).

HNO_2 ($pK_a = 3.39$) is the stronger acid

$C_2H_3O_2^-$ ($pK_a = 4.75$) is the stronger base

Copyright © Sterling Test Prep.

Acid–Base Ionization

Strength of acids and bases

A *strong acid* ionizes entirely in aqueous solutions.

Strong acids include $HClO_4$, HCl, H_2SO_4, HNO_3, HBr, and HI. For these strong acids, the equilibrium lies *far to the right*.

Weak acids partially ionize with ionization equilibriums *far to the left*. Weak acids include:

$$HC_2H_3O_2, \quad HNO_2, \quad H_2SO_3, \quad H_3PO_4, \text{ and } HClO.$$

For example, consider the reversible process when an acid dissolves in water:

$$HA \ (aq) + H_2O \leftrightarrow H_3O^+ \ (aq) + A^- \ (aq)$$

In the forward reaction above, the acid HA donates a proton to the water molecule to form a hydronium ion (H_3O^+) and conjugate base (A^-).

Water acts as a Brønsted-Lowry base, and the acid strength is measured by its degree of ionization (or dissociation) in water. If an acid ionizes completely, it is a strong acid.

A *strong acid* has a *weak conjugate base* (i.e., the conjugate base loses its proton to water readily). Cl^-, Br^-, I^-, ClO_4^-, HSO_4^-, and NO_3^- are weak conjugate bases.

A strong acid is less able to compete with water for a proton.

Ionization constant K_a

The ionization constant K_a for a *strong acid is exceptionally large*, and the equilibrium shifts far to the right (i.e., dissociated proton and stable anion).

A weak acid does not readily give up its proton to water, and it has a strong conjugate base.

Strong conjugate bases include:

$$C_2H_3O_2^-, \ F^-, \ CN^-, \ NO_2^-, \ HSO_3^-, \ SO_3^{2-}, \ H_2PO_4^-, \ HPO_4^{2-}, \ PO_4^{3-}.$$

The weaker the acid, the stronger its conjugate base.

The ionization equilibrium for weak acids shifts far to the left.

An aqueous solution of a strong acid has hydronium ions (H_3O^+) and the acid's conjugate base.

For example, there are H_3O^+ and Cl^- ions in an aqueous HCl solution, but virtually no HCl molecules.

An aqueous solution of a weak acid, acetic acid, contains mainly undissociated molecules, $HC_2H_3O_2$, with about ~1% H_3O^+ and $C_2H_3O^-$ ions.

Any duplication (copies, uploads, PDFs) is illegal.

Bond strength and polarity for acids

Factors determine the strength of acids, such as the *strength* and *polarity* of the X−H bond in the molecule and the hydration energy of the ionic species in an aqueous solution.

For inorganic binary acids, H−X bond strength decreases down the group:

HF, HCl, HBr, HI),

The weaker the bond, the easier it ionizes in an aqueous solution.

The stronger bond is the acid; strength increases down the group:

HF < HCl < HBr < HI

Relative ionization strength

Among the hydrohalic acids, HF is the weak acid; the others are strong acids. The relative strength of HCl, HBr and HI cannot be differentiated in an aqueous solution because each of them dissociates almost completely.

Less polar solvents are used to determine their relative strength.

Strong bases, like strong acids (e.g., NaOH, KOH, and $Ba(OH)_2$), ionize completely when dissolved in water. A strong base tends to accept a proton.

Weak bases do not ionize completely when dissolved in water and show a minimal tendency to accept a proton.

HCl, HBr, and HI ionize partially in acetone or methanol, which have a weaker ionizing strength than water.

The ionization of HCl in acetone:

$(CH_3)_2CO$ (*l*) (*acetone*) $+ HCl \rightleftarrows (CH_3)_2COH^+ + Cl^-$

The degree of ionization in acetone increases as:

HCl < HBr < HI

Copyright © Sterling Test Prep.

Oxoacids produce H+

Oxoacids contain one or more ~OH groups covalently bonded to a central atom, a metal, or a nonmetal.

The ~OH group ionizes entirely or partially in an aqueous solution, producing hydrogen ions.

Examples of oxoacids: H_2CO_3, HNO_3, H_3PO_4, H_2SO_4, $HClO_4$, $HC_2H_3O_2$

H_2SO_4

H_3PO_4

$HClO_4$

Any duplication (copies, uploads, PDFs) is illegal.

Notes for active learning

Copyright © Sterling Test Prep.

Periodic Trends of Acids and Bases

Periodic trends for acidity

The acidity of hydrogen halides increases down a group; trend of relative acidity for hydrides of Group 15:

$$H_2O < H_2S < H_2Se < H_2Te$$

For the same period hydrides, relative acidity increases from left to right:

$$CH_4 < NH_3 < H_2O < HF$$

$$PH_3 < H_2S \ll HCl$$

Water is a stronger acid than ammonia, and in an acid-base reaction, H_2O acts as a Brønsted-Lowry acid, which donates a proton to NH_3:

$$H_2O + NH_3\ (aq) \rightleftarrows NH_4^+\ (aq) + OH^-\ (aq)$$

In reaction with HF, water acts as a Brønsted-Lowry base, which accepts a proton:

$$HF\ (aq) + H_2O \rightleftarrows H_3O^+\ (aq) + F^-\ (aq)$$

Electronegativity for acid strength

Relative acid strength depends on the central atoms' *electronegativity*. The more electronegative, the more polarized the O$-$H bond, and the *more readily it ionizes* in an aqueous solution to release the H^+ ion.

For example, N, S, and Cl are more electronegative than P; HNO_3, H_2SO_4, and $HClO_4$ are stronger acids, whereas H_3PO_4 is weak.

The order of electronegativity for relative acid strength:

$$Cl \approx N > S > P$$

Acid strength:

$$HClO_4 > HNO_3 > H_2SO_4 > H_3PO_4$$

Any duplication (copies, uploads, PDFs) is illegal.

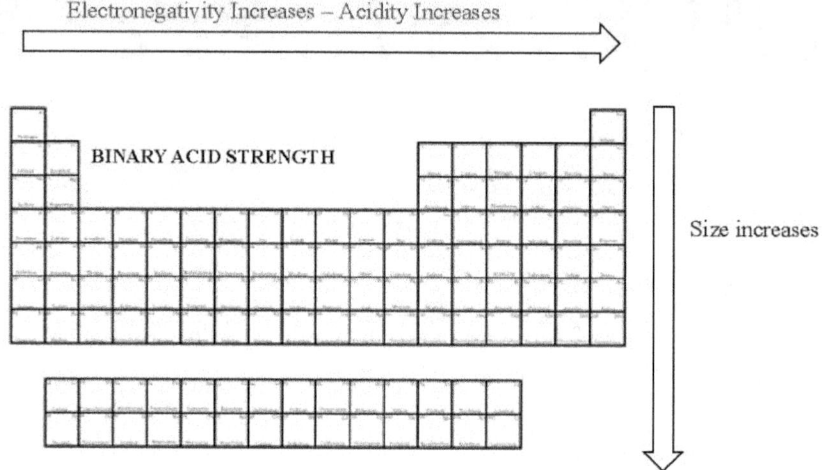

Acidity increases with increases in electronegativity and atomic size

Induction and acid strength

For oxoacids that have the central atoms with elements of the same group in the periodic table, relative strength decreases from top to bottom (as the electronegativity of the central atom decreases):

$$HOCl > HOBr > HOI \qquad HClO_2 > HBrO_2 > HIO_2 \qquad HClO_4 > HBrO_4 > HIO_4$$

For oxoacids containing identical central atoms, acidity increases as oxygen atoms bond.

Acidity increases as follows:

$$HOCl < HClO_2 < HClO_3 < HClO_4$$

$$H_2SO_3 < H_2SO_4$$

$$HNO_2 < HNO_3$$

When oxygen atoms bond to the central atom, the O–H bond in the molecule becomes highly polarized (due to the inductive electronegative effect).

The O–H bond ionizes readily to release an H^+ ion.

Acetic acid (CH_3COOH) is an organic acid containing the carboxyl (~COOH) group. In an aqueous solution, ionization of an acetic acid involves breaking the carboxyl group's O–H bond, but not the C–H bonds in the methyl group (CH_3).

However, if one or more of the hydrogen atoms in the methyl group is substituted with a more electronegative atom, the inductive effect causes the electron cloud to be drawn away from the carbonyl group. The O–H bond becomes more polarized and ionizes readily, increasing the acidity.

Copyright © Sterling Test Prep.

The following K_a values illustrate the effect on acetic acid's acidity and its derivatives when methyl hydrogens are substituted with electronegative atoms.

The stronger the acid, the higher the K_a value:

$$CH_3COOH\ (aq)\ (acetic\ acid) + H_2O \rightleftharpoons CH_3COO^-\ (aq) + H_3O^+\ (aq)$$

$$K_a = \mathbf{1.8 \times 10^{-5}}$$

$$ClCH_2COOH\ (aq)\ (chloroacetic\ acid) + H_2O \rightleftharpoons ClCH_2COO^-\ (aq) + H_3O^+\ (aq)$$

$$K_a = \mathbf{1.4 \times 10^{-3}}$$

$$FCH_2COOH\ (aq)\ (fluoroacetic\ acid) + H_2O \rightleftharpoons FCH_2COO^-\ (aq) + H_3O^+\ (aq)$$

$$K_a = \mathbf{2.6 \times 10^{-3}}$$

$$CCl_3COOH\ (aq)\ (trichloroacetic\ acid) + H_2O \rightleftharpoons CCl_3COO^-\ (aq) + H_3O^+\ (aq)$$

$$K_a = \mathbf{3.0 \times 10^{-1}}$$

Monoprotic and polyprotic acids

The *monoprotic acids* include HCl, HF, HOCl, HNO_2, and $HC_2H_3O_2$ because each molecule contains a single ionizable hydrogen ion.

The *polyprotic acids* contain more than one ionizable hydrogen, and examples include:

$$H_2SO_4,\ H_2SO_3,\ H_2C_2O_4\ and\ H_3PO_4$$

The hydrogen ionizes in stages with ionization constants, such as for H_3PO_4.

$$H_3PO_4\ (aq) \rightleftharpoons H^+\ (aq) + H_2PO_4^-\ (aq); \qquad K_{a1} = 7.5 \times 10^{-3}$$

$$H_2PO_4^-\ (aq) \rightleftharpoons H^+\ (aq) + HPO_4^{2-}\ (aq); \qquad K_{a2} = 6.2 \times 10^{-8}$$

$$HPO_4^{2-}\ (aq) \rightleftharpoons H^+\ (aq) + PO_4^{3-}\ (aq); \qquad K_{a3} = 4.8 \times 10^{-13}$$

From above, acid strength decreases in order:

$$H_3PO_4 \gg H_2PO_4^- \gg HPO_4^{2-}$$

Any duplication (copies, uploads, PDFs) is illegal.

Ionization of polyprotic acids

Sulfuric acid (H_2SO_4) is a strong acid, but only the first hydrogen ionizes completely:

$$H_2SO_4\,(aq) \rightarrow H^+\,(aq) + HSO_4^-\,(aq) \qquad K_{a1} = \text{exceptionally large}$$

The second H does not often dissociate, and HSO_4^- is a weak acid:

$$HSO_4^-\,(aq) \rightleftharpoons H^+\,(aq) + SO_4^{2-}\,(aq) \qquad K_{a2} = 1.2 \times 10^{-2}$$

Hydroxide bases

Weak bases include NH_3 (or NH_4OH), NH_2OH, $Mg(OH)_2$, and hydroxides and oxides as slightly soluble in water.

Hydroxides of Group I metals (LiOH, NaOH, KOH, ROH, and CsOH) are strong bases, but only NaOH and KOH are commercially important and commonly used laboratory bases.

Hydroxide bases are *soluble* in water, and they dissociate entirely in an aqueous solution, producing a high concentration of hydroxide ions.

A moderately dilute solution of NaOH contains [$^-$OH], [Na+], and [NaOH].

$Ba(OH)_2$ is a relatively strong base among the alkaline earth metal hydroxides.

The other metal hydroxides are sparingly soluble in water, limiting their basicity.

A saturated solution of these hydroxides contains a low concentration of $^-$OH.

These hydroxides react with strong acids:

$$Na_2O\,(s) + H_2O \rightarrow 2\,NaOH\,(aq)$$

$$BaO\,(s) + H_2O \rightarrow Ba(OH)_2\,(aq)$$

$$MgO\,(s) + 2\,HCl\,(aq) \rightarrow MgCl_2\,(aq) + H_2O$$

Hydroxides of some metals (e.g., $Al(OH)_3$, $Cr(OH)_3$, $Zn(OH)_2$, $Sn(OH)_2$ and $Pb(OH)_2$) exhibit amphoteric properties (i.e., act as an acid or a base).

For example:

$$Al(OH)_3\,(s) + OH^-\,(aq) \rightleftharpoons Al(OH)_4^-\,(aq)$$

$$Al(OH)_3\,(s) + 3\,H_3O^+\,(aq) \rightleftharpoons [Al(H_2O)_6]^{3+}\,(aq)$$

Copyright © Sterling Test Prep.

Hydride bases

Hydrides of reactive metals (e.g., NaH, MgH_2, CaH_2) form strongly basic solutions if dissolved in water.

The hydride ion reacts with water to produce hydroxide ions and hydrogen gas:

$$H^- \, (aq) + H_2O \rightarrow H_2 \, (g) + OH^- \, (aq)$$

The oxide ion O^{2-} has an extraordinarily strong affinity for protons (H^+), reacting with H_2O to form hydroxide ions.

$$O^{2-} \, (aq) + H_2O \rightarrow 2 \, OH^- \, (aq)$$

Oxides of nonmetals are acidic, forming acidic solutions when dissolved in water.

$$CO_2 \, (g) + H_2O \rightleftharpoons H_2CO_3 \, (aq) \rightleftharpoons H^+ \, (aq) + HCO_3^- \, (aq)$$

$$SO_2 \, (g) + H_2O \rightleftharpoons H_2SO_3 \, (aq) \rightleftharpoons H^+ \, (aq) + HSO_3^- \, (aq)$$

Ammonia is the only weak base that is of commercial importance.

Ammonia does not contain hydroxide ions, but it reacts with water and ionizes as follows:

$$NH_3 \, (aq) + H_2O \rightleftharpoons NH_4^+ \, (aq) + OH^- \, (aq)$$

The base dissociation constant K_b is given by the expression:

$$K_b = [NH_4^+] \cdot [^-OH] / [NH_3]$$

$$K_b = 1.8 \times 10^{-5}$$

Any duplication (copies, uploads, PDFs) is illegal.

Notes for active learning

Copyright © Sterling Test Prep.

Percent Dissociation

Degree of ionization

The *degree of ionization* (or *percent dissociation*) of a weak acid is:

Percent dissociation = [acid ionized] / [initial acid] $\times$ 100%

For strong acids, the percent dissociation at equilibrium is $\sim$ 100%.

The percent dissociation for weak acids depends on the acid's K_a and initial concentration.

For example, the percent dissociation of acetic acid (Use $HC_2H_3O_2$ $K_a = 1.8 \times 10^{-5}$) at 0.10 M concentration is:

$(1.3 \times 10^{-3}$ M / 0.10 M$) \times$ 100% = 1.3 %

The *stronger* the acid, the larger is the K_a, and the *greater* is the percent ionization.

The *percent ionization of a weak acid* depends on the K_a and the extent of dilution.

The *more dilute* an acid solution, the *higher* is the percentage of ionization.

For example, consider a 0.010 M acetic acid solution and its ionization products.

Use an ICE table (**I**nitial, **C**hange, **E**quilibrium) to simplify calculations in reversible equilibrium reactions.

Once the equilibrium row is completed (by summing the initial and change rows), its contents can be substituted into the equilibrium constant expression to solve for K_a.

$CH_3COOH\ (aq) \rightleftarrows H^+\ (aq) + CH_3COO^-\ (aq)$

Initial [ ], M:	0.010	0.00	0.00
Change, Δ[ ], M:	$-x$	$+x$	$+x$
Equilibrium [ ], M :	$(0.010 - x)$	x	x

Acid ionization constant K_a

The acid ionization constant, K_a, is given by:

$K_a = [H_3O^+] \cdot [C_2H_3O_2^-] / [CH_3COOH]$

$K_a = x^2 / (0.010 - x)$

$K_a = 1.8 \times 10^{-5}$

Any duplication (copies, uploads, PDFs) is illegal.

Since $K_a \ll 0.010$, approximate that $x \ll 0.010$, and $(0.010 - x) \sim 0.010$

Then

$$K_a = x^2 / (0.010 - x)$$

$$K_a \approx x^2 / 0.010$$

$$K_a = 1.8 \times 10^{-5}$$

$$x^2 = 1.8 \times 10^{-7}$$

and

$$x = \sqrt{(1.8 \times 10^{-7})}$$

$$x = 4.2 \times 10^{-4}$$

$$x = [H_3O^+]$$

$$x = 4.2 \times 10^{-4} \text{ M}$$

The *degree of ionization* of acetic acid at this concentration is

$$(4.2 \times 10^{-4} \text{ M} / 0.010 \text{ M}) \times 100\% = 4.2\%$$

The degree of ionization of the acid increases as the solution is diluted. For 0.1 M acetic acid, the degree of ionization is 0.42%, 10-fold lower than in 0.010 M acid solution.

The percent dissociation of weak bases depends on the base solution's K_b value and dilution.

Larger K_b and greater dilution result in a higher percent dissociation.

Hydrolysis of salts

The presence of salt affects the dissociation of acids and bases due to salts' *hydrolysis*.

For example, CH_3COOH dissociates less in a solution containing CH_3COONa salt, while NH_4OH dissociates less in a solution containing NH_4Cl salt.

When salts (or ionic compounds) dissolve in water, it is assumed that they dissociate entirely into separate ions.

Some of these ions react with water and behave as acids or bases.

The acidic or basic nature of a salt solution depends on whether it is a product of a:

    1) strong acid-strong base reaction,

    2) a weak acid-strong base reaction,

    3) a strong acid-weak base reaction, or

    4) a weak acid-weak base reaction.

                                                        Copyright © Sterling Test Prep.

Salts of Acid–Base Reactions

Salts

Salts are the products of acid-base reactions.

The general reaction to produce salt:

$$\text{Acid} + \text{Base} \rightarrow \text{Salt} + \text{Water}$$

The acid and base are neutralized; H^+ and OH^- form water.

The acid's nonmetallic ions and metal ions of the base form the salt.

For example, NaCl is a product of the following acid-base reaction:

$$HCl\,(aq) + NaOH\,(aq) \rightarrow NaCl\,(aq) + H_2O$$

In the chemical formula of salt, the cation is contributed by the base (e.g., Na^+), while the acid contributes the anion (e.g., Cl^-).

Four types of salts

The four types of salts are:

- **Salts of strong acid-strong base** reactions (e.g., $NaCl$, KNO_3, $NaClO_4$)

- **Salts of weak acid-strong base** reactions (e.g., $NaC_2H_3O_3$, K_2CO_3, KCN, $NaCHO_2$)

- **Salts of strong acid-weak base** reactions (e.g., NH_4Cl, NH_4NO_3, $HONH_3Cl$)

- **Salts of weak acid-weak base** reactions (e.g., $NH_4C_2H_3O_2$, NH_4CN, NH_4HS)

When dissolved in water, these salts produce acidic, basic, or neutral solutions.

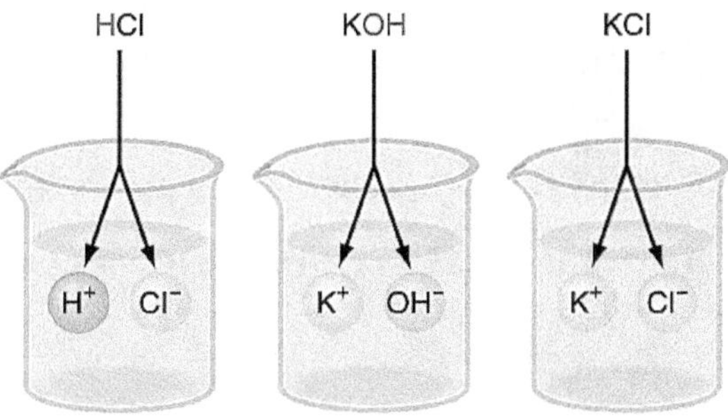

The acidic dissociation constant increases with stable anions

Any duplication (copies, uploads, PDFs) is illegal.

*Salts of **Strong Acid-Strong Base** Reactions*: (e.g., NaCl, NaNO$_3$, KBr)

- Salts form a neutral solution because neither the cation nor anion reacts with water and offsets the equilibrium concentrations of H$_3$O$^+$ and OH$^-$ in the solution.

*Salts of **Weak Acid-Strong Base** Reactions*: (e.g., NaF, NaNO$_2$, NaC$_2$H$_3$O$_2$)

- Salts form basic solutions when dissolved in H$_2$O.

  The anions of such salts react with H$_3$O increasing [$^-$OH].

- Sodium acetate (NaC$_2$H$_3$O$_2$) is a product of the reaction between acetic acid (HC$_2$H$_3$O$_2$), which is a weak acid, and a strong base (NaOH).

$$\text{HC}_2\text{H}_3\text{O}_2 \ (aq) + \text{NaOH} \ (aq) \rightarrow \text{NaC}_2\text{H}_3\text{O}_2 \ (aq) + \text{H}_2\text{O}$$

*Salts of **Strong Acid-Weak Base** Reactions*: (e.g., NH$_4$Cl, NH$_4$NO$_3$, (CH$_3$)$_2$NH$_2$Cl, C$_5$H$_5$NHCl)

- Aqueous solutions of salts are acidic.

- The cations react with H$_2$O and increase [H$_3$O$^+$].

- Example: NH$_4$Cl is produced when HCl (strong acid) reacts with NH$_3$ (a weak base):

$$\text{HCl} \ (aq) + \text{NH}_3 \ (aq) \rightarrow \text{NH}_4\text{Cl} \ (aq) \rightarrow \text{NH}_4^+ \ (aq) + \text{Cl}^- \ (aq)$$

In an aqueous solution, NH$_4^+$ equilibrium increases [H$_3$O$^+$] and creates an acidic solution:

$$\text{NH}_4^+ \ (aq) + \text{H}_2\text{O} \rightleftarrows \text{H}_3\text{O}^+ \ (aq) + \text{NH}_3 \ (aq)$$

$$K_a = [\text{H}_3\text{O}^+] \cdot [\text{NH}_3] \ / \ [\text{NH}_4^+]$$

While in an NH$_3$ solution, the following equilibrium occurs:

$$\text{NH}_3 \ (aq) + \text{H}_2\text{O} \rightleftarrows \text{NH}_4^+ \ (aq) + \text{OH}^- \ (aq)$$

$$K_b = [\text{NH}_4^+] \cdot [^-\text{OH}] \ / \ [\text{NH}_3]$$

$$K_a \times K_b = \{[\text{H}_3\text{O}^+] \cdot [\text{NH}_3] \ / \ [\text{NH}_4^+]\} \times \{[\text{NH}_4^+] \cdot [^-\text{OH}] \ / \ [\text{NH}_3]\}$$

$$K_a \times K_b = K_w = [\text{H}_3\text{O}^+] \cdot [^-\text{OH}]$$

$$K_w = 1.0 \times 10^{-14}$$

Copyright © Sterling Test Prep.

For NH_4^+ :

$$K_a = K_w \, / \, K_b \text{ (for } NH_3\text{)}$$

$$K_a = (1.0 \; x \; 10^{-14}) \, / \, (1.8 \; x \; 10^{-5})$$

$$K_a = 5.6 \times 10^{-10}$$

An aqueous solution of NH_4^+ has a $K_a = 5.6 \times 10^{-10}$ at 25 °C (which is $> K_w$).

A 0.10 M solution of NH_4Cl or NH_4NO_3 has $[H_3O^+] \approx 7.5 \times 10^{-6}$ M and pH $\approx$ 5.1

*Salts of **Weak Acid-Weak Base** Reactions*: (e.g., $NH_4C_2H_3O_2$, NH_4CN, NH_4NO_2)

Aqueous salts can be neutral, acidic, or basic, depending on the relative magnitude of the K_a of the weak acid and the K_b of the weak base.

If $K_a \approx K_b$, salt forms an approximately neutral solution.

For example:

$$K_a \text{ of } HC_2H_3O_2 = 1.8 \times 10^{-5}$$

$$K_b \text{ of } NH_3 = 1.8 \times 10^{-5}$$

When $NH_4C_2H_3O_2$ dissolves in water and dissociates, the following equilibria exist:

$$NH_4C_2H_3O_2 \, (aq) \rightarrow NH_4^+ \, (aq) + C_2H_3CO_2^- \, (aq)$$

$$NH_4^+ \, (aq) + H_2O \rightleftarrows H_3O^+ \, (aq) + NH_3 \, (aq) \qquad\qquad K_a = 5.6 \times 10^{-10}$$

$$C_2H_3O_2^- \, (aq) + H_2O \rightleftarrows HC_2H_3O_2 \, (aq) + OH^- \, (aq) \qquad\qquad K_b = 5.6 \times 10^{-10}$$

K_a (for NH_4^+) = K_b (for $C_2H_3O_2^-$), at equilibrium $[H_3O^+] = [^-OH]$ and the $NH_4C_2H_3O_2$ solution is neutral.

If $\boldsymbol{K_a} > \boldsymbol{K_b}$, the salt solution is acidic.

For NH_4NO_2:

$$K_a \, (HNO_2) = 4.0 \times 10^{-4}$$

$$K_b \, (NH_3) = 1.8 \times 10^{-5}$$

$K_a > K_b \rightarrow$ acidic solution because hydrolysis forms a solution with $[H_3O^+] > [^-OH]$.

If $\boldsymbol{K_a} < \boldsymbol{K_b}$, the salt solution is basic.

Any duplication (copies, uploads, PDFs) is illegal.

For NH_4CN in solution,

$$K_a \text{ (HCN)} = 6.2 \times 10^{-10}$$

$$K_b \text{ (NH}_3\text{)} = 1.8 \times 10^{-5}$$

The following equilibria exist:

$$NH_4CN \ (aq) \rightarrow NH_4^+ \ (aq) + CN^- \ (aq)$$

$$NH_4^+ \ (aq) + H_2O \rightleftharpoons H_3O^+ \ (aq) + NH_3 \ (aq)$$

$$K_a = 5.6 \times 10^{-10}$$

$$CN^- \ (aq) + H_2O \rightleftharpoons HCN \ (aq) + OH^- \ (aq)$$

$$K_b = 1.6 \times 10^{-5}$$

$K_b \text{ (CN}^-) > K_b \text{ (NH}_4^+)$, at equilibrium $[^-OH] > [H_3O^+]$; an aqueous solution of NH_4CN is basic.

Copyright © Sterling Test Prep.

pH Calculations for Acid–Base Solutions

Calculating pH of strong acid and strong base solutions

Strong acids are assumed to ionize completely in an aqueous solution.

For monoprotic acids (i.e., acids with a single ionizable hydrogen) such as HCl and HNO_3, the [hydronium ion; H_3O^+] in solution is the same as the molar concentration of the acid:

$$[H_3O^+] = [HX]$$

For example, consider 0.10 M HCl (*aq*)

$$[H_3O^+] = 0.10 \text{ M}$$

$$pH = -\log(0.10)$$

$$pH = 1.00$$

A strong base such as NaOH has [$^-$OH] equal to the molar concentration of dissolved NaOH.

A solution of 0.10 M NaOH (*aq*) has:

$$[^-OH] = 0.10 \text{ M}$$

$$pOH = -\log[^-OH]$$

$$-\log[^-OH] = -\log(0.10)$$

$$pOH = 1.00$$

$$pH = 14.00 - 1.00$$

$$pH = 13.00$$

A strong base such as $Ba(OH)_2$ produces twice the concentration of $^-$OH as the molar concentration of $Ba(OH)_2$ in solution:

$$Ba(OH)_2 \, (aq) \; \rightarrow \; Ba^{2+} \, (aq) + 2 \, OH^- \, (aq)$$

$$[^-OH] = 2 \times [Ba(OH)_2]$$

In a solution of 0.010 M $Ba(OH)_2$

$$[^-OH] = 0.020 \text{ M}$$

$$pOH = 1.70, \text{ and } pH = 12.30$$

Any duplication (copies, uploads, PDFs) is illegal.

Calculating pH of weak acid solutions

Unlike strong acids, weak acids do not ionize completely.

At equilibrium, $[H^+]$ is less than the concentration of the acid.

The concentration of H^+ in a weak acid solution depends on the initial acid concentration and the K_a of the acid.

Determine $[H^+]$ of a weak acid using the "ICE" table as follows:

For example, consider a 0.10 M acetic acid solution and its ionization products.

$$HC_3H_3O_2\,(aq) \rightleftarrows H^+\,(aq) + C_2H_3O_2^-\,(aq)$$

Initial [ ], M:	0.10	0.00	0.00
Change, Δ[ ], M:	$-x$	$+x$	$+x$
Equilibrium [ ], M :	$(0.10 - x)$	x	x

The acid ionization constant, K_a, is given by the expression:

$$K_a = [H_3O^+]\cdot[C_2H_3O_2^-] / [CH_3COOH]$$

$$K_a = x^2 / (0.010 - x)$$

$$K_a = 1.8 \times 10^{-5}$$

Since,

$$K_a \ll 0.10$$

approximate that $x \ll 0.10$, and $(0.10 - x) \sim 0.10$

$$K_a = x^2 / (0.10 - x)$$

$$K_a \approx x^2 / 0.10$$

$$K_a = 1.8 \times 10^{-5}$$

$$x^2 = 1.8 \; x \; 10^{-6}$$

$$x = \sqrt{(1.8 \; x \; 10^{-6})}$$

$$x = 1.3 \times 10^{-3}$$

Copyright © Sterling Test Prep.

Note that

$$x = [H_3O^+]$$

$$[H_3O^+] = 1.3 \times 10^{-3} \text{ M}$$

$$pH = -\log(1.3 \times 10^{-3})$$

$$pH = 2.89$$

Calculating pH of weak base solutions

The concentration of ^-OH in a weak base, such as $NH_3\,(aq)$, depends on its K_b value and the initial concentration of the base.

To determine $[^-OH]$ and pH of 0.10 M $NH_3\,(aq)$, set the following "ICE" table:

$$NH_3\,(aq) + H_2O \rightleftharpoons NH_4^+\,(aq) + {}^-OH\,(aq)$$

Initial [ ], M:	0.10	0.00	0.00
Change, Δ[ ], M:	$-x$	$+x$	$+x$
Equilibrium [ ], M:	$(0.10 - x)$	x	x

$$K_b = [NH_4^+] \cdot [^-OH] \,/\, [NH_3]$$

$$K_b = x^2 \,/\, (0.10 - x)$$

$$K_b = 1.8 \times 10^{-5}$$

Using the approximation method:

$$K_b = x^2 \,/\, (0.10 - x)$$

$$K_b \approx x^2 \,/\, 0.10$$

$$K_b = 1.8 \times 10^{-5}$$

$$x^2 = 1.8 \times 10^{-6}$$

$$x = \sqrt{(1.8 \times 10^{-6})}$$

$$x = 1.3 \times 10^{-3}$$

continued...

Any duplication (copies, uploads, PDFs) is illegal.

Where

$$x = [^-OH]$$

$$x = 1.3 \times 10^{-3}\ M$$

$$pOH = -\log(1.3 \times 10^{-3})$$

$$pOH = 2.87$$

$$pH = 11.13$$

Calculating pH of basic or acidic salt solutions

1. For example, consider a solution of 0.050 M sodium acetate, which dissociates completely and establishes the following equilibrium:

$$NaC_2H_3O_2\,(aq) \rightarrow Na^+\,(aq) + C_2H_3O_2^-\,(aq)$$

The acetate ion establishes the equilibrium in aq solution:

$$C_2H_3O_2^-\,(aq) + H_2O \rightleftarrows HC_2H_3O_2\,(aq) + {}^-OH\,(aq)$$

$$K_b = [HC_2H_3O_2]\cdot[^-OH] / [C_2H_3O_2^-]$$

$$K_b = 5.6 \times 10^{-10}$$

By approximation

$$[^-OH] = \sqrt{(K_b[C_2H_3O_2^-])}$$

$$[^-OH] = \sqrt{\{(5.6 \times 10^{-10})\cdot(0.050)\}}$$

$$[^-OH] = 5.3 \times 10^{-6}\ M$$

$$pOH = -\log(5.3 \times 10^{-6})$$

$$pOH = 5.28$$

$$pH = 8.72 \text{ (solution is basic)}$$

2. For example, consider a solution of 0.050 M NH_4Cl, which dissociates and establishes the following equilibrium:

$$NH_4Cl\,(aq) \rightarrow NH_4^+\,(aq) + Cl^-\,(aq)$$

$$NH_4^+\,(aq) + H_2O \rightleftarrows H_3O^+\,(aq) + NH_3\,(aq)$$

$$K_a = [H_3O^+]\cdot[NH_3] / [NH_4^+]$$

$$K_a = 5.6 \times 10^{-10}$$

continued...

Copyright © Sterling Test Prep.

By approximation:

$$[H_3O^+] = \sqrt{(K_a[NH_4^+])}$$

$$[H_3O^+] = \sqrt{\{(5.6 \times 10^{-10})\cdot(0.050)\}}$$

$$[H_3O^+] = 5.3 \times 10^{-6} \text{ M}$$

$$pH = -\log(5.3 \times 10^{-6})$$

$$pH = 5.28 \text{ (solution is acidic)}$$

Notes for active learning

Notes for active learning

Copyright © Sterling Test Prep.

Buffered Solutions

Buffers

A *buffer* is a solution that maintains pH (little or no change) even when a small amount of strong acid or strong base is added. A buffer solution contains a weak acid and the "salt" of its conjugate base or a weak base and the "salt" of its conjugate acid.

The *buffering capacity* is the amount of H^+ or OH^- ion the buffer absorbs without significantly altering the pH.

A buffer with large concentrations of buffering components absorbs significant quantities of strong acid or strong base, with little change in its pH, has a large buffering capacity.

A buffer is effective within a pH range, typically about ± 1 of the pK_a of its acid component.

Some *common buffer systems*:

Buffer	pK_a	pH Range
$HCHO_2 - NaCHO_2$	3.74	2.75–4.75
$HC_2H_3O_2 - NaC_2H_3O_2$	4.74	3.75–5.75
$KH_2PO_4 - K_2HPO_4$	7.21	6.20–8.20 (a blood buffer)
$CO_2/H_2O - NaHCO_3$	6.37	5.40–7.40 (a blood buffer)
$NH_3 - NH_4Cl$	9.25	8.25–10.25

When 0.01 mol of HCl is added to 1 L of pure water, $[H^+]$ increases from 10^{-7} to 10^{-2} M, and the pH changes from about 7 to 2. This change in pH indicates that water is not a buffer.

When the same HCl amount is added to a solution containing a mixture of 1 M acetic acid ($HC_2H_3O_2$) and 1 M sodium acetate ($NaC_2H_3O_2$), the solution's pH changes little – it goes from 4.74 to 4.66. A solution composed of acetic acid and sodium acetate is a buffer solution.

For example, consider a buffered solution composed of KH_2PO_4 and K_2HPO_4. The species present in the solution are primarily K^+, $H_2PO_4^-$ and HPO_4^{2-}. K^+ is a spectator ion and not involved in the buffering reaction.

If a small amount of strong acid is added to this solution, the H^+ ions from the acid react with the base component of the buffer (HPO_4^{2-}):

$$H^+ (aq) + HPO_4^{2-} (aq) \rightarrow H_2PO_4^- (aq) \ldots \text{(buffering reaction 1)}$$

Any duplication (copies, uploads, PDFs) is illegal.

When a strong base such as NaOH is added, the $^-$OH reacts with the buffer's acid component ($HC_2H_3O_2$).

$$^-OH\,(aq) + HC_2H_3O_2\,(aq) \rightarrow H_2O + C_2H_3O_2^-\,(aq) \ldots \text{(buffering reaction 2)}$$

Reactions (1) and (2) are critical buffering reactions that maintain the solution's pH.

Biological buffers

Buffered solutions are vital to living organisms. Metabolic reactions are controlled or accelerated by biological catalysts called *enzymes*, often proteins that function within a narrow pH range.

The human body fluid must be at a specific (narrow) pH range. Human blood is maintained at the pH range of 7.30 - 7.40. A drop below pH 7 or rise above pH 7.5 can be fatal.

Two buffer systems – the phosphate buffer ($H_2PO_4^-$-HPO_4^{2-}) and carbonic acid-bicarbonate buffer (H_2CO_3-HCO_3^-) are essential to maintaining the normal blood pH.

The buffering reactions of bicarbonate buffer are:

$$H^+\,(aq) + HCO_3^-\,(aq) \rightarrow H_2O + CO_2\,(aq)$$

$$^-OH\,(aq) + CO_2\,(aq) + H_2O \rightarrow 2\,HCO_3^-\,(aq)$$

Buffers resist changes in pH

The *capacity of a buffer* is determined by the sizes of [AB] and [B$^-$].

Consider a buffer made up of acetic acid and sodium acetate, in which the significant species present in the solution are $HC_2H_3O_2$ and $C_2H_3O_2^-$.

For example, if a small amount of HCl (aq) is added to this solution, most H^+ (from HCl) is absorbed by the conjugate base, $C_2H_3O_2^-$.

$$H^+\,(aq) + C_2H_3O_2^-\,(aq) \rightarrow HC_2H_3O_2^-\,(aq)$$

Since $C_2H_3O_2^-$ is present in a larger quantity than the added H^+, the reaction shifts almost entirely to the right. This buffering reaction prevents a significant increase in [H^+] and minimizes the pH change.

For example, if a strong base, such as NaOH (aq), is added, most $^-$OH ions (from NaOH) react with the buffer's acidic component.

$$^-OH\,(aq) + HC_2H_3O_2\,(aq) \rightarrow H_2O + C_2H_3O_2^-\,(aq)$$

Because of the larger concentration of $HC_2H_3O_2$ compared to $^-$OH, this reaction goes almost to completion. This buffering reaction prevents a significant increase in [$^-$OH] and minimizes pH changes.

                                        Copyright © Sterling Test Prep.

For example, for a buffer containing the weak acid HB and the salt NaB, such that B^- is the conjugate base to the acid, the concentration $[H^+]$ and pH of the buffer depending on the dissociation constant, K_a, of the acid component, and the concentration ratio $[B^-] / [HB]$ in the buffer solution.

For example, consider the equilibrium:

$$HB\ (aq) \leftrightarrows H^+\ (aq) + B^-\ (aq)$$

$$K_a = [H^+]{\cdot}[B^-] / [HB]$$

Rearranging the expression:

$$[H^+] = K_a \times ([HB] / [B^-])$$

$$pH = pK_a + \log([B^-] / [HB])$$

Calculating pH of buffered solutions

The *pH of a buffered solution* is determined by the ratio $[B^-] / [HB]$.

The following example illustrates how to calculate the change in pH and the buffering capacity of a buffered solution after a strong acid is added.

Calculate the change in pH when 0.010 mol of HCl adds to 1.0 L of each of the buffers:

Buffer A: 1.0 M $HC_2H_3O_2$ + 1.0 M $NaC_2H_3O_2$

Buffer B: 0.020 M $HC_2H_3O_2$ + 0.020 M $NaC_2H_3O_2$

Henderson-Hasselbalch equation

The expression $pH = pK_a + \log([B^-] / [HB])$ is the *Henderson-Hasselbalch equation*, which is useful for calculating the pH of solutions when the K_a and the ratio $[B^-] / [HB]$ are known.

For buffers, the *Henderson-Hasselbalch equation* calculates pH:

$$pH = pK_a + \log([B^-] / [HB])$$

$$pH = pK_a + \log([C_2H_3O_2^-] / [HC_2H_3O_2])$$

$$pH = -\log(1.8 \times 10^{-5}) + \log(1)$$

$$pH = 4.74$$

Any duplication (copies, uploads, PDFs) is illegal.

For example, 0.010 mol HCl is added to Buffer A with the following reaction:

$$H^+(aq) + C_2H_3O_2^-(aq) \rightarrow HC_2H_3O_2(aq)$$

[ ] before reaction: 0.010 M   1.0 M   1.0 M

[ ] after reaction: 0   0.99 M   1.01 M

The new pH:

$$pH = pK_a + \log([B^-] / [HB])$$

$$pH = 4.74 + \log(0.99 / 1.01)$$

$$pH = 4.74 - 0.010$$

$$pH = 4.73 \; (pH \text{ is changed} \approx 0.21\%)$$

0.010 mol HCl is added to Buffer B with the following reaction:

$$H^+(aq) + C_2H_3O_2^-(aq) \rightarrow HC_2H_3O_2(aq)$$

[ ] before rxn: 0.010 M   0.020 M   0.020 M

[ ] after rxn: 0   0.010 M   0.030 M

The expression to calculate pH:

$$pH = pK_a + \log([B^-] / [HB])$$

$$pH = 4.74 + \log(0.010 / 0.030)$$

$$pH = 4.74 - 0.48$$

$$pH = 4.26 \; (pH \text{ decreases by } 10\%)$$

From above, Buffer A, which contains larger quantities of buffering components, has a higher buffering capacity than Buffer B.

For Buffer A to decrease its pH by 0.48 units (or 10%), it must absorb the equivalent of 0.50 mol of HCl.

Copyright © Sterling Test Prep.

Summary of buffer solutions

1. The solution contains a weak acid HX and its conjugate base X^-, or a weak base B and its conjugate acid BH^+ in appreciable amounts.

2. A buffer solution maintains its pH by absorbing H^+ or OH^- produced by a strong acid or strong base, so these ions do not accumulate.

3. The buffering reaction involves H^+ with the conjugate base **X^-** in the buffer, or the reaction of ^-OH with the acid component (**HX**) of the buffer:

$$H^+ (aq) + \mathbf{X^-} (aq) \rightarrow HX (aq)$$

$$^-OH (aq) + \mathbf{HX} (aq) \rightarrow H_2O + X^-$$

These two reactions prevent a significant increase in $[H^+]$ or $[^-OH]$ in the solution.

4. The *buffering capacity* of a solution implies the amount of H^+ or ^-OH it absorbs without significantly changing its pH. This depends on the concentration of the weak acid and its conjugate base in the solution.

5. The *buffering range* of a solution depends on the pK_a of the buffer's acid component.

   A buffer is most effective when the pH range $= pK_a \pm 1$.

Any duplication (copies, uploads, PDFs) is illegal.

Notes for active learning

Copyright © Sterling Test Prep.

Ionic Equations and Neutralization

Molecular, ionic and net ionic equations

Examples of molecular, complete (or total) ionic, and net ionic equations for strong acid-strong base reactions:

Molecular:

$$HCl\,(aq) + NaOH\,(aq) \rightarrow H_2O\,(l) + NaCl\,(aq)$$

Complete ionic:

$$H^+\,(aq) + Cl^-\,(aq) + Na^+\,(aq) + OH^-\,(aq) \rightarrow H_2O\,(l) + Na^+\,(aq) + Cl^-\,(aq)$$

Net ionic:

$$H^+\,(aq) + OH^-\,(aq) \rightarrow H_2O\,(l)$$

Spectator ions:

$$Na^+ \text{ and } Cl^-$$

Weak acids and weak bases ionize partially, and most weak acids and bases remain in the molecular form in solution.

Therefore, weak acids and bases should NOT be written in ionized forms when writing the ionic equations.

The three equations for the reaction between acetic acid (a weak acid) and sodium hydroxide (a strong base) are:

Molecular:

$$HC_2H_3O_2\,(aq) + NaOH\,(aq) \rightarrow H_2O\,(l) + NaC_2H_3O_2\,(aq)$$

Complete ionic:

$$HC_2H_3O_2\,(aq) + Na^+\,(aq) + OH^-\,(aq) \rightarrow H_2O\,(l) + Na^+\,(aq) + C_2H_3O_2^-\,(aq)$$

Net ionic:

$$HC_2H_3O_2\,(aq) + OH^-\,(aq) \rightarrow H_2O\,(l) + C_2H_3O_2^-\,(aq)$$

Spectator ion:

$$Na^+$$

Any duplication (copies, uploads, PDFs) is illegal.

Neutralization

Neutralization is a chemical reaction where an acid and a base react quantitatively, resulting in no excess of hydrogen (H^+) or hydroxide ions (OH^-) in the aqueous solution.

The products of acid-base reactions are salt and water.

$$HCl \, (aq) + NaOH \, (aq) \rightarrow H_2O \, (l) + NaCl \, (aq)$$

$$HClO_4 \, (aq) + KOH \, (aq) \rightarrow H_2O \, (l) + KClO_4 \, (aq)$$

$$HC_2H_3O_2 \, (aq) + NaOH \, (aq) \rightarrow H_2O \, (l) + NaC_2H_3O_2 \, (aq)$$

Note that the substances involved are subject to dissociation.

Stoichiometry of acid-base neutralization

For example, how many mL of 0.1725 M NaOH (aq) are needed to neutralize 25.00 mL of 0.2040 M HCl (aq)? According to the equation:

$$HCl \, (aq) + NaOH \, (aq) \rightarrow H_2O \, (l) + NaCl \, (aq)$$

Moles of NaOH needed = Moles of HCl present

$$(\text{Liters NaOH} \times \text{Molarity NaOH}) = (\text{Liters HCl} \times \text{Molarity HCl})$$

$$(\text{Liters of NaOH} \times 0.1725 \text{ mol/L}) = (0.02500 \text{ L} \times 0.2040 \text{ mol/L})$$

Divide each side by 0.1725 mol/L:

$$\text{Liters of NaOH} = \frac{0.02500 \text{ L} \times 0.2040 \text{ mol/L}}{0.1725 \text{ mol/L}}$$

$$\text{Liters of NaOH} = 0.02957 \text{ L} = 29.57 \text{ mL}$$

For example, if 10.00 mL of acetic acid $HC_2H_3O_2$ of an unknown concentration requires 38.64 mL of 0.2250 M KOH to neutralize, what is the molarity of the acetic acid?

From the reaction:

$$HC_2H_3O_2 \, (aq) + NaOH \, (aq) \rightarrow H_2O \, (l) + NaC_2H_3O_2 \, (aq)$$

Moles of acetic acid = Moles of NaOH

$$(\text{Liter of acid} \times \text{Molarity of acid}) = (\text{Liter of base} \times \text{Molarity of base})$$

$$(0.01000 \text{ L} \times \text{Molarity } HC_2H_3O_2) = (0.03864 \text{ L} \times 0.2250 \text{ M NaOH})$$

Divide each side by 0.01000 L:

$$\text{Molarity of } HC_2H_3O_2 = \frac{0.03864 \text{ L} \times 0.2250 \text{ M}}{0.01000 \text{ L}}$$

$$HC_2H_3O_2 = 0.8694 \text{ M}$$

 Copyright © Sterling Test Prep.

Titration

Molar concentration determination

Titration (or *volumetric analysis*) is a standard laboratory procedure to determine an *unknown concentration.*

Titration (*volumetric analysis*) is an essential application of neutralization reactions. A reagent of known concentration is slowly added to a sample of unknown concentration until the neutralization reaction (i.e., a reaction between an acid and a base) is complete.

Titration determines the molar concentration of a solution (analyte) using the volume and concentration of another (titrant) by adding an exact amount of the *titrant* from a buret to another reactant (*analyte*) in a flask.

The *analyte* is the substance (i.e., sample) of an unknown concentration being analyzed.

The *titrant* is the analytical reagent of known concentration added to the sample. It is carefully added from a buret until the *equivalence point* (i.e., the point of neutralization) is reached.

If the titrant's volume and concentration are known, its number of moles can be calculated.

An *indicator* is added to the reaction to determine the endpoint of a redox titration (i.e., when sample molecules have been depleted).

The *endpoint* is marked by an indicator's color change when a solution transitions from acidic to slightly basic. A redox indicator undergoes a color change at the neutralization reaction's equivalence point.

In a useful titration, the *equivalence point* should match the *endpoint.*

The number of moles of the analyte and its concentration can be determined from reaction stoichiometry.

Titration experimental factors

1. The reaction between titrant and analyte occurs rapidly.

2. The balanced equation is known.

3. The endpoint occurs precisely at or close to the equivalence point.

4. The volume of titrant to reach the equivalence point is accurately measurable.

Stoichiometry determines unknown concentrations

An acid (e.g., HCl (*aq*)) is titrated with aqueous NaOH as follows:

$$HCl\,(aq) + NaOH\,(aq) \rightarrow NaCl\,(aq) + H_2O$$

For example, the stoichiometric ratio of HCl to NaOH is 1 mole HCl to 1 mole NaOH. The standard solution's volume and concentration (acid or base) are known in titration. However, only the volume of the other solution (acid or base), whose concentration is to be determined, is known.

The above stoichiometry enables calculating unknown concentrations.

For example, suppose that 25.00 mL of 0.2250 M HCl (*aq*) is required to neutralize 27.45 mL of aqueous NaOH solution, whose concentration is unknown. Calculate the moles of each reactant and the concentration of NaOH.

No. of mol of HCl reacted:

$$25.0\text{ mL} \times (1\text{ L}/1{,}000\text{ mL}) \times (0.2250\text{ mol/L}) = 0.005625\text{ mol}$$

Since HCl and NaOH react in a 1:1 ratio,

$$\text{No. of mol of NaOH} = \text{No. of mol of HCl}$$

No. of mol of NaOH = 0.005625 mol

$$\text{Molarity of NaOH} = (0.005625\text{ mol} / 0.02745\text{ L})$$

$$\text{Molarity of NaOH} = 0.2049\text{ M}$$

The stoichiometric ratio may not be 1 to 1, as below.

For example, suppose 20.0 mL H_2SO_4 of unknown concentration requires 32.0 mL of 0.205 M NaOH. Calculate the concentration of the H_2SO_4 solution.

$$H_2SO_4\,(aq) + 2\,NaOH\,(aq) \rightarrow Na_2SO_4\,(aq) + 2\,H_2O$$

The stoichiometric ratio is 1 mole of H_2SO_4 to 2 moles of NaOH.

$$\text{No. of mol of NaOH reacted} = \frac{0.205\text{ mol NaOH}}{1\text{ L solution}} \times 32.0\text{ mL} \times \frac{1\text{ L}}{1{,}000\text{ mL}}$$

No. of mol of NaOH reacted = 0.00656 mol

No. of mol of H_2SO_4 = Mol NaOH $\times$ (1 mol H_2SO_4 / 2 mol NaOH)

No. of mol of H_2SO_4 = 0.00656 mol NaOH $\times$ (1 mol H_2SO_4 / 2 mol NaOH)

Copyright © Sterling Test Prep.

No. of mol of H_2SO_4 = 0.00328 mol

Molarity of H_2SO_4 = $\dfrac{0.00328 \text{ mol } H_2SO_4}{0.0200 \text{ L}}$

Molarity of H_2SO_4 = 0.164 M

Strong acid–strong base titrations

For example, consider the strong acid-strong base titration of 20.0 mL of 0.100 M HCl (*aq*) with 0.100 M NaOH (*aq*) solution. Calculate the pH of the acid solution:

(a) before NaOH is added,

(b) after 15.0 mL of NaOH is added,

(c) after 19.5 mL of NaOH is added,

(d) after 20.0 mL of NaOH is added,

(e) after 21.0 mL of NaOH is added and

(f) after 25.0 mL of NaOH is added.

(a) Before titration:

$[H^+]$ = 0.100 M

pH = 1.000

(b) When 15.0 mL of NaOH is added, the following reaction occurs:

$H^+ (aq) + OH^- (aq) \rightarrow H_2O$

[ ] before mixing:

0.100 M   0.100 M

[ ] after mixing, but before reaction:

0.0571 M   0.0429 M

[ ] after reaction:

0.0142 M   0 M

$[H^+]$ = 0.0142 M

pH = $-\log(0.0142)$

pH = 1.848

Any duplication (copies, uploads, PDFs) is illegal.

(c) After 19.5 mL of NaOH is added, the calculation of $[H^+]$ is:

$$H^+ (aq) + {}^-OH (aq) \rightarrow H_2O$$

[ ] before mixing:

$$0.100 \text{ M} \qquad 0.100 \text{ M}$$

[ ] after mixing, but before reaction:

$$0.0506 \text{ M} \qquad 0.0494 \text{ M}$$

[ ] after reaction:

$$0.0012 \text{ M} \qquad 0 \text{ M}$$

$$[H^+] = 0.0012 \text{ M}$$

$$pH = -\log(0.0012)$$

$$pH = 2.92$$

Before the equivalent point, $[H^+]$ can be calculated:

$$[H^+] = \frac{(\text{initial mol of } H^+ - \text{mol of } {}^-OH \text{ added})}{(\text{L of HCl titrated} + \text{L of NaOH added})}$$

$$[H^+] = \frac{(0.00200 \text{ mol } H^+ - 0.00195 \text{ mol } {}^-OH)}{(0.0200 \text{ L of HCl} + 0.0195 \text{ L NaOH})}$$

$$[H^+] = (0.000050 \text{ mol} / 0.0395 \text{ L})$$

$$[H^+] = 0.0013 \text{ M}$$

$$pH = 2.90$$

(d) When 20.0 mL of 0.100 M NaOH has been added,

$$H^+ (aq) + {}^-OH (aq) \rightarrow H_2O$$

[ ] after mixing, but before reaction:

$$0.0500 \text{ M} \qquad 0.0500 \text{ M}$$

[ ] after reaction:

$$0 \qquad 0$$

Copyright © Sterling Test Prep.

This point of the titration is the equivalence point, whereby only Na^+ and Cl^- occur in the solution.

Since neither reacts with water, the solution has a pH = 7.00.

(e) When 21.0 mL of 0.100 M NaOH has been added, there is excess OH^-:

$$H^+ (aq) + {}^-OH (aq) \rightarrow H_2O$$

[ ] after mixing, but before reaction:

        0.0488 M   0.0512 M

[ ] after reaction:

        0   0.0024 M

$[^-OH] = 0.0024$ M

pOH = 2.62

pH = 11.38

(f) When 25.0 mL of NaOH is added,

$$H^+ (aq) + {}^-OH (aq) \rightarrow H_2O$$

[ ] after mixing, but before reaction:

        0.0444 M   0.0556 M

[ ] after reaction:

        0   0.0112 M

$[^-OH] = 0.0112$ M

pOH = 1.953

pH = 12.047

In strong acid-strong base titrations, an abrupt change from about pH 3 to 11 occurs within ±0.5 mL of NaOH, added near the equivalent point.

Any duplication (copies, uploads, PDFs) is illegal.

Weak acid–strong base titrations

An example of a weak acid-strong base titration follows.

General reaction:

$$HA \; (aq) + {}^-OH \; (aq) \; \rightarrow \; H_2O + A^- \; (aq)$$

When acetic acid (weak acid) is titrated with sodium hydroxide (strong acid), the net reaction is:

$$HC_2H_3O_2 \; (aq) + {}^-OH \; (aq) \; \rightarrow \; C_2H_3O_2{}^- \; (aq) + H_2O$$

For example, consider the titration of 20.0 mL of 0.100 M $HC_2H_3O_2$ (aq) with 0.100 M NaOH (aq) solution.

Calculate the pH of the solution:

    (a) before the NaOH is added;

    (b) after 10.0 mL of NaOH is added;

    (c) after 15.0 mL of NaOH is added;

    (d) after 20.0 mL of NaOH is added;

    (e) after 25.0 mL of NaOH is added.

(a) Before titration:

$$[H^+] = \sqrt{(0.100 \text{ M}) \cdot (1.8 \times 10^{-5})}$$

$$[H^+] = 1.34 \times 10^{-3} \text{ M}$$

$$pH = -\log(1.34 \times 10^{-3})$$

$$pH = 2.873$$

(b) After adding 10.0 mL of 0.100 M NaOH, $[H^+]$ is calculated as follows:

$$HC_2H_3O_2 \; (aq) + \; OH^- \; (aq) \; \rightarrow \; C_2H_3O_2{}^- \; (aq) + H_2O$$

Before mixing:

        0.100 M   0.1000 M   0.0000 M

After mixing (before rxn):

        0.0667 M   0.0333 M   0.0000 M

After reaction:

        0.0333 M   0.0000 M   0.0333 M

Copyright © Sterling Test Prep.

Using the Henderson-Hasselbalch equation,

$$pH = pK_a + \log([B^-] / [HB])$$

$$[H^+] = K_a \times \frac{[HC_2H_3O_2]}{[C_2H_3O_2^-]}$$

$$[H^+] = 1.8 \times 10^{-5} \times (0.0333 \text{ M} / 0.0333 \text{ M})$$

$$[H^+] = 1.8 \times 10^{-5} \text{ M}$$

$$pH = pK_a + \log([C_2H_3O_2^-] / [HC_2H_3O_2])$$

$$pH = 4.74 + \log(1)$$

$$pH = 4.74$$

When a weak acid is half-neutralized, 50% of the acid is converted to its conjugate base.

Halfway to the equivalence point,

$$[C_2H_3O_2^-] = [HC_2H_3O_2]$$

Under this condition,

$$[H^+] = K_a, \text{ and } pH = pK_a$$

(c) After adding 15.0 mL of 0.100 M NaOH, $[H^+]$ is calculated as follows:

$$HC_2H_3O_2\,(aq) + {}^-OH\,(aq) \rightarrow C_2H_3O_2^-\,(aq) + H_2O$$

Before mixing:	0.100 M	0.100 M	0.000 M
After mixing (before rxn):	0.0571 M	0.0429 M	0.000
After reaction:	0.0142 M	0.000 M	0.0429 M

Using the Henderson-Hasselbalch equation:

$$pH = pK_a + \log([B^-] / [HB])$$

$$[H^+] = K_a \times [HC_2H_3O_2] / [C_2H_3O_2^-]$$

$$[H^+] = 1.8 \times 10^{-5} \times (0.0142 \text{ M} / 0.0429 \text{ M})$$

$$[H^+] = 6.0 \times 10^{-6} \text{ M}$$

$$pH = pK_a + \log([C_2H_3O_2^-] / [HC_2H_3O_2])$$

$$pH = 4.74 + \log(3.02)$$

$$pH = 4.74 + 0.48$$

$$pH = 5.22$$

Any duplication (copies, uploads, PDFs) is illegal.

(d) At the equivalent point, when 20.0 mL of 0.100 M NaOH has been added, acid has been reacted and converted to its conjugate base.

The latter undergoes hydrolysis (reacts with water) as follows:

$$C_2H_3O_2^- \ (aq) + H_2O \ (l) \leftrightharpoons HC_2H_3O_2 \ (aq) + {}^-OH \ (aq)$$

Initial [ ], M:

| 0.0500 | 0.000 | 0.000 |

Change, Δ[ ], M:

| $-x$ | $+x$ | $+x$ |

Equilibrium [ ], M:

| $(0.0500 - x)$ | x | x |

By approximation:

$$[{}^-OH] = x = \sqrt{(K_b \times [C_2H_3O_2^-]_0)}$$

$$[{}^-OH] = \sqrt{(5.6 \times 10^{-10}) \cdot (0.0500)}$$

$$[{}^-OH] = 5.3 \times 10^{-6} \ M$$

$$pOH = -\log(5.3 \times 10^{-6})$$

$$pOH = 5.28$$

$$pH = 8.72$$

Since the conjugate base of a weak acid undergoes hydrolysis (reacts with water), the pH of the solution at the equivalence point is greater than 7.00.

In the case of acetic-NaOH titration, the pH at the equivalence point is about 8.72.

For weak acids with a larger K_a, the pH at the equivalence point is closer to neutral pH; for a smaller K_a, the pH at the equivalence point is higher than neutral pH.

Copyright © Sterling Test Prep.

Indicators

Titration endpoint

The *titrant* is a solution of known concentration that is added (titrated) to another solution to determine the concentration of a second chemical species (i.e., analyte).

The *analyte* is the substance whose quantity or concentration is to be determined.

An *indicator* is a substance that changes color to mark a titration's endpoint. Indicators exhibit one color in acid (or protonated form, HIn) and another in base (or deprotonated form, In^-). Most indicators used in acid-base titration are weak organic acids.

Each indicator has a range of $pH = pK_a \pm 1$, where the change of colors occurs.

A suitable indicator gives the *endpoint* corresponding to the titration's equivalence point. A pH range falls within the sharp increase (or decrease) of pH changes in the titration curves. The endpoint approximates the equivalence point with the known concentration of the titrant to calculate the amount or concentration of the analyte (i.e., unknown quantity or concentration).

Indicator equilibrium

Like a weak acid, an indicator has the following equilibrium in aqueous solution:

$$HIn\ (aq) \rightleftharpoons H^+\ (aq) + In^-\ (aq)$$

$$K_a = \frac{[H^+] \cdot [In^-]}{[HIn]}$$

Rearranging

$$[H^+] = K_a \times ([HIn] / [In^-])$$

$$pH = pK_a + \log([In^-] / [HIn]$$

when

$$[HIn] = 10 \times [In^-]$$

$$pH = pK_a + \log([In^-] / 10[In^-])$$

$$pH = pK_a - 1.0;\ \text{the indicator assumes the color of the acid form.}$$

when

$$[In^-] = 10 \times [HIn]$$

$$pH = pK_a + \log(10[HIn] / [HIn])$$

$$pH = pK_a + 1.0;\ \text{the indicator assumes the color of base form.}$$

Any duplication (copies, uploads, PDFs) is illegal.

Common acid–base indicator

Phenolphthalein, the most common acid-base indicator, has $K_a \sim 10^{-9}$. Its acid form (HIn) is colorless, and the conjugate base form (In⁻) is pink.

It is colorless when the solution's pH ≤ 8 when 90% or more of the species are in the acid form (HIn) and pink at pH ≥ 10 when 90% or more of the species are in the conjugate base form (In⁻).

The pH range at which an indicator changes depends on the K_a.

For phenolphthalein, which has $K_a \sim 10^{-9}$, its color changes in the pH range 8–10.

It is a suitable indicator for strong acid-strong base titrations and weak acid-strong base titrations.

pH of common indicators

Indicators	Acid color	Base color	pH Range	Type of Titrations
Methyl orange	orange	yellow	3.2–4.5	strong acid-strong base strong acid-weak base
Bromocresol green	yellow	blue	3.8–5.4	strong acid-strong strong acid-weak base
Methyl red	red	yellow	4.5–6.0	strong acid-strong base strong acid-weak base
Bromothymol blue	yellow	blue	6.0–7.6	strong acid-strong base
Phenol Red	orange	red	6.8–8.2	strong acid-strong base weak acid-strong base

Copyright © Sterling Test Prep.

Titration Curves

Interpreting titration curves

A *pH curve* is a graph of the pH of the solution *vs.* the volume of titrant in an acid-base titration.

The *equivalence point* is the point in a titration at which the amount of titrant added is just enough to neutralize the analyte solution completely.

At the equivalence point in an acid-base titration, moles of base = moles of acid, and the solution only contains salt and water.

The *buffering region* pH is usually $pK_a \pm 1$ (or $14 - pK_b \pm 1$), which resists pH changes before the curve's smooth (horizontal) portion is reached.

The data used to plot a pH curve may be obtained by computation *or* measuring a pH directly with the pH meter during titration.

The acid is incrementally added to the alkali (how titrations usually are performed).

Strong acid–strong base titration

The following titration curve is for the reaction:

$$NaOH\ (aq) + HCl\ (aq) \rightarrow NaCl\ (aq) + H_2O\ (l)$$

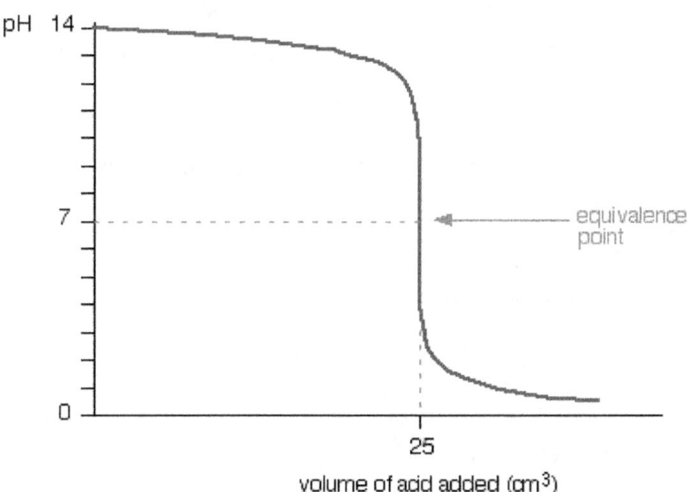

The pH falls a minimal amount until near the equivalence point, then a steep drop.

After the *equivalence point*, it is like the end of the strong acid-strong base reaction.

The buffering region's pH is usually $pK_a \pm 1$ (or $14 - pK_b \pm 1$), which resists pH changes before the curve's smooth (horizontal) portion is reached.

Strong acid–weak base titration

The following titration curve is for the reaction:

$$NH_3\ (aq) + HCl\ (aq) \rightarrow NH_4Cl\ (aq)$$

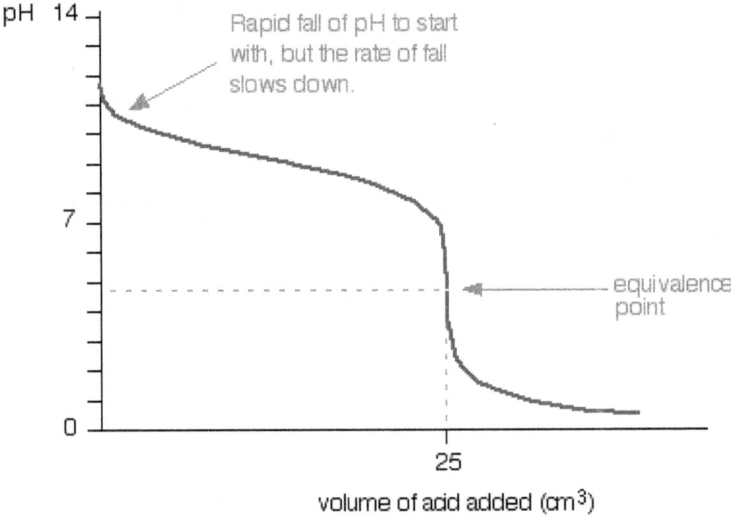

Since it is a weak base rather than a strong base, the titration curve's beginning is vastly different.

The pH decreases steeply as the acid is added, but the curve soon becomes less steep because a buffer solution is created composed of excess ammonia and formed ammonium chloride.

The equivalence point is now acidic (pH ~ 5) but is on the curve's steepest part.

The *equivalence point* is the point in a titration at which the amount of titrant added is just enough to neutralize the analyte solution completely.

At the equivalence point in an acid-base titration, moles of base = moles of acid, and the solution only contains salt and water.

After the *equivalence point*, it is like the end of the strong acid-weak base reaction.

The buffering region's pH is usually $pK_a \pm 1$ (or $14 - pK_b \pm 1$), which resists pH changes before the curve's smooth (horizontal) portion is reached.

 Copyright © Sterling Test Prep.

Weak acid–strong base titration

The following titration curve is for the reaction:

$$CH_3COOH\ (aq) + NaOH\ (aq) \rightarrow CH_3COONa\ (aq) + H_2O\ (aq)$$

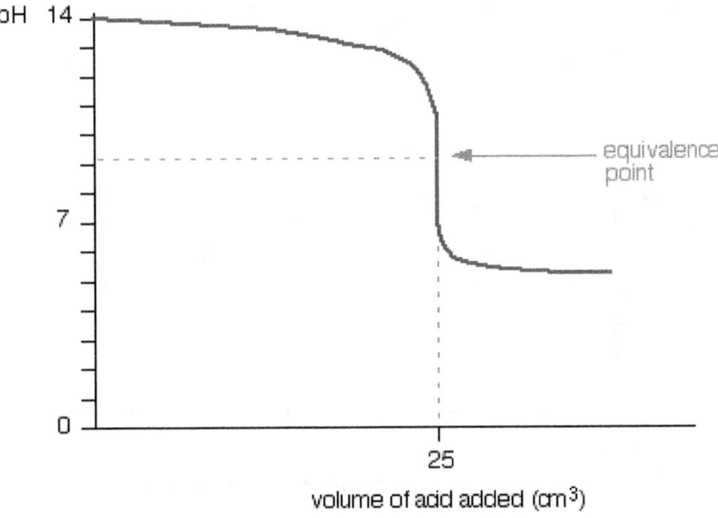

In the first part of the graph, there is an excess of sodium hydroxide, and this part of the curve is like the strong acid-strong base titration.

The *equivalence point* is the point in a titration at which the amount of titrant added is just enough to neutralize the analyte solution completely.

At the equivalence point in an acid-base titration, moles of base = moles of acid, and the solution only contains salt and water.

After the *equivalence point*, it is like the end of the weak acid-strong base reaction.

Once the acid is in excess, there is a difference between forming a buffer solution containing sodium ethanoate and ethanoic acid.

The buffering region's pH is usually $pK_a \pm 1$ (or $14 - pK_b \pm 1$), which resists pH changes before the curve's smooth (horizontal) portion is reached.

Weak acid–weak base titration

The following titration curve is for the reaction:

$$CH_3COOH\ (aq) + NH_3\ (aq) \rightarrow CH_3COONH_4\ (aq)$$

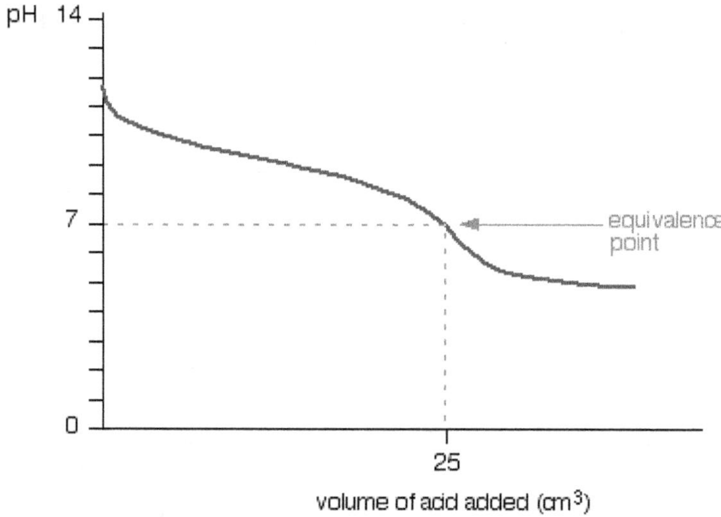

The acid and the base are equally weak, so the equivalence point is around pH 7.

This titration curve is essentially a combination of the two previous graphs.

The *equivalence point* is the point in a titration at which the amount of titrant added is just enough to neutralize the analyte solution completely.

At the equivalence point in an acid-base titration, moles of base = moles of acid, and the solution only contains salt and water.

After the *equivalence point*, it is like the end of the weak acid-weak base reaction.

There is no steep portion of this graph; the lack of a steep portion is an important identifying factor of a weak acid-weak base titration curve.

The buffering region's pH is usually $pK_a \pm 1$ (or $14 - pK_b \pm 1$), which resists pH changes before the curve's smooth (horizontal) portion is reached.

Copyright © Sterling Test Prep.

Four titration curves comparison

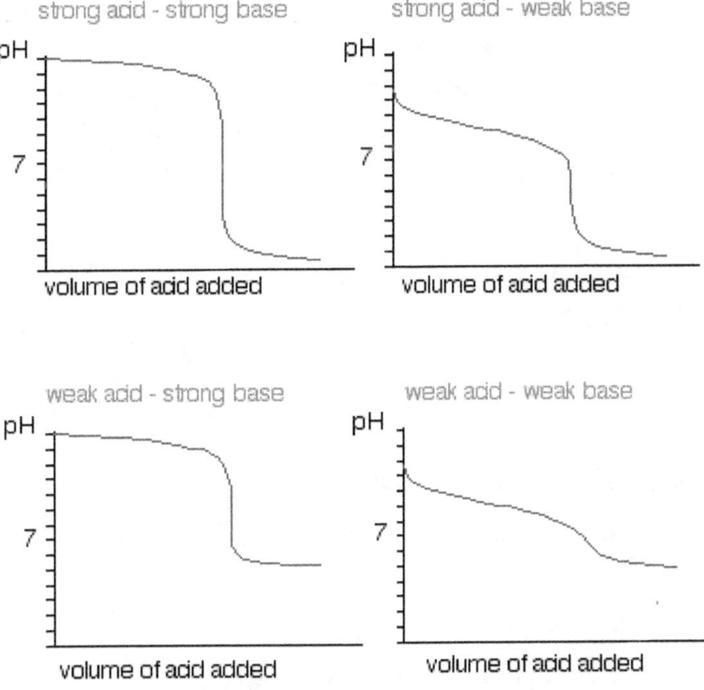

Redox titration transfers electrons

A *redox titration* is a type of titration based on the transfer of electrons.

In redox titration, standard solutions of oxidizing agents are used because solutions of reducing agents may react with oxygen in the air.

The analyzed species must be in a single oxidation state before titration.

For example, when iron ore is dissolved in hydrochloric acid, Fe^{2+} and Fe^{3+}, as iron(II) and iron(III), are present.

However, since it must be in a single oxidation state before titration with a standard $KMnO_4$ solution, iron(III) is reduced to iron(II) by reaction with excess zinc:

$$2\ Fe^{3+}\ (aq) + Zn\ (s) \rightarrow 2\ Fe^{2+}\ (aq) + Zn^{2+}\ (aq)$$

$$MnO_4^-\ (aq) + 5\ Fe^{2+}\ (aq) + 8\ H^+\ (aq) \rightarrow Mn^{2+}\ (aq) + 5\ Fe^{3+}\ (aq) + 4\ H_2O\ (l)$$

$KMnO_4$ and $K_2Cr_2O_7$ are the most used oxidizing agents in redox titration because the solutions change color during reduction; this color change serves as the titration indicator.

$KMnO_4$ is a purple solution but becomes colorless when MnO_4^- is reduced to Mn^{2+}.

The bright orange of $K_2Cr_2O_7$ changes to purplish-blue when $Cr_2O_7^{2-}$ is reduced to Cr^{3+}.

Any duplication (copies, uploads, PDFs) is illegal.

Sample redox titration

For example, a solution contains iron(II) and iron(III) ions. A 50.0 mL solution sample is titrated with 35.0 mL of 0.00280 M $KMnO_4$, which oxidizes Fe^{2+} to Fe^{3+}. The permanganate ion is reduced to manganese(II) ion. Another 50.0 mL solution sample is treated with zinc metal, which reduces Fe^{3+} to Fe^{2+}. The resulting solution is again titrated with 0.00280 M $KMnO_4$ and 48.0 mL is required. What are the concentrations of Fe^{2+} and Fe^{3+} in the solution?

For this problem, several steps are required.

1) The stoichiometric relationship of Fe(II) to permanganate is five to one:

$$MnO_4^-\,(aq) + 5\ Fe^{2+}\,(aq) + 8\ H^+\,(aq) \rightarrow Mn^{2+}\,(aq) + 5\ Fe^{3+}\,(aq) + 4\ H_2O\,(l)$$

2) Calculate the moles of Fe(II) reacted:

$(0.00280\ mol/L)\,(0.0350\ L) = 0.000098\ mol\ of\ MnO_4^-$

$(0.0000980\ mol\ Mn)\cdot(5\ mol\ Fe\ /\ 1\ mol\ Mn) = 0.000490\ mol\ Fe(II)$

3) Determine the total iron content:

$(0.00280\ mol/L)\cdot(0.0480\ L) = 0.0001344\ mol\ of\ MnO_4^-$

$(0.0000980\ mol\ Mn)\cdot(5\ mol\ Fe\ /\ 1\ mol\ Mn) = 0.000672\ mol\ total\ Fe$

4) Determine Fe(III) in solution and its molarity:

$0.000672\ mol - 0.000490\ mol = 0.000182\ mol$

$(0.000182\ mol)\,/\,(0.050\ L) = 0.00364\ M\ Fe(III)$

5) Determine the molarity of Fe(II):

$(0.000490\ mol)\,/\,(0.050\ L) = 0.0098\ M\ Fe(II)$

Copyright © Sterling Test Prep.

Practice Questions

1. Which of the following indicators is green at pH 7?

    I. phenolphthalein   II. bromothymol blue   III. methyl red

    A. I only
    B. II only

    C. III only
    D. I and II only

2. Which of the following is NOT a strong acid?

    A. $HClO_3$
    B. HF

    C. HBr
    D. HCl

3. Which of the following compounds is the strongest base?

    A. ClO_3^-
    B. NH_3

    C. ClO^-
    D. ClO_2^-

4. A Brønsted-Lowry acid is defined as a substance that:

    A. acts as a proton donor in any system
    B. acts as a proton acceptor in any system

    C. increases $[H^+]$ when placed in water
    D. decreases $[H^+]$ when placed in water

5. Which of the following acts as the best buffer solution?

    A. Strong acids or bases
    B. Strong acids and their salts

    C. Salts
    D. Weak acids or bases and their salts

6. Which of the following is true if a drain cleaner solution is a strong electrolyte?

    I. Slightly reactive   II. Highly reactive   III. Highly ionized

    A. I only
    B. II only

    C. III only
    D. I and III only

7. Which of the following statements describes a basic solution?

    A. $[H_3O^+] \times [^-OH] \neq 1 \times 10^{-14}$
    B. $[H_3O^+] / [^-OH] = 1 \times 10^{-14}$

    C. $[H_3O^+] > [^-OH]$
    D. $[H_3O^+] < [^-OH]$

Any duplication (copies, uploads, PDFs) is illegal.

8. Which of the following is a general property of an acidic solution?

       I. Neutralizes bases   II. Tastes sour   III. Turns litmus paper red

A. I only

B. II only

C. III only

D. I, II and III

9. Compared to a solution with a higher pH, a solution with a lower pH has a(n):

A. decreased K_a

B. increased [$^-$OH]

C. increased pK_a

D. increased [H^+]

10. Which of the following statements correctly define an Arrhenius acid?

A. Decreases [H^+] when placed in aqueous solutions

B. Increases [H^+] when placed in aqueous solutions

C. Acts as a proton acceptor in any system

D. Acts as a proton donor in any system

11. A Brønsted-Lowry base is a(n):

A. electron acceptor

B. proton acceptor

C. electron donor

D. proton donor

12. Which of the following is the conjugate base of water?

A. $^-$OH (*aq*)

B. H^+ (*aq*)

C. H_2O (*l*)

D. H_3O^+ (*aq*)

13. Which of the following explains why distilled H_2O is neutral?

A. [H^+] = [OH^-]

B. Distilled H_2O has no OH^-

C. Distilled H_2O has no H^+

D. Distilled H_2O has no ions

14. In the following reaction, what does the symbol $\rightleftharpoons$ indicate?

$$^-OH + NH_4^+ \rightleftharpoons H_2O + NH_3$$

A. The rate of the reverse reaction is the same as the forward reaction

B. The forward reaction does not proceed

C. The reaction does not produce an equilibrium

D. The reverse reaction does not proceed

Copyright © Sterling Test Prep.

15. What is the $[H_3O^+]$ of a solution with a pH $= 2.34$?

    A. 1.3×10^1 M   **C.** 4.6×10^{-3} M

    B. 2.3×10^{-10} M   **D.** 2.4×10^{-3} M

Any duplication (copies, uploads, PDFs) is illegal.

Notes for active learning

Copyright © Sterling Test Prep.

Detailed Explanations

1. B is correct.

Bromothymol blue is a pH indicator often used for solutions with neutral pH near 7 (e.g., managing the pH of pools and fish tanks).

Bromothymol blue acts as a weak acid in a solution that can be protonated or deprotonated.

Bromothymol blue appears yellow when protonated (lower pH), blue when deprotonated (higher pH), and bluish-green in neutral solution.

Phenolphthalein is an indicator for acid-base titrations.

A weak acid dissociates protons (H^+ ions) in the solution.

The phenolphthalein molecule is colorless, and the phenolphthalein ion is pink.

It turns colorless in acidic solutions and pink in basic solutions.

With basic conditions, the phenolphthalein (neutral) $\rightleftharpoons$ ions (pink) equilibrium shifts to the right, leading to more ionization as H^+ ions are removed.

Methyl red has a pK_a of 5.1 and is a pH indicator dye that changes color in acidic solutions: it turns red in pH under 4.4, orange in pH 4.4-6.2, and yellow in pH over 6.2.

2. B is correct.

Strong acids dissociate a proton to produce the weakest conjugate base (i.e., most stable anion).

Weak acids dissociate a proton to produce the strongest conjugate base (i.e., least stable anion).

Hydrofluoric (HF) acid has a pK_a of about 3.8 and is weak because it does not dissociate completely.

The F^- anion is the least stable of the halogen anions (due to its small valence shell).

3. C is correct.

Perchloric acid ($HClO_4$) is the strongest acid listed and the weakest conjugate base (i.e., most stable anion).

Hypochlorite (ClO^-) is the strongest base (i.e., least stable anion).

In *oxyacids* (or *oxoacids*), more oxygens increase the acid strength.

The weakest conjugate bases of oxyacids (containing more than one oxygen) are stabilized by resonance.

Any duplication (copies, uploads, PDFs) is illegal.

4. A is correct.

The Brønsted-Lowry acid-base theory focuses on the ability to accept and donate protons (H^+).

A Brønsted-Lowry acid is a substance that donates a proton in an acid-base reaction, while a Brønsted-Lowry base is a substance that accepts the proton.

5. D is correct.

A *buffer* is an aqueous solution that consists of a weak acid and its conjugate base, or vice versa.

Buffered solutions resist changes in pH and are often used to keep the pH at a nearly constant value in many chemical applications. It does this by readily absorbing or releasing protons (H^+) and ^-OH.

When acid is added to the solution, the buffer releases ^-OH and accepts H^+ ions from the acid.

Weak acids or bases and their salts are the best buffer systems.

6. C is correct.

An *electrolyte* is a substance that dissociates into cations (i.e., positive ions) and anions (i.e., negative ions) when placed in a solution.

7. D is correct.

In a basic solution, the concentration of ^-OH is higher than H^+ or H_3O^+.

8. D is correct.

An acid can neutralize a base (i.e., a substance with a pH above 7) to form a salt.

Acids have a sour taste (e.g., lemon juice has a pH of about 2). Acids are known to have a sour taste (e.g., lemon juice) because the sour taste receptors on the tongue detect the dissolved hydrogen (H^+) ions.

An acid is a chemical substance with a pH of less than 7.

Litmus paper is red under acidic conditions and blue under basic conditions.

A pH greater than 7 and feels slippery are qualities of bases, not acids. An acid is a chemical substance with a pH of less than 7, producing H^+ ions in water.

An acid can be neutralized by a base (i.e., a substance with a pH above 7) to form a salt.

Litmus paper is red under acidic conditions and blue under basic conditions.

However, acids are not known to have a slippery feel; this is a characteristic of bases.

Bases feel slippery because they dissolve the fatty acids and oils from the skin and reduce the friction between the skin cells.

Copyright © Sterling Test Prep.

9. D is correct.

Acidic solutions contain hydronium ions (H_3O^+).

These ions are in aqueous form because they are dissolved in water.

Although chemists often write H^+ (*aq*), referring to a single hydrogen nucleus (a proton), it exists as the hydrogen atom.

10. B is correct.

An Arrhenius acid increases the concentration of H^+ ions in an aqueous solution.

Since pH is the negative log of the activity of H^+ ions in an aqueous solution, acids have a low pH.

An acid can act as a proton donor, but this is the Brønsted-Lowry definition of an acid, not Arrhenius.

11. B is correct.

Brønsted-Lowry acids are proton donors (e.g., HCl, H_2SO_4).

Brønsted-Lowry bases are proton acceptors (e.g., HSO_4^-, NO_3^-).

Lewis acids are electron-pair acceptors, whereas Lewis bases are electron-pair donors.

12. A is correct.

By the Brønsted-Lowry acid-base theory:

An acid (reactant) dissociates a proton to become the conjugate base (product).

A base (reactant) gains a proton to become the conjugate acid (product).

The definition is expressed as an equilibrium expression

$$acid + base \leftrightarrow conjugate\ base + conjugate\ acid.$$

13. A is correct.

If the $[H_3O^+] = [^-OH]$, it has a pH of 7, and the solution is neutral.

14. A is correct.

The symbol $\rightleftharpoons$ indicates that the reaction is in equilibrium, whereby the *forward reaction rate* equals the *reverse reaction rate*.

Equilibrium refers to the rate and not the relative magnitude of the remaining reactants or the formed products during the reaction.

Any duplication (copies, uploads, PDFs) is illegal.

15. C is correct.

The formula for pH is:

$$pH = -\log[H_3O^+]$$

Rearrange to solve for $[H_3O^+]$:

$$[H_3O^+] = 10^{-pH}$$

$$[H_3O^+] = 10^{-2.34}$$

$$[H_3O^+] = 4.6 \times 10^{-3} \text{ M}$$

Copyright © Sterling Test Prep.

Notes for active learning

Notes for active learning

Copyright © Sterling Test Prep.

CHAPTER 9

Electrochemistry

Electrochemistry is **the study of chemical processes that cause electrons to move**. This movement of electrons is called electricity, which can be generated by movements of electrons from one element to another in a reaction known as an oxidation-reduction ("redox") reaction.

Electrochemistry studies the interconversion of electrical and chemical energy. It utilizes spontaneous oxidation-reduction reactions for electrical energy to drive nonspontaneous reactions. Electrochemistry involves an oxidation-reduction process. Some electrochemistry applications include batteries, corrosion control, metallurgy, and electrolysis.

- Electrochemical Cells

- Electrolysis

- Galvanic (Voltaic) Cells

- Reduction Potential

- Electrical Potential Difference

- Electron Flow

- Batteries

- Practice Questions & Detailed Explanations

Copyright © Sterling Test Prep.

Electrochemical Cells

Galvanic and electrolytic cells

The two types of electrochemical cells are *galvanic* (or Voltaic) and *electrolytic.*

Galvanic (or Voltaic) cells convert chemical energy to electrical energy by *spontaneous* redox reactions. A galvanic cell consists of two different metals (electrodes) connected through a conducting solution (an electrolyte) and externally, completing a circuit. A galvanic (voltaic) cell has a resistor or voltmeter in place of the battery.

Galvanic (Voltaic) cells have a positive cell potential; thus, no electrical input is required.

Electrolytic cells use electrical energy to drive a non-spontaneous redox reaction, often used to decompose chemical compounds via electrolysis. Electrolytic cells convert electrical energy to chemical energy. Electrolytic cells involve non-spontaneous reactions and thus require an external electron source (e.g., battery).

For example, the electrolysis of sodium chloride (NaCl, common salt) forms sodium Na^+ metal and chlorine Cl^- gas; an electric current supplies the energy required to make the reaction proceed.

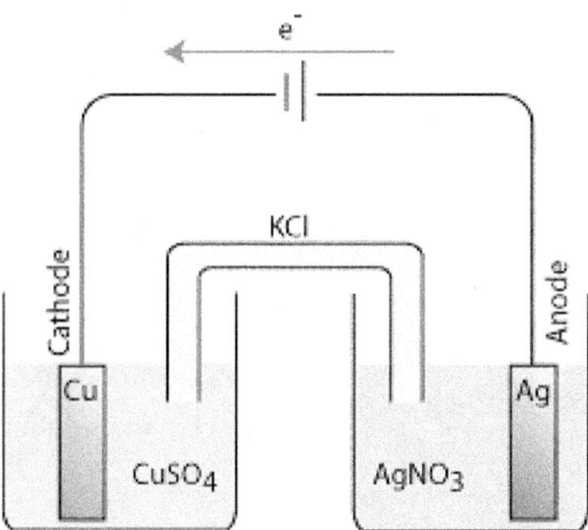

Electrolytic cells use electrical energy to drive non-spontaneous redox reactions

Anode and cathode reactions

Oxidation occurs at the *anode* for an electrochemical cell, and reduction occurs at the *cathode.*

To designate anode and cathode, write a balanced net ionic equation for the spontaneous cell reaction.

- The oxidizing agent (more positive or less negative reduction potential $E°$) is the cathode.

- The oxidizing agent (less positive or more negative reduction potential $E°$) is the anode.

- Oxidation occurs in the anode half-cell and reduction in the cathode half-cell.

- The anode is negative (-), and the cathode is positive (+).

Mnemonic:　　**An Ox** = **AN**ode **Ox**idation　　　**Red Cat** = **RED**uction **CAT**hode

When drawing a galvanic cell, the anode is on the left by convention, with cathode on the right.

Use the mnemonic **ABC** as convention (**A**node / **B**ridge / **C**athode).

Galvanic (or Voltaic) cells	Electrolytic cells
Cell converts chemical energy into electrical energy.	Cell converts electrical energy into chemical energy.
The redox reaction is *spontaneous* and produces electrical energy.	The redox reaction is *not spontaneous*, and electrical energy must be supplied to initiate the reaction.
The electrons are supplied by the species getting oxidized. The electrons migrate from the anode to the cathode in the external circuit.	An *external battery* supplies the electrons which enter through the cathode and come out through the anode
The two half-cells occur in different chambers and are connected through a salt bridge.	Both electrodes are placed in the same vessel in the solution.
The anode is *negative*, and the cathode is the positive electrode.	The anode is *positive*, and the cathode is the negative electrode.
Oxidation occurs at the anode, while reduction occurs at the cathode.	Oxidation occurs at the anode, while reduction occurs at the cathode.

Comparison between galvanic and electrolytic cells

　　　　　　　　　　　　　　Copyright © Sterling Test Prep.

Oxidation and reduction at electrodes

Oxidation occurs at the anode for electrolytic cells, while reduction occurs at the cathode.

Electrons originate from the anode because oxidation (loss of electrons) occurs.

$$M \rightarrow M^+ + e^-$$

Electrons travel to the cathode, where they join cations on the cathode surface, and reduction (gain of electrons) occurs.

$$M^+ + e^- \rightarrow M$$

Mnemonic: OIL RIG: **O**xidation **I**s **L**osing e^-, **R**eduction **I**s **G**aining e^-.

Oxidation is an *increase in charge* (more positive) due to the *loss* of e^-.

Reduction is a *decrease in charge* (more negative) due to the *gain* of e^-.

Concentration cells

A *concentration cell* is an electrochemical cell in which the half-cells are of the same type but with different electrolyte concentrations.

The following cell notations are examples of concentration cells:

$$Cu|Cu^{2+}(aq, 0.0010 \text{ M})\|Cu^{2+}(aq, 1.0 \text{ M})|Cu$$

$$Ag|Ag^+(aq, 0.0010 \text{ M})\|Ag^+(aq, 0.10 \text{ M})|Ag$$

In concentration cells, the half-cell with the lower electrolyte concentration serves as an anode half-cell, and one with the higher electrolyte concentration is the cathode half-cell.

At the anode half-cell, an oxidation reaction occurs to increase the electrolyte concentration, and at the cathode half-cell, a reduction reaction occurs to decrease its electrolyte concentration.

Any duplication (copies, uploads, PDFs) is illegal.

Notes for active learning

Copyright © Sterling Test Prep.

Electrolysis

Forced current

Electrolysis forces a current through a cell to produce a chemical change, negative cell potential. The electrolysis process extracts metals from molten salts, and a sustained current is applied to the molten salt.

Electrolysis requires potential (voltage) input.

The potential (voltage) input + the cell potential must be > 0 for reactions to occur. Electrolytic cells have negative cell potentials, so a potential input greater than the cell potential's magnitude is needed for electrolysis.

A battery is a contained unit that produces electricity. Commercial batteries are galvanic cells using solids or pastes as reactants to maximize the electrical output per unit mass. A *fuel cell* is a galvanic cell requiring constant reactants to generate electricity.

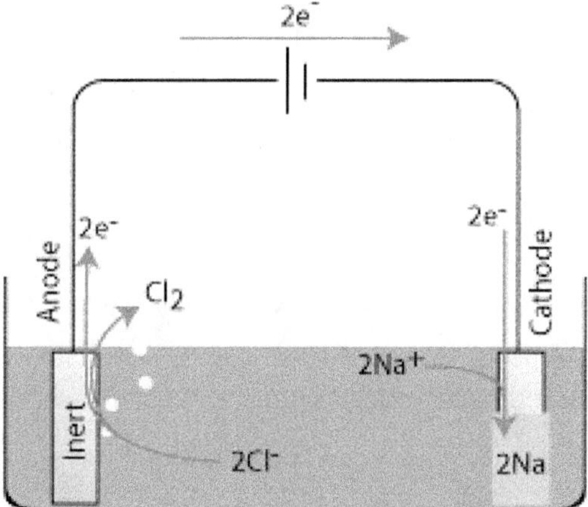

Salt electrolysis charge flow with arrows as how the battery forces electron flow

For example, the following spontaneous reaction occurs in a galvanic (voltaic) cell consisting of $Zn|Zn^{2+}$ *(aq)* and $Cu|Cu^{2+}$ *(aq)* half-cells:

$$Zn\ (s) + Cu^{2+}\ (aq)\ \rightarrow\ Zn^{2+}\ (aq) + Cu\ (s); \qquad E°_{cell} = 1.10\ V$$

In an electrolytic cell, the reverse reaction occurs if a voltage greater than the E° cell is applied, which causes the reverse reaction, and the excess voltage is *overpotential* or *overvoltage*.

Electrolytic cell reactions

Three types of reactions occur in electrolytic cells if sufficient voltage is applied:

- Solute ions or molecules may be oxidized or reduced.

- The solvent can be oxidized or reduced.

- Metal electrode that forms the anode can be oxidized.

Which reactions occur depends on the thermodynamic and kinetic properties of each reaction.

Electric current

An *electric current* flowing through a cell is measured in *ampere* (A), equal to the charge in *coulomb* (C) flowing per second.

$$1 \text{ ampere} = 1 \text{ coulomb} / 1 \text{ second}$$

$$\text{Ampere} = \text{Coulomb/second} (1 \text{ A} = 1 \text{ C/s})$$

$$\text{Coulomb} = \text{Ampere} \times \text{time in sec} (1 \text{ C} = 1 \text{ A·s}) = \text{total charge}$$

$$\text{Joule} = \text{Coulomb} \times \text{Volt} (1 \text{ J} = 1 \text{ C.V})$$

Electrolytes are ions that conduct electricity by motion. Without electrolytes, electricity cannot propagate.

Electrolysis of water

The *decomposition of water* is a *nonspontaneous* process with an amount of energy of about 400 kJ/mol of water. The decomposition of water can be effectuated by electrolysis, where O_2 forms at the *anode* and H_2 at the *cathode*. Assumes an electrolytic cell with:

$$[H^+] = [OH^-] = 1 \text{ M and } P_{H2} = P_{O2} = 1 \text{ atm}$$

For the electrolysis of pure water, where $[H^+] = [OH^-] = 10^{-7}$ M, the potential for the overall process is −1.23 V.

Anode reaction:	$2\,H_2O \rightarrow O_2 + 4\,H^+\,(aq) + 4\,e^-$	$E° = -1.23$ V
Cathode reaction:	$4\,H_2O + 4\,e^- \rightarrow 2\,H_2 + 4\,OH^-\,(aq)$	$E° = -0.83$ V

Net reaction:	$6\,H_2O \rightarrow 2\,H_2 + O_2 + 4\,(H^+ + OH^-)$	$E° = -2.06$ V
	$2\,H_2O \rightarrow 2\,H_2 + O_2$	

A voltage of 1.23 V is enough for the electrolysis of water. However, an additional voltage (overpotential or overvoltage) of about 1 V is needed to force the electrolysis.

Copyright © Sterling Test Prep.

Electrolysis of solution containing mixtures of ions

The metal with the smallest potential is the first to be formed.

For example, consider a solution of Cu^{2+}, Zn^{2+} and Ag^+ electrolyzed with a current with sufficient voltage to reduce the cations.

The reduction potentials of these elements are as follows:

$$Ag^+ (aq) + e^- \rightarrow Ag\ (s) \qquad E° = 0.80 \text{ V}$$

$$Cu^{2+} (aq) + 2\ e^- \rightarrow Cu\ (s) \qquad E° = 0.34 \text{ V}$$

$$Zn^{2+} (aq) + 2\ e^- \rightarrow Zn\ (s) \qquad E° = -0.76 \text{ V}$$

Since the reduction of Ag^+ to Ag has the most positive potential, Ag deposits at the cathode before other metals, followed by Cu and Zn, respectively. If the voltage is controlled, starting with the lowest voltage, separating the three metals using electrolysis is possible.

For example, how many grams of Ag is deposited in the cathode of an electrolytic cell if a current of 3.50 A is applied to a solution of $AgNO_3$ for 12 minutes? (Use the molecular mass of Ag = 107.86 g/mol and the conversion of 1 mol $e^- = 9.65 \times 10^4$ C)

Balanced equation:

$$Ag^+ + e^- \rightarrow Ag\ (s)$$

Formula to calculate deposit mass:

mass of deposit = (atomic mass × current × time) / 96,500 C

mass of deposit = [107.86 g × 3.50 A × (12 min × 60 s/min)] / 96,500 C

mass of deposit = 2.82 g

Stoichiometry of electrolysis

The number of substances formed at the anode or cathode is calculated from the current's magnitude (amperes) and time (sec).

For example, if a current of 1.50 A flows through an aqueous solution of $CuSO_4$ for 25.0 minutes, the amount of *charge passing* through the solution is:

charge = (1.50 C/s)·(25.0 minutes)·(60 sec/min)

charge = 2,250 C

The number of *moles of electrons* passing through the solution:

moles of electrons = (2,250 C) × (1 mol e^-)

moles of electrons = 0.0233 mol

Any duplication (copies, uploads, PDFs) is illegal.

Reduction of Cu^{2+} requires 2 mol e^- per mole of Cu:

$$Cu^{2+} + 2\,e^- \rightarrow Cu$$

Cu formed at cathode:

$$(0.0233\ \text{mol } e^-) \times \frac{(1\ \text{mol Cu})}{(2\ \text{mol } e^-)} \times \frac{(63.55\ \text{g Cu})}{(1\ \text{mol Cu})}$$

Cu formed at cathode = 0.741 g Cu

If the process uses a cell with 3.0 V:

energy consumed = 3.0 V × 2250 C

energy consumed = 6,750 C.V

6,750 C.V = 6,800 J

6,800 J = 6.8 kJ

Faraday's Law relates the number of elements

In 1833, Michael Faraday (1791-1867) proposed quantitative laws to express magnitudes of electrolytic effects. The mass of elements deposited at an electrode is directly proportional to the charge (A·sec or Coulombs).

Faraday constant is the electric charge of 1 mole of electrons, a conversion factor between moles and coulombs. Faraday constant (F):

$$F = [6.02 \times 10^{23}\ \text{electrons/mol } e^-] / [6.24 \times 10^{18}\ e^-/C]$$

$$F = 96,485\ \text{C/mol } e^-$$

1 mol electrons is 6.022×10^{23} electrons, and a coulomb is the (negative) charge of 6.241×10^{18} electrons. The Faraday constant is the quotient of these.

For electrolysis, divide the charge in coulombs by the Faraday constant to calculate the number of moles that underwent electrolysis.

Current I = coulombs of charge per second

$$I = q\,/\,t$$

Faraday's constant:

$$F = \text{coulombs of charge per mol of electron}$$

$$F = \text{total charge}\,/\,\text{the total moles of electrons}$$

$$F = q\,/\,n$$

$$q = It \text{ and } q = nF$$

Copyright © Sterling Test Prep.

Thus:

$$It = nF$$

Current × time = moles of e^- × Faraday's constant.

Using this equation, solve for moles of electrons (n). Using the half-equation stoichiometry, determine how many moles of the element are made for every e^- transferred.

For example, 1 mol of Cu is deposited for every 2 moles of electrons for the half-reaction, using Faraday's constant = 96,485 C/mo.

$$Cu^{2+} + 2\ e^- \rightarrow Cu$$

For example, if 1 amp of current passes a cathode for 10 minutes, how much Zn (s) forms in the following reaction? (Use the molecular mass of Zn = 65 g/mole)

$$Zn^{2+} + 2\ e^- \rightarrow Zn\ (s)$$

Calculate the mass of metal deposited at the cathode:

*Step 1: **Calculate total charge using current and time***

$$Q = \text{current} \times \text{time}$$

$$Q = 1\ A \times (10\ \text{minutes} \times 60\ \text{s/minute})$$

$$Q = 600\ A \cdot s = 600\ C$$

*Step 2: **Calculate moles of electron that has the same amount of charge***

$$\text{moles } e^- = Q\ /\ 96{,}500\ \text{C/mol}$$

$$\text{moles } e^- = 600\ C\ /\ 96{,}500\ \text{C/mol}$$

$$\text{moles } e^- = 6.22 \times 10^{-3}\ \text{mol}$$

*Step 3: **Calculate moles of metal deposit***

Half-reaction of zinc ion reduction:

$$Zn^{2+}\ (aq) + 2\ e^- \rightarrow Zn\ (s)$$

$$\text{moles of Zn} = (\text{coefficient Zn}\ /\ \text{coefficient } e^-) \times \text{moles } e^-$$

$$\text{moles of Zn} = (½) \times 6.22 \times 10^{-3}\ \text{mol}$$

$$\text{moles of Zn} = 3.11 \times 10^{-3}\ \text{mol}$$

Any duplication (copies, uploads, PDFs) is illegal.

*Step 4: **Calculate mass of metal deposit***

$$\text{mass Zn} = \text{moles Zn} \times \text{molecular mass of Zn}$$

$$\text{mass Zn} = 3.11 \times 10^{-3} \text{ mol} \times (65 \text{ g/mol})$$

$$\text{mass Zn} = 0.20 \text{ g}$$

For example, how long would it take to deposit 4.00 grams of Cu from a $CuSO_4$ solution if a current of 2.5 A is applied? (Use molecular mass of Cu = 63.55 g/mol and conversion of 1 mol $e^- = 9.65 \times 10^4$ C and 1 C = A·s)

Balanced equation:

$$Cu^{2+} + 2 \, e^- \rightarrow Cu \, (s)$$

Formula to calculate deposit mass:

$$\text{mass of deposit} = \text{atomic mass} \times \text{moles of electron}$$

Since Cu has 2 electrons per atom, multiply the moles by 2:

$$\text{mass of deposit} = \text{atomic mass} \times (2 \times \text{moles of electron})$$

$$\text{mass of deposit} = \text{atomic mass} \times (2 \times \text{current} \times \text{time}) \, / \, 96{,}500 \text{ C}$$

$$4.00 \text{ g} = 63.55 \text{ g} \times (2 \times 2.50 \text{ A} \times \text{time}) \, / \, 96{,}500 \text{ C}$$

$$\text{time} = 4{,}880 \text{ s}$$

$$\text{time} = (4{,}880 \text{ s} \times 1 \text{ min/60s} \times 1 \text{ hr/60 min})$$

$$\text{time} = 1.36 \text{ hr}$$

Copyright © Sterling Test Prep.

Galvanic (Voltaic) Cells

Spontaneous oxidation-reduction reactions

A *galvanic cell* is a typical application of electrochemistry as batteries. A galvanic cell (i.e., *voltaic cell*) uses a spontaneous oxidation-reduction reaction to produce an electric current. Galvanic cells produce voltage and were invented between 1780-1799 by Italian physicists Luigi Galvani (1737-1798) and Alessandro Volta (1745-1827).

In galvanic cells, species with more positive reduction potential serve as a cathode, and those with less positive or more negative reduction potential serve as the anode half-cell.

Galvanic cells:

- Oxidation occurs at the anode.

- Reduction occurs at the cathode.

- Salt bridges or porous disks allow ions to flow without extensive mixing of the solutions.

Salt bridge contains a strong electrolyte held in a gel-like matrix.

Porous disk contains tiny passages for the hindered flow of ions.

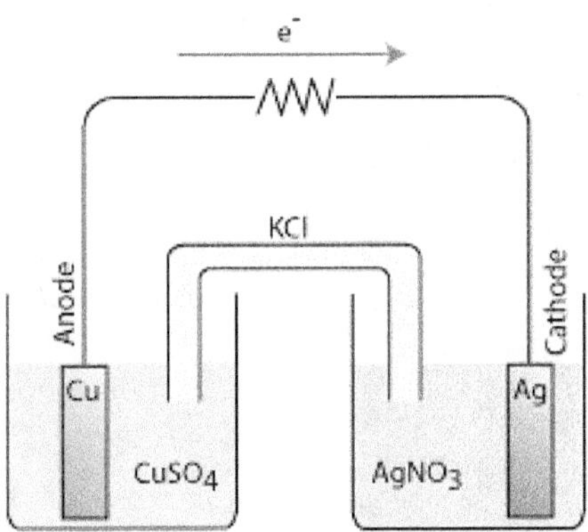

Galvanic (Voltaic) Cell

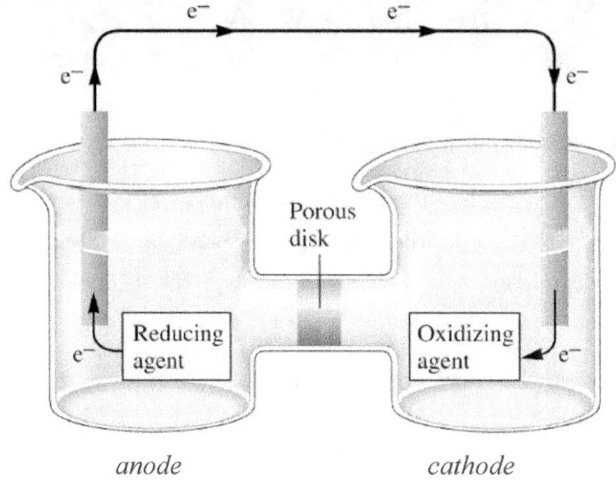

Oxidation-reduction reaction examples

For example, spontaneous oxidation-reduction reactions:

$$Zn\,(s) + CuSO_4\,(aq) \rightarrow Cu\,(s) + ZnSO_4\,(aq)$$

$$Zn\,(s) + Cu\cancel{SO_4}\,(aq) \rightarrow Cu\,(s) + Zn\cancel{SO_4}\,(aq)$$

$$Zn\,(s) + Cu^{2+}\,(aq) \rightarrow Cu\,(s) + Zn^{2+}\,(aq)$$

$$Cr\,(s) + 3\,AgNO_3\,(aq) \rightarrow Cr(NO_3)_3\,(aq) + 3\,Ag\,(s)$$

$$Cr\,(s) + 3\,Ag\cancel{NO_3}\,(aq) \rightarrow Cr\cancel{(NO_3)_3}\,(aq) + 3\,Ag\,(s)$$

$$Cr\,(s) + 3\,Ag^{+}\,(aq) \rightarrow Cr^{3+}\,(aq) + 3\,Ag\,(s)$$

$$MnO_4^{-}\,(aq) + 5\,Fe^{2+}\,(aq) + 8\,H^{+}\,(aq) \rightarrow Mn^{2+}\,(aq) + 5\,Fe^{3+}\,(aq) + 4\,H_2O\,(l)$$

$$Zn\,(s) + 2\,MnO_2\,(s) \rightarrow ZnO\,(s) + Mn_2O_3\,(s)$$

Redox half-reactions

Oxidation half-reaction is the species losing electrons (increases charge).

Oxidation example,

$$Cu \rightarrow Cu^{2+} + 2\,e^{-}$$

Reduction half-reaction is the species gaining electrons (decreases charge).

Reduction example,

$$2\,Ag^{+} + 2\,e^{-} \rightarrow 2\,Ag$$

Copyright © Sterling Test Prep.

Ten steps for balancing redox half-reactions

1. Split the equation into two half-reactions.

2. Balance atoms other than O or H.

3. Balance O using H_2O. Add water as needed to balance oxygens on the side deficient in O.

4. If in an acidic solution, balance hydrogens using H^+.

    Add H^+ ions as needed to the side deficient in H.

5. In basic solution, add OH^- ions equal to the number of H^+ ions to each side of the chemical equation.

6. The mass is balanced. To balance charge, determine the charge on each side of the reaction and add needed electrons to the positive side, so charge is the same.

7. Repeat steps 1 through 4 for each half-reaction.

8. If electrons lost do not equal the number gained, multiply each half-reaction by the necessary factor.

9. Sum half-reactions to obtain the balanced net ionic reaction. Inspect H^+ ions and H_2O molecules to cancel.

10. Check the final equation for mass and charge balance.

Important: If expressing H^+ as H_3O^+, change the number of H^+ into the same number of H_3O^+.

Add that same number of H_2O to the opposite side of the equation.

Redox reactions can be separated into two half-reactions:

Example 1:

$$\textit{Oxidation half-reaction:}\quad Zn\,(s) \rightarrow Zn^{2+}\,(aq) + 2\ e^-$$

$$\textit{Reduction half-reaction:}\quad Cu^{2+}\,(aq) + 2\ e^- \rightarrow Cu\,(s)$$

$$\text{Overall reaction:}\quad Zn\,(s) + Cu^{2+}\,(aq) \rightarrow Cu\,(s) + Zn^{2+}\,(aq)$$

Example 2:

Oxidation half-reaction:

$$5\ Fe^{2+}(aq) \rightarrow 5\ Fe^{3+}(aq) + 5\ e^-$$

Reduction half-reaction:

$$MnO_4^-(aq) + 8\ H^+(aq) + 5\ e^- \rightarrow Mn^{2+}(aq) + 4\ H_2O\ (l)$$

Overall reaction:

$$MnO_4^-(aq) + 5\ Fe^{2+}(aq) + 8\ H^+(aq) \rightarrow Mn^{2+}(aq) + 5\ Fe^{3+}(aq) + 4\ H_2O\ (l)$$

Galvanic cell half-reactions

Redox reactions are split into two half-reactions in *galvanic cells*, each in two separate compartments as *half-cells*. The chemical energy drives electrons through the external circuit connecting the two half-cells, thus producing an electric current.

For example, consider the reaction:

$$Zn\ (s) + CuSO_4\ (aq) \rightarrow Cu\ (s) + ZnSO_4\ (aq)$$

If zinc metal is placed in the $CuSO_4$ solution, an exothermic reaction occurs, producing heat. In a galvanic cell, zinc metal is placed in a $ZnSO_4$ solution in one container, a copper metal in $CuSO_4$ solution in another container. A wire connects two metals to complete the circuit; the two solutions are connected by a salt bridge containing strong electrolytes such as $KCl\ (aq)$ or $K_2SO_4\ (aq)$ that allow ions to flow between the two half-cells. The reactants – zinc metal and copper ions – are not allowed direct contact.

Due to the potential difference between the two half-cells, electrons are forced to flow from the zinc electrode (the anode) to the copper electrode (the cathode).

At the interface between the zinc electrode and $ZnSO_4$ solution, Zn atoms are oxidized to Zn^{2+} ions; at the interface of the copper electrode and the $CuSO_4$ solution, Cu^{2+} ions combine with incoming electrons reduced to Cu atoms.

These oxidation and reduction processes as half-reaction equations:

Anode-reaction (oxidation):

$$Zn\ (s) \rightarrow Zn^{2+}(aq) + 2\ e^-$$

Cathode-reaction (reduction):

$$Cu^{2+}(aq) + 2\ e^- \rightarrow Cu\ (s)$$

Copyright © Sterling Test Prep.

Cell notation

The two half-cells are represented by notations $Zn|Zn^{2+}$ and $Cu^{2+}|Cu$.

In a galvanic cell, the more readily oxidized metal serves as an anode and the other as a cathode.

Determine the species' tendency to oxidize from the standard reduction potentials table.

Zinc is more easily oxidized than copper, serving as the anode, while copper forms the cathode.

> Oxidation half-reaction occurs at the anode

> Reduction half-reaction at the cathode.

This galvanic cell is represented using the following cell notation:

> $Zn|ZnSO_4\,(aq)||CuSO_4\,(aq)|Cu$

Cell notation has the anode on the left side, with the cathode on the right.

At the anode half-cell, Zn^{2+} ions are continuously formed, creating an excess of positive ions.

Cu2+ ions are continuously reduced to Cu at the cathode half-cell, causing a decrease in cation concentration.

To maintain electrically neutral solutions in half-cells, anions flow into the anode half-cell, and cations flow into the cathode half-cell. Thus, an electric current is the flow of charged particles. Electrons flow from anode to cathode through the wire; cations and anions flow in the opposite direction through the salt bridge.

Any duplication (copies, uploads, PDFs) is illegal.

Notes for active learning

Copyright © Sterling Test Prep.

Reduction Potential

Standard reduction potential

The *standard half-cell potential* (or *reduction potential*) is determined by connecting the half-cell (under standard conditions) to the *standard hydrogen electrode* (SHE) as a reference half-cell.

This reference half-cell consists of a Pt-electrode in a solution containing 1 M H^+, into which H_2 gas is purged at a constant pressure of 1 atm.

Since the reference half-cell is assigned zero potential, the cell potential measured under these conditions is the substance's standard half-cell potential.

For example, when a $Zn|Zn^{2+}(aq, 1 M)$ half-cell is connected to the reference half-cell (SHE), the voltage measured at 25 °C is found to be 0.763 V.

Electron flows from $Zn|Zn^{2+}$ to SHE; relative to SHE, the reduction potential for $Zn|Zn^{2+}$ half-cell is the negative value of the measured voltage. ($E°$ = reduction potential)

$$Zn^{2+}(aq) + 2\ e^- \rightarrow Zn\ (s) \qquad E° = -076\ V$$

$$2\ H^+(aq) + 2\ e^- \rightarrow H_2\ (s) \qquad E° = 0.000\ V$$

$$Cu^{2+}(aq) + 2\ e^- \rightarrow Cu\ (s) \qquad E° = 0.34\ V$$

Since values are obtained against the standard hydrogen potential, species with positive reduction potentials are easier to reduce than H^+ ions.

In contrast, those with negative reduction potentials are more difficult to reduce.

Therefore, relative to H^+, Cu^{2+} is easier to reduce, whereas Zn^{2+} is challenging. When $Zn|Zn^{2+}$ is connected to $Cu|Cu^{2+}$ half-cells, the spontaneous process is the flow of e^- from $Zn|Zn^{2+}$ to $Cu|Cu^{2+}$ half-cell.

Half-reactions

A half-reaction is the oxidation or reduction reaction component of a redox reaction. A half-reaction is derived from the change in oxidation states of individual substances in the redox reaction.

Balancing redox reactions requires splitting the equation into the two half-reactions of reduction and oxidation.

In general, the half-reactions with the most positive (or the least negative) reduction potential occurs before others.

Any duplication (copies, uploads, PDFs) is illegal.

The *kinetic factor* drives the reaction; if the difference in the standard reduction potentials between two half-cell reactions is small, the concentration may determine the outcome of electrolysis.

For example, the following reactions are possible for the electrolysis of 1 M NaCl (*aq*):

At anode:	$2\,Cl^-\,(aq) \rightarrow Cl_2\,(g) + 2\,e^-$	$E° = -1.36\ V$
	$2\,H_2O \rightarrow O_2\,(g) + 4\,H^+\,(aq) + 4\,e^-$	$E° = -1.23\ V$
At cathode:	$2\,H_2O + 2\,e^- \rightarrow H_2\,(g) + 2\,OH^-\,(aq)$	$E° = -0.83\ V$
	$Na^+\,(aq) + e^- \rightarrow Na\,(s)$	$E° = -2.71\ V$

In the electrolysis of aqueous sodium chloride, Cl_2 forms at the anode instead of O_2, although $E°$ (-1.23 V) for the formation of O_2 from H_2O is less negative than for the formation of Cl_2 from Cl^- (-1.36 V).

The formation of O_2 involves higher activation energy (overvoltage); thus, kinetically less favorable.

Therefore, H_2 is formed at the cathode since it requires less voltage than the reduction of $Na^+(aq)$.

For example, suppose a solution containing metal cations that require lower potential (or one with a positive reduction potential) is electrolyzed.

The metal deposits on the cathode, and no hydrogen are produced. The electrolysis of aqueous copper(II) chloride solution yields Cl_2 gas at the cathode's anode and copper metal deposits.

If the anode is copper, it is oxidized because of the lower voltage requirements relative to water.

Anode reaction:	$Cu\,(s) \rightarrow Cu^{2+}\,(aq) + 2\,e^-$	$E° - -0.34\ V$
Cathode reaction:	$Cu^{2+}\,(aq) + 2\,e^- \rightarrow Cu\,(s)$	$E° = +0.34\ V$

Copyright © Sterling Test Prep.

Electrical Potential Difference

Cell potential

Electrons flow from one electrode to another in a galvanic cell because there is an electrical potential difference between the two half-cells, called the *cell potential.*

The magnitude of cell potential depends on the two half-cells' nature, the concentration of electrolytes in each half-cell, and temperature.

> Cell potential for galvanic (voltaic) cells is *positive* because the voltaic cell generates potential.

> Cell potential for electrolytic cells is *negative* because it requires potential input.

The *standard cell potential*, $E°_{cell}$, is measured under standard conditions (1 atm pressure for gas, 1 M of electrolytes, and 25 °C). The parameters for a standard cell require the species in solution to have a concentration of 1 M, the partial pressure of gases to be 1 atm, and the temperature to be 25 °C (298 K).

> Reduction potential = potential of the reduction half-reaction

> Oxidation potential = potential of the oxidation half-reaction

> Cell potential = reduction potential + oxidation potential

In the zinc-copper cell, electrons flow from $Zn|Zn^{2+}$ half-cell to $Cu|Cu^{2+}$ half-cell. The $Zn|Z^{2+}$ half-cell has a higher electrical potential than $Cu|Cu^{2+}$.

Using these conditions makes it possible to describe a standard electrode potential (E°) in volts and calculate the cell's electrical potential. The superscript ° denotes standard conditions.

Cell potential and Gibbs free energy

Cell potential measures the potential difference between half-cells. It indicates if the cell reaction is spontaneous or nonspontaneous. Spontaneity for a process is often associated with its Gibbs free energy change (ΔG).

The mathematical relationship between cell potential and ΔG is:

> $\Delta G° = -nF E°$ under standard conditions

> $\Delta G = -nF E$ under nonstandard conditions

where ΔG = change in Gibbs free energy in Joules, n = the moles of electrons transferred, the Faraday's constant $F = 96,485$ C/mol e⁻, E = cell potential

Any duplication (copies, uploads, PDFs) is illegal.

Maximum work produced

Faraday's constant (C/mol e$^-$) and volts (V) for cell potential gas ΔG units of coulomb-volts.

Coulomb-volts are equivalent to Joules.

A potential difference of 1 V is equivalent to 1 Joule of *work done per coulomb of charge* that flows between two points in the circuit. (1 V = 1 J/C or 1 J = 1 C·V)

Maximum work produced = charge × maximum potential

$$w_{max} = -q E_{max} = \Delta G$$

Electrical charge

$$q = nF$$

$$\Delta G = -nF E_{cell}$$

or

$$\Delta G° = -nF E°_{cell}$$

where n = mole of electrons transferred or that flow through circuit, Faraday's constant F = 96,485 C/mole e$^-$

For example:

$$Zn\,(s) + Cu^{2+}\,(aq) \rightarrow Zn^{2+}\,(aq) + Cu\,(s) \qquad E°_{cell} = 1.10 \text{ V}$$

$$\Delta G° = -2 \text{ mol e}^- \times (96{,}485 \text{ C / mol e}^-) \times 1.10 \text{ V}$$

$$\Delta G° = -2.12 \times 10^5 \text{ J}$$

2.12×10^2 kJ is the maximum work per mole of Zn reacted by Cu^{2+}.

Net cell potential

The *net cell potential* is the sum of the two half-cell potentials.

For the Zn-copper cell, the two half-cell potentials:

Anode half-cell reaction:

$$Zn\,(s) \rightarrow Zn^{2+}\,(aq) + 2 \text{ e}^- \qquad E°_{Zn \rightarrow Zn2+} = 0.76 \text{ V}$$

Cathode half-cell reaction:

$$Cu^{2+}\,(aq) + 2 \text{ e}^- \rightarrow Cu\,(s) \qquad E°_{Cu2+ \rightarrow Cu} = 0.34 \text{ V}$$

Copyright © Sterling Test Prep.

Overall cell reaction:

$$Zn\ (s) + Cu^{2+}\ (aq) \rightarrow\ Zn^{2+}\ (aq) + Cu\ (s)$$

$$E°_{cell} = E°_{Zn \rightarrow Zn2+} + E°_{Cu2+ \rightarrow Cu}$$

$$E°_{cell} = 0.76\ V + 0.34\ V$$

$$E°_{cell} = 1.10\ V$$

For $Cr|Cr^{3+}$ and $Ag|Ag^+$ half-cells, cell notation and potential:

$$Cr|Cr^{3+}\ (aq, 1\ M)\|Ag^+\ (aq, 1\ M)|Ag$$

$$\text{(anode)} \qquad \text{(cathode)}$$

Anode half-cell reaction:

$$Cr\ (s)\ \rightarrow Cr^{3+}\ (aq) + 3\ e^- \qquad E°_{Cr \rightarrow Cr3+} = 0.73\ V$$

Cathode half-cell reaction:

$$Ag^+\ (aq) + e^- \rightarrow\ Ag\ (s) \qquad E°_{Ag+ \rightarrow Ag} = 0.80\ V$$

Overall cell reaction:

$$Cr\ (s) + 3\ Ag^+\ (aq) \rightarrow\ Cr^{3+}\ (aq) + 3\ Ag\ (s)$$

$$E°_{cell} = E°_{Cr \rightarrow Cr3+} + E°_{Ag+ \rightarrow Ag}$$

$$E°_{cell} = 0.73\ V + 0.80\ V$$

$$E°_{cell} = 1.53\ V$$

The standard electrode potentials are usually listed as reduction potentials relative to a standard hydrogen electrode (SHE) with 0 volts.

For example, use the following metal ion/metal reaction potentials, calculate the standard cell potential:

$$Co\ (s) + Cu^{2+}\ (aq) \rightarrow Co^{2+}\ (aq) + Cu\ (s)$$

$$Cu^{2+}\ (aq)|Cu\ (s) \quad Ag^+\ (aq)|Ag\ (s) \quad Co^{2+}\ (aq)|Co\ (s) \quad Zn^{2+}\ (aq)|Zn\ (s)$$

$$+0.34\ V \qquad +0.80\ V \qquad -0.28\ V \qquad -0.76\ V$$

The cell reaction is:

$$Co\ (s) + Cu^{2+}\ (aq) \rightarrow Co^{2+}\ (aq) + Cu\ (s)$$

Separate it into half reactions:

$$Cu^{2+}\ (aq) \rightarrow Cu\ (s) \qquad\qquad Co\ (s) \rightarrow Co^{2+}\ (aq)$$

Any duplication (copies, uploads, PDFs) is illegal.

The potentials provided are written in this format:

$$Cu^{2+} (aq) \mid Cu (s)$$

$$+0.34 \text{ V}$$

It means that for the reduction reaction:

$$Cu^{2+} (aq) \rightarrow Cu (s), \text{ the potential is } +0.34 \text{ V}$$

The reverse reaction or oxidation reaction is:

$$Cu (s) \rightarrow Cu^{2+} (aq) \text{ has opposing potential value: } -0.34 \text{ V}$$

Obtain the potential values for each half-reaction.

Reverse sign for potential values of oxidation reactions:

$$Cu^{2+} (aq) \rightarrow Cu (s) = +0.34 \text{ V} \quad \text{(reduction)}$$

$$Co (s) \rightarrow Co^{2+} (aq) = +0.28 \text{ V} \quad \text{(oxidation)}$$

Determine the standard cell potential:

Standard cell potential = sum of half-reaction potential

Standard cell potential = 0.34 V + 0.28 V

Standard cell potential = 0.62 V

Nernst equation

The Nernst Equation is used if the temperature is standard (298 K) and uses the natural logarithm (ln) or common logarithm, base 10 (log).

The Nernst equation is used for a cell not under standard conditions:

$$E_{cell} = E_{cell}^{\circ} - (RT / nF) \ln Q$$

where E_{cell} = nonstandard cell potential in volts; E°_{cell} = standard cell potential in volts; R = constant, 8.314 J/mol·K, T = temperature in K; n = moles of electrons transferred; F = Faraday's constant = 96,485 C/mol e^- (C represents coulombs) and Q = the reaction quotient expression.

$$E_{cell} = E^{\circ}_{cell} - [(0.0257 \text{ V}) / n] \ln Q$$

or

$$E_{cell} = E^{\circ}_{cell} - [(0.0592 \text{ V}) / n] \log Q$$

Copyright © Sterling Test Prep.

Modifying one of the previous cells, as follows:

$$\text{Zn (s)} \mid \text{Zn}^{2+} (aq), \, 3.00\text{M}; \, \text{SO}_4^{2-} (3.00\text{M}) \parallel \text{Cu}^{2+} (aq), \, 0.0015\text{M}; \, \text{SO}_4^{2-} (0.0015\text{M}) \mid \text{Cu (s)}$$

$$E^{\circ}_{cell} = +1.10\text{V}$$

For standard temperature (298 K), the Nernst equation expression:

$$E_{cell} = E^{\circ}_{cell} - [(0.0257 \text{ V}) / n] \ln [\text{Zn}^{2+}] / [\text{Cu}^{2+}]$$

$$E_{cell} = +1.10 \text{ V} - (0.0257 \text{ V} / 2 \text{ mol e}^-) \ln (3.00 \text{ M} / 0.0015 \text{ M})$$

Evaluating the above expression gives:

$$E_{cell} = +1.10 \text{ V} - 0.098 \text{ V}$$

$$E_{cell} = 1.00 \text{ V}$$

Table of standard reduction potentials

Reduction potential measures the tendency of a chemical species to acquire electrons and become reduced (i.e., the gain of electrons is a reduction).

Reduction potential is measured in volts (V) or millivolts (mV).

The more positive the E°, the higher the tendency for the species to be reduced (a strong oxidizing agent).

The more negative the E°, the higher the tendency for the species to be oxidized (i.e., strong reducing agent).

Standard Reduction Potentials in Aqueous Solution at 25 °C *	
Acidic Solutions	**E° (V)**
$\text{F}_2 (g) + 2 \text{ e}^- \rightarrow 2 \text{ F}^- (aq)$	+2.87
$\text{Co}^{3+} (aq) + \text{e}^- \rightarrow \text{Co}^{2+} (aq)$	+1.82
$\text{Pb}^{4+} (aq) + 2 \text{ e}^- \rightarrow \text{Pb}^{2+} (aq)$	+1.8
$\text{H}_2\text{O}_2 (aq) + 2 \text{ H}^+ (aq) + 2 \text{ e}^- \rightarrow 2 \text{ H}_2\text{O}$	+1.77
$\text{NiO}_2 (s) + 4 \text{ H}^+ (aq) + 2 \text{ e}^- \rightarrow \text{Ni}^{2+} (aq) + 2 \text{ H}_2\text{O}$	+1.7
$\text{PbO}_2 (s) + \text{SO}_4^{2-} (aq) + 4 \text{ H}^+ (aq) + 2 \text{ e}^- \rightarrow \text{PbSO}_4 (s) + 2 \text{ H}_2\text{O}$	+1.69
$\text{Au}^+ (aq) + \text{e}^- \rightarrow \text{Au} (s)$	+1.68
$2 \text{ HClO} (aq) + 2 \text{ H}^+ (aq) + 2 \text{ e}^- \rightarrow \text{Cl}_2 (g) + 2 \text{ H}_2\text{O}$	+1.63
$\text{Ce}^{4+} (aq) + \text{e}^- \rightarrow \text{Ce}^{3+} (aq)$	+1.61
$\text{NaBiO}_3 (s) + 6 \text{ H}^+ (aq) + 2 \text{ e}^- \rightarrow \text{Bi}^{3+} (aq) + \text{Na}^+ (aq) + 3 \text{ H}_2\text{O}$	+1.6

Any duplication (copies, uploads, PDFs) is illegal.

$MnO_4^- (aq) + 8 H^+ (aq) + 5 e^- \rightarrow Mn^{2+} (aq) + 4 H_2O$	+1.51
$Au^{3+} (aq) + 3 e^- \rightarrow Au (s)$	+1.5
$ClO_3^- (aq) + 6 H^+ (aq) + 5 e^- \rightarrow 1/2 Cl_2 (g) + 3 H_2O$	+1.47
$BrO_3^- + 6 H^+ (aq) + 6 e^- \rightarrow Br^- (aq) + 3 H_2O$	+1.44
$Cl_2 (g) + 2 e^- \rightarrow 2 Cl^- (aq)$	+1.36
$Cr_2O_7^{2-} + 14 H^+ (aq) + 6 e^- \rightarrow 2 Cr^{3+} (aq) + 7 H_2O$	+1.33
$N_2H_5^+ (aq) + 3 H^+ (aq) + 2 e^- \rightarrow 2 NH_4^+ (aq)$	+1.24
$MnO_2 (s) + 4 H^+ (aq) + 2 e^- \rightarrow Mn^{2+} (aq) + 2 H_2O$	+1.23
$O_2 (g) + 4 H^+ (aq) + 4 e^- \rightarrow 2 H_2O$	+1.23
$Pt^{2+} (aq) + 2 e^- \rightarrow Pt (s)$	+1.2
$IO_3^- (aq) + 6 H^+ (aq) + 5 e^- \rightarrow \frac{1}{2} I_2 (aq) + 3 H_2O$	+1.19
$ClO_4^- (aq) + 2 H^+ (aq) + 2 e^- \rightarrow ClO_3^- (aq) + H_2O$	+1.19
$Br_2 (l) + 2 e^- \rightarrow 2 Br^- (aq)$	+1.07
$AuCl_4^- + 3 e^- \rightarrow Au (s) + 4 Cl^- (aq)$	+1
$Pd^{2+} (aq) + 2 e^- \rightarrow Pd (s)$	+0.99
$NO_3^- (aq) + 4 H^+ (aq) + 3 e^- \rightarrow NO (g) + 2 H_2O$	+0.96
$NO_3^- (aq) + 3 H^+ (aq) + 2 e^- \rightarrow HNO_2 (aq) + H_2O$	+0.94
$2 Hg^{2+} (aq) + 2 e^- \rightarrow Hg_2^{2+} (aq)$	+0.92
$Hg^{2+} (aq) + 2 e^- \rightarrow Hg (l)$	+0.86
$Ag^+ (aq) + e^- \rightarrow Ag (s)$	+0.79
$Hg_2^{2+} (aq) + 2 e^- \rightarrow 2 Hg (l)$	+0.79
$Fe^{3+} (aq) + e^- \rightarrow Fe^{2+} (aq)$	+0.77
$SbCl_6^- (aq) + 2 e^- \rightarrow SbCl_4^- (aq) + 2 Cl^- (aq)$	+0.75
$[PtCl_4]^{2-} (aq) + 2 e^- \rightarrow Pt (s) + 4 Cl^- (aq)$	+0.73
$O_2 (g) + 2 H^+ (aq) + 2 e^- \rightarrow H_2O_2 (aq)$	+0.68
$[PtCl_6]^{2-} (aq) + 2 e^- \rightarrow [PtCl_4]^{2-} (aq) + 2 Cl^- (aq)$	+0.68
$H_3AsO_4 (aq) + 2 H^+ (aq) + 2 e^- \rightarrow H_3AsO_3 (aq) + H_2O$	+0.58
$I_2 (s) + 2 e^- \rightarrow 2 I^- (aq)$	+0.54
$TeO_2 (s) + 4 H^+ (aq) + 4 e^- \rightarrow Te (s) + 2 H_2O$	+0.53
$Cu^+ (aq) + e^- \rightarrow Cu (s)$	+0.521

Copyright © Sterling Test Prep.

Reaction	$E°$
$[RhCl_6]^{3-} (aq) + 3\ e^- \rightarrow Rh\ (s) + 6\ Cl^- (aq)$	+0.44
$Cu^{2+} (aq) + 2\ e^- \rightarrow Cu\ (s)$	+0.34
$HgCl_2 (s) + 2\ e^- \rightarrow 2\ Hg\ (l) + 2\ Cl^- (aq)$	+0.27
$AgCl\ (s) + e^- \rightarrow Ag\ (s) + Cl^- (aq)$	+0.22
$SO_4^{2-} (aq) + 4\ H^+ (aq) + 2\ e^- \rightarrow SO_2 (g) + 2\ H_2O$	+0.2
$SO_4^{2-} (aq) + 4\ H^+ (aq) + 2\ e^- \rightarrow H_2SO_3 (g) + H_2O$	+0.17
$Cu^{2+} (aq) + e^- \rightarrow Cu^+ (aq)$	+0.15
$Sn^{4+} (aq) + 2\ e^- \rightarrow Sn^{2+} (aq)$	+0.15
$S\ (s) + 2\ H^+ (aq) + 2\ e^- \rightarrow H_2S\ (aq)$	+0.14
$AgBr\ (s) + e^- \rightarrow Ag\ (s) + Br^- (aq)$	+0.07
$2\ H^+ (aq) + 2\ e^- \rightarrow H_2 (g)$ (reference electrode)	0.0
$N_2O\ (g) + 6\ H^+ (aq) + H_2O + 4\ e^- \rightarrow 2\ NH_3OH^+ (aq)$	−0.05
$Pb^{2+} (aq) + 2\ e^- \rightarrow Pb\ (s)$	−0.13
$Sn^{2+} (aq) + 2\ e^- \rightarrow Sn\ (s)$	−0.14
$AgI\ (s) + e^- \rightarrow Ag\ (s) + I^- (aq)$	−0.15
$Sn^{4+} (aq) + 2\ e^- \rightarrow Sn^{2+} (aq)$	+0.15
$S\ (s) + 2\ H^+ (aq) + 2\ e^- \rightarrow H_2S\ (aq)$	+0.14
$AgBr\ (s) + e^- \rightarrow Ag\ (s) + Br^- (aq)$	+0.07
$2\ H^+ (aq) + 2\ e^- \rightarrow H_2 (g)$ (reference electrode)	0.0
$N_2O\ (g) + 6\ H^+ (aq) + H_2O + 4\ e^- \rightarrow 2\ NH_3OH^+ (aq)$	−0.05
$Pb^{2+} (aq) + 2\ e^- \rightarrow Pb\ (s)$	−0.13
$Sn^{2+} (aq) + 2\ e^- \rightarrow Sn\ (s)$	−0.14
$AgI\ (s) + e^- \rightarrow Ag\ (s) + I^- (aq)$	−0.15
$[SnF_6]^{2-} (aq) + 4\ e^- \rightarrow Sn\ (s) + 6\ F^- (aq)$	−0.25
$Ni^{2+} (aq) + 2\ e^- \rightarrow Ni\ (s)$	−0.25
$Co^{2+} (aq) + 2\ e^- \rightarrow Co\ (s)$	−0.28
$Tl^+ (aq) + e^- \rightarrow Tl\ (s)$	−0.34
$PbSO_4 (s) + 2\ e^- \rightarrow Pb\ (s) + SO_4^{2-} (aq)$	−0.36
$Se\ (s) + 2\ H^+ (aq) + 2\ e^- \rightarrow H_2Se\ (aq)$	−0.4
$Cd^{2+} (aq) + 2\ e^- \rightarrow Cd\ (s)$	−0.40

Any duplication (copies, uploads, PDFs) is illegal.

Reaction	$E°$ (V)
$Cr^{3+}(aq) + e^- \rightarrow Cr^{2+}(aq)$	−0.41
$Fe^{2+}(aq) + 2\,e^- \rightarrow Fe\,(s)$	−0.44
$2\,CO_2\,(g) + 2\,H^+(aq) + 2\,e^- \rightarrow (COOH)_2\,(aq)$	−0.49
$Ga^{3+}(aq) + 3\,e^- \rightarrow Ga\,(s)$	−0.53
$HgS\,(s) + 2\,H^+(aq) + 2\,e^- \rightarrow Hg\,(l) + H_2S\,(g)$	−0.72
$Cr^{3+}(aq) + 3\,e^- \rightarrow Cr\,(s)$	−0.74
$Zn^{2+}(aq) + 2\,e^- \rightarrow Zn\,(s)$	−0.76
$2\,H_2O\,(l) + 2\,e^- \rightarrow H_2\,(g) + 2\,OH^-(aq)$	−0.83
$Cr^{2+}(aq) + 2\,e^- \rightarrow Cr\,(s)$	−0.91
$Mn^{2+}(aq) + 2\,e^- \rightarrow Mn\,(s)$	−1.18
$V^{2+}(aq) + 2\,e^- \rightarrow V\,(s)$	−1.18
$Zr^{4+}(aq) + 4\,e^- \rightarrow Zr\,(s)$	−1.53
$Al^{3+}(aq) + 3\,e^- \rightarrow Al\,(s)$	−1.66
$H_2\,(g) + 2\,e^- \rightarrow 2\,H^-(aq)$	−2.25
$Mg^{2+}(aq) + 2\,e^- \rightarrow Mg\,(s)$	−2.37
$Na^+(aq) + e^- \rightarrow Na\,(s)$	−2.71
$Ca^{2+}(aq) + 2\,e^- \rightarrow Ca\,(s)$	−2.87
$Sr^{2+}(aq) + 2\,e^- \rightarrow Sr\,(s)$	−2.89
$Ba^{2+}(aq) + 2\,e^- \rightarrow Ba\,(s)$	−2.9
$Rb^+(aq) + e^- \rightarrow Rb\,(s)$	−2.93
$K^+(aq) + e^- \rightarrow K\,(s)$	−2.93
$Li^+(aq) + e^- \rightarrow Li\,(s)$	−3.05
Basic Solutions	**$E°$ (V)**
$ClO^-(aq) + H_2O + 2\,e^- \rightarrow Cl^-(aq) + 2\,OH^-(aq)$	+0.89
$OOH^-(aq) + H_2O + 2\,e^- \rightarrow 3\,OH^-(aq)$	+0.88
$2\,NH_2OH\,(aq) + 2\,e^- \rightarrow N_2H_4\,(aq) + 2\,OH^-(aq)$	+0.74
$ClO_3^-(aq) + 3\,H_2O + 6\,e^- \rightarrow Cl^-(aq) + 6\,OH^-(aq)$	+0.62
$MnO_4^-(aq) + 2\,H_2O + 3\,e^- \rightarrow MnO_2\,(s) + 4\,OH^-(aq)$	+0.59
$MnO_4^-(aq) + e^- \rightarrow MnO_4^{2-}(aq)$	+0.56
$NiO_2\,(s) + 2\,H_2O + 2\,e^- \rightarrow Ni(OH)_2\,(s) + 2\,OH^-(aq)$	+0.49

Copyright © Sterling Test Prep.

$Ag_2CrO_4\,(s) + 2\,e^- \;\rightarrow\; 2\,Ag\,(s) + CrO_4^{2-}\,(aq)$	+0.45
$O_2\,(g) + 2\,H_2O + 4\,e^- \;\rightarrow\; 4\,OH^-\,(aq)$	+0.4
$ClO_4^-\,(aq) + H_2O + 2\,e^- \;\rightarrow\; ClO_3^-\,(aq) + 2\,OH^-\,(aq)$	+0.36
$Ag_2O\,(s) + H_2O + 2\,e^- \;\rightarrow\; 2\,Ag\,(s) + 2\,OH^-\,(aq)$	+0.34
$2\,NO_2^-\,(aq) + 3\,H_2O + 4\,e^- \;\rightarrow\; N_2O\,(g) + 6\,OH^-\,(aq)$	+0.15
$N_2H_4\,(aq) + 2\,H_2O + 2\,e^- \;\rightarrow\; 2\,NH_3\,(aq) + 2\,OH^-\,(aq)$	+0.1
$[Co(NH_3)_6]^{3+}\,(aq) + e^- \;\rightarrow\; [Co(NH_3)_6]^{2+}\,(aq)$	+0.1
$HgO\,(s) + H_2O + 2\,e^- \;\rightarrow\; Hg\,(l) + 2\,OH^-\,(aq)$	+0.09
$O_2\,(g) + H_2O + 2\,e^- \;\rightarrow\; OOH^-\,(aq) + OH^-\,(aq)$	+0.08
$NO_3^-\,(aq) + H_2O + 2\,e^- \;\rightarrow\; NO_2^-\,(aq) + 2\,OH^-\,(aq)$	+0.01
$MnO_2\,(s) + 2\,H_2O + 2\,e^- \;\rightarrow\; Mn(OH)_2\,(s) + 2\,OH^-\,(aq)$	−0.05
$CrO_4^{2-}\,(aq) + 4\,H_2O + 3\,e^- \;\rightarrow\; Cr(OH)_3\,(s) + 5\,OH^-\,(aq)$	−0.12
$Cu(OH)_2\,(s) + 2\,e^- \;\rightarrow\; Cu\,(s) + 2\,OH^-\,(aq)$	−0.36
$Fe(OH)_3\,(s) + e^- \;\rightarrow\; Fe(OH)_2\,(s) + OH^-\,(aq)$	−0.56
$2\,H_2O + 2\,e^- \;\rightarrow\; H_2\,(g) + 2\,OH^-\,(aq)$	−0.83
$2\,NO_3^-\,(aq) + 2\,H_2O + 2\,e^- \;\rightarrow\; N_2O_4\,(g) + 4\,OH^-\,(aq)$	−0.85
$Fe(OH)_2\,(s) + 2\,e^- \;\rightarrow\; Fe\,(s) + 2\,OH^-\,(aq)$	−0.88
$SO_4^{2-}\,(aq) + H_2O + 2\,e^- \;\rightarrow\; SO_3^{2-}\,(aq) + 2\,OH^-\,(aq)$	−0.93
$N_2\,(g) + 4\,H_2O + 4\,e^- \;\rightarrow\; N_2H_4\,(aq) + 4\,OH^-\,(aq)$	−1.15
$[Zn(OH)_4]^{2-}\,(aq) + 2\,e^- \;\rightarrow\; Zn\,(s) + 4\,OH^-\,(aq)$	−1.22
$Zn(OH)_2\,(s) + 2\,e^- \;\rightarrow\; Zn\,(s) + 2\,OH^-\,(aq)$	−1.25
$[Zn(CN)_4]^{2-}\,(aq) + 2\,e^- \;\rightarrow\; Zn\,(s) + 4\,CN^-\,(aq)$	−1.26
$Cr(OH)_3\,(s) + 3\,e^- \;\rightarrow\; Cr\,(s) + 3\,OH^-\,(aq)$	−1.3
$SiO_3^{2-}\,(aq) + 3\,H_2O + 4\,e^- \;\rightarrow\; Si\,(s) + 6\,OH^-\,(aq)$	−1.7

**table is for reference only, and values are approximate*

Any duplication (copies, uploads, PDFs) is illegal.

Notes for active learning

Copyright © Sterling Test Prep.

Electron Flow

Direction of electron flow

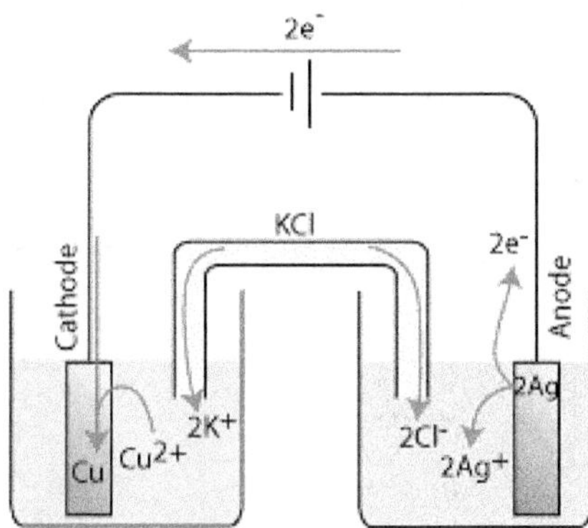

Electrolytic charge flow

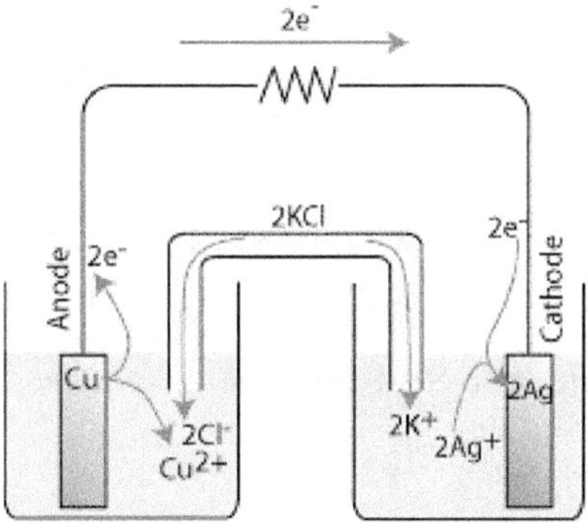

Galvanic (voltaic) charge flow

Electrons flow from anode to cathode.

Mnemonic: *A to C in alphabetical order*.

Alternatively, AC power as *the A comes first and stands for Anode*.

Oxidation (at the anode) produces electrons (and cations) and releases electrons toward the cathode.

The cathode receives those electrons and uses them for reduction.

The species with the highest oxidation potential (lowest reduction potential) is the anode, and the species with the highest reduction potential is the cathode.

In the diagram above, the galvanic (voltaic) cell shows a natural flow because Cu (higher oxidation potential/lower reduction potential) is the anode, and Ag (higher reduction potential) is the cathode.

The electrolytic cell shows the opposite; a battery drives the reaction to force Cu to be the cathode and Ag to be the anode. Electrons flow in wires and electrodes, while ions flow in an electrolyte solution, thus creating a completed circuit.

The oxidation-reduction reaction continues until the electrolyte concentrations in the half-cells become equal.

At *anode half-cell*:

$$Cu\ (s)\ \rightarrow\ Cu^{2+}\ (aq) + 2\ e^-\ (\text{in } 0.0010\ M\ Cu^{2+})$$

At *cathode half-cell*:

$$Cu^{2+}\ (aq) + 2\ e^-\ \rightarrow\ Cu\ (s)\ (\text{in } 0.50\ M\ Cu^{2+})$$

Any duplication (copies, uploads, PDFs) is illegal.

Electromotive force and voltage

Electromotive force (emf), or cell potential (E_{cell}), is the driving force that enables electrons to flow from one electrode to the other, which has the unit volt (V).

A volt is one Joule/coulomb, where the coulomb is the unit of charge.

Voltage is the driving force for reactions, often called the EMF (electromotive force).

The higher the voltage, the greater the driving force.

A voltmeter must have its negative terminal connected to the anode and positive terminal to the cathode for the correct polarity ($\pm$ sign).

Voltage is positive for a spontaneous reaction.

Copyright © Sterling Test Prep.

Batteries

Batteries as galvanic cells

Batteries are galvanic cells connected in series, where the battery potential equals the cells' potentials.

Three types of batteries:

Primary batteries are not rechargeable.

Secondary batteries are rechargeable.

Fuel cells need an ample supply of fuel to provide the energy.

Primary batteries include common (acidic) dry, alkaline, and mercury batteries. A new dry cell battery has a potential of about 1.5 V regardless of size, but the amount of energy a battery can deliver depends on its size. A D battery delivers more current (greater amperes) than an AAA battery.

Standard dry batteries use aqueous NH_4Cl paste as an electrolyte and are acidic batteries due to the ionization:

$$NH_4^+ \, (aq) \rightarrow NH_3 \, (aq) + H^+ \, (aq)$$

Acidic dry batteries consist of Zinc casing as an anode, the reducing agent; graphite rod as an (inert) cathode; aqueous NH_4Cl paste as an electrolyte, and MnO_2 powder as the oxidizing agent.

$$Zn|Zn^{2+},NH_4^+,NH_3 \, (aq)\|Mn_2O_3,MnO_2|C \, (s)$$

Anode reaction:

$$Zn \, (s) \rightarrow Zn^{2+} \, (aq) + 2 \, e^-$$

Cathodic reaction:

$$2 \, MnO_2 \, (s) + 2 \, NH_4^+ \, (aq) + 2 \, e^- \rightarrow Mn_2O_3 \, (s) + 2 \, NH_3 \, (aq) + H_2O \, (l)$$

Net reaction:

$$Zn \, (s) + 2 \, MnO_2 \, (s) + 2 \, NH_4^+ \, (aq) \rightarrow Zn^{2+} \, (aq) + Mn_2O_3 \, (s) + 2 \, NH_3 \, (aq) + H_2O \, (l)$$

The reverse reaction is prevented by forming $[Zn(NH_3)_4]^{2+}$ ions.

Any duplication (copies, uploads, PDFs) is illegal.

Alkaline batteries use zinc (reducing agent) and MnO_2 (oxidizing agent), but an aqueous KOH paste (not NH_4Cl) as the electrolyte.

Reaction at anode: $\quad Zn\,(s) + 2\,OH^-\,(aq) \rightarrow ZnO\,(s) + H_2O\,(l) + 2\,e^-$

Reaction a cathode: $\quad 2\,MnO_2\,(s) + H_2O\,(l) + 2\,e^- \rightarrow Mn_2O_3\,(s) + 2\,OH^-\,(aq)$

Net reaction: $\quad Zn\,(s) + 2\,MnO_2\,(s) \rightarrow ZnO\,(s) + Mn_2O_3\,(s)$

Since the reactants involved are solid, alkaline batteries deliver a constant voltage until the limiting reactant is exhausted. They last longer because zinc metal corrodes more slowly under basic conditions.

Mercury batteries are used in calculators and watches.

The following reactions occur at the anode and cathode sections of the cell:

Anode reaction: $\quad Zn\,(s) + 2\,OH^-\,(aq) \rightarrow ZnO\,(s) + H_2O + 2\,e^-$

Cathode reaction: $HgO\,(s) + H_2O + 2\,e^- \rightarrow Hg\,(l) + 2\,OH^-\,(aq)$

Net cell reaction: $\quad Zn\,(s) + HgO\,(s) \rightarrow ZnO\,(s) + Hg\,(l)$

Lead-storage batteries

Lead-storage batteries are used in automobiles and contain sulfuric acid as an electrolyte. Each battery cell contains several lead alloy grids with alternating grids packed with lead metal and lead(IV) oxide, PbO_2. The standard 12-V car battery contains six cells connected in series.

Each set of grids, which are electrodes, is connected in parallel, enabling the cell to deliver more current. The amount of current delivered depends on the electrode's surface area. Each cell in a lead storage battery has a potential of about 2.01 V.

Spontaneous reactions occurring in a lead storage battery:

Anode reaction:

$$Pb\,(s) + HSO_4^-\,(aq) \rightarrow PbSO_4\,(s) + H^+\,(aq) + 2\,e^-$$

Cathode reaction:

$$PbO_2\,(s) + 3\,H^+\,(aq) + HSO_4^-\,(aq) + 2\,e^- \rightarrow PbSO_4\,(s) + 2\,H_2O\,(m)$$

Net reaction:

$$Pb\,(s) + PbO_2\,(s) + 2\,H^+\,(aq) + 2\,HSO_4^-\,(aq) \rightarrow 2\,PbSO_4\,(s) + 2\,H_2O\,(l) \rightarrow \text{discharging}$$

Copyright © Sterling Test Prep.

The greatest advantage of lead-storage batteries (i.e., secondary batteries) is being rechargeable. However, while the car is driven, the battery obtains energy from the motor through the alternator. Starting the engine causes the discharge reaction to occur.

The following re-charging reaction occurs:

$$2\ PbSO_4\ (s) + 2\ H_2O\ (l)\ \rightarrow\ Pb\ (s) + PbO_2\ (s) + 2\ H^+\ (aq) + 2\ HSO_4^-\ (aq)$$

Lead-storage batteries have a longer lifetime than other batteries and deliver a large amount of current and electrical energy within a short time.

The major disadvantages of lead-storage batteries are:

1) they are heavy, bulky, and relatively non-portable

2) lead is a toxic metal, and disposal is an environmental problem

3) the battery must be kept upright because H_2SO_4 is corrosive

Nickel-cadmium batteries

Nickel-cadmium batteries are rechargeable batteries used in small appliances such as cordless phones. They contain cadmium as the anode and hydrated nickel oxide as the cathode.

The electrolyte is an aqueous KOH paste.

Anode reaction:

$$Cd\ (s) + 2\ OH^-\ (aq) \rightarrow Cd(OH)_2\ (s) + 2\ e^-$$

Cathode reaction:

$$2\ NiO(OH)\ (s) + 2\ H_2O\ (l) + 2\ e^- \rightarrow 2\ Ni(OH)_2\ (s) + 2\ OH^-\ (aq)$$

Net cell reaction:

$$Cd\ (s) + 2\ NiO(OH)\ (s) + 2\ H_2O\ (l) \rightarrow Cd(OH)_2\ (aq) + 2\ Ni(OH)_2\ (aq)$$

Any duplication (copies, uploads, PDFs) is illegal.

Notes for active learning

Copyright © Sterling Test Prep.

Practice Questions

1. Nickel-cadmium batteries are used in rechargeable electronic calculators. Given the following reaction for a discharging NiCad battery, which substance is reduced?

$$Cd\,(s) + NiO_2\,(s) + 2H_2O\,(l) \rightarrow Cd(OH)_2\,(s) + Ni(OH)_2\,(s)$$

A. $Cd(OH)_2$ **C.** NiO_2

B. H_2O **D.** Cd

2. How long must a current of 2 amps run for to liberate 3 moles of H_2O when electrolyzed in the following reaction? (Use Faraday's constant = 96,500C/mol and the conversion factor of 1 amp = 1 coulomb/sec)

$$H_2O_2 + 2\,H^+ + 2\,e^- \rightleftarrows 2\,H_2O$$

A. 4.83×10^4 sec   **C.** 1.45×10^5 sec

B. 4.83×10^5 sec   **D.** 2.41×10^5 sec

3. Which is NOT true for the following redox reaction occurring in an electrolytic cell?

$$Cd\,(s) + Zn(NO_3)_2\,(aq) \xrightarrow{\text{Electricity}} Zn\,(s) + Cd(NO_3)_2\,(aq)$$

A. Cd^{2+} is produced at the anode   **C.** Oxidation half-reaction: $Cd \rightarrow Cd^{2+} + 2\,e^-$

B. Zn metal is produced at the anode   **D.** Reduction half-reaction: $Zn^{2+} + 2\,e^- \rightarrow Zn$

4. Which statements is true of the following redox reaction occurring in a nonspontaneous electrolytic cell?

$$Br_2\,(l) + 2\,NaCl\,(aq) \xrightarrow{\text{Electricity}} Cl_2\,(g) + 2\,NaBr\,(aq)$$

A. Br_2 liquid is produced at the cathode

B. Cl_2 gas is produced at the cathode

C. Reduction half–reaction: $2\,Cl^- \rightarrow Cl_2 + 2\,e^-$

D. Cl_2 gas is produced at the anode

Any duplication (copies, uploads, PDFs) is illegal.

5. Which is true regarding the redox reaction occurring in a spontaneous electrochemical cell?

$$\text{Sn}\,(s) + \text{Cu}^{2+}\,(aq) \rightarrow \text{Cu}\,(s) + \text{Sn}^{2+}\,(aq)$$

A. Electrons flow from the Sn electrode to the Cu electrode

B. Anions in the salt bridge flow from the Sn half-cell to the Cu half-cell

C. Sn is oxidized at the cathode

D. Cu^{2+} is reduced at the anode

6. Which of the following is a unit of electrical charge?

A. joule

B. volt

C. coulomb

D. ampere

7. If a galvanic cell has two electrodes, which is correct?

A. Oxidation occurs at the negatively charged cathode

B. Oxidation occurs at the positively charged cathode

C. Oxidation occurs at the positively charged anode

D. Oxidation occurs at the negatively charged anode

8. A battery operates by:

A. oxidation

B. reduction

C. both oxidation and reduction

D. neither oxidation nor reduction

9. What is the term for the electrode in an electrochemical cell where oxidation occurs?

A. oxidation electrode

B. reduction electrode

C. anode

D. cathode

10. What is the general term for an apparatus containing two solutions with electrodes in separate compartments connected by a wire and salt bridge?

A. electrolytic cell

B. voltaic cell

C. battery

D. electrochemical cell

11. The electrode with the standard reduction potential of 0 V is assigned as the standard reference electrode and uses the half-reaction:

A. $2\,\text{NH}_4^+\,(aq) + 2\,\text{e}^- \rightleftarrows \text{H}_2\,(g) + 2\,\text{NH}_3\,(g)$

B. $2\,\text{H}^+\,(aq) + 2\,\text{e}^- \rightleftarrows \text{H}_2\,(g)$

C. $\text{Cu}^{2+}\,(aq) + 2\,\text{e}^- \rightleftarrows \text{Cu}\,(s)$

D. $\text{Zn}^{2+}\,(aq) + 2\,\text{e}^- \rightleftarrows \text{Zn}\,(s)$

Copyright © Sterling Test Prep.

12. In which type of cell does the following reaction occur when electrons are forced into a system by applying an external voltage?

$$Fe^{2+} + 2\,e^- \rightarrow Fe\,(s) \qquad E° = -0.44\ V$$

A. electrolytic cell

B. battery

C. electrochemical cell

D. galvanic cell

13. What is the term for a reaction representing separate oxidation or reduction processes?

A. reduction reaction

B. redox reaction

C. oxidation reaction

D. half-reaction

14. What is the term for providing electricity to a nonspontaneous redox process to cause a reaction?

A. hydrolysis

B. protolysis

C. electrochemistry

D. electrolysis

15. How is electrolysis different from the chemical process inside a battery?

A. Pure compounds cannot be generated *via* electrolysis

B. Electrolysis only uses electrons from a cathode

C. Electrolysis does not use electrons

D. They are the same process in reverse

16. How many electrons are needed to balance the following half-reaction $H_2S \rightarrow S_8$ in an acidic solution?

A. 16 electrons to the right side

B. 6 electrons to the right side

C. 12 electrons to the left side

D. 8 electrons to the right side

17. How are photovoltaic cells different from many other forms of solar energy?

A. Light is reflected, and the coolness of the shade provides a temperature differential

B. Light is passively converted into heat

C. Light is converted directly to electricity

D. Light is converted into heat and then into electricity

Any duplication (copies, uploads, PDFs) is illegal.

18. Which is true regarding the following redox reaction occurring in a spontaneous electrochemical cell?

$$\text{Sn } (s) + \text{Cu}^{2+} (aq) \rightarrow \text{Cu } (s) + \text{Sn}^{2+} (aq)$$

A. Anions in the salt bridge flow from the Cu half-cell to the Sn half-cell

B. Electrons flow from the Cu electrode to the Sn electrode

C. Cu^{2+} is oxidized at the cathode

D. Sn is reduced at the anode

19. What is the term for converting chemical energy to electrical energy from redox reactions?

A. Redox chemistry

B. Electrochemistry

C. Cell chemistry

D. Battery chemistry

Copyright © Sterling Test Prep.

Detailed Explanations

1. C is correct.

Determine the *oxidation number* of each species.

The reduced substance has a decrease in the oxidation number.

In NiO_2, the oxidation number of Ni = +4.

In $Ni(OH)_2$, the oxidation number of Ni = +2.

2. C is correct.

3 moles of H_2O need 3 moles of e^-.

Calculate time needed for 3 moles of H_2O:

$$(3 \text{ moles } e^-) \times (96{,}500 \text{ coulomb} / 1 \text{ mole } e^-) \times (1 \text{ sec} / 2 \text{ coulomb})$$
$$\text{time (for 3 moles of } H_2O) = 1.45 \times 10^5 \text{ sec}$$

3. B is correct.

An *electrolytic cell* is a nonspontaneous electrochemical cell requiring electrical energy to be supplied (e.g., battery) to initiate the reaction.

The *anode* is positive, and the *cathode* is the negative electrode.

For electrolytic and galvanic cells, *oxidation occurs at the anode and reduction at the cathode.*

Therefore, Zn metal is produced at the cathode because it is a reduction product (from an oxidation number of +3 on the left to 0 on the right). Zn will not be produced at the anode.

4. D is correct.

In electrochemical (i.e., galvanic) cells, oxidation occurs at the anode and reduction at the cathode.

Cl_2 gas is produced at the anode because it is an oxidation product (from -1 on the left to 0 on the right).

5. A is correct.

A *spontaneous* electrochemical cell is known as a *galvanic cell.*

Any duplication (copies, uploads, PDFs) is illegal.

6. C is correct.

A *coulomb* is a unit of electrical charge.

7. D is correct.

Oxidation (i.e., loss of electrons) occurs at the *anode*.

8. C is correct.

A battery is an electrochemical (i.e., galvanic) cell with reduction and oxidation reactions to generate electricity.

9. C is correct.

Oxidation (i.e., loss of electrons) occurs at the anode.

10. D is correct.

The salt bridge should be the most apparent indication of an electrochemical cell – electrolytic cells do not have salt bridges.

11. B is correct.

By convention, the reference standard for potential is hydrogen reduction.

12. A is correct.

Electrolytic cells do not occur spontaneously; the reaction requires *external electrical energy*.

13. D is correct.

A redox reaction, or oxidation-reduction reaction, involves transferring electrons between two reacting substances.

An oxidation reaction refers explicitly to the substance that is losing electrons, and a reduction reaction refers explicitly to the substance that is gaining reactions.

The oxidation and reduction reactions alone are *half-reactions* because they occur together to form a whole reaction.

Therefore, half-reaction is the answer because it can represent a separate oxidation or reduction process.

Copyright © Sterling Test Prep.

14. D is correct.

Electrolysis uses a direct electric current to provide a nonspontaneous redox process to drive the reaction.

The direct electric current must be passed through an ionic substance or solution that contains electrolytes.

15. D is correct.

Electrolysis is the same process in reverse for the chemical process inside a battery.

Electrolysis is used to separate elements.

16. A is correct.

Half-reaction:

$$H_2S \rightarrow S_8$$

Balancing half-reaction in acidic conditions:

Step 1: Balance atoms except for H and O

$$8\ H_2S \rightarrow S_8$$

Step 2: To balance oxygen, add H_2O to the side with fewer oxygen atoms

There is no oxygen at all, so skip this step.

Step 3: To balance hydrogen, add H^+:

$$8\ H_2S \rightarrow S_8 + 16\ H^+$$

Balance charges by adding electrons to the side with a greater/more positive total charge

Total charge on left side: 0

Total charge on right side: $16(+1) = +16$

Add 16 electrons to right side:

$$8\ H_2S \rightarrow S_8 + 16\ H^+ + 16\ e^-$$

17. C is correct.

The operation of a photovoltaic (PV) cell has the following requirements:

1) Light is absorbed, which excites electrons.

2) Separation of charge carries opposite types.

3) The separated charges are transferred to an external circuit.

Any duplication (copies, uploads, PDFs) is illegal.

In contrast, solar panels supply heat by absorbing sunlight.

A photoelectrolytic (or photoelectrochemical) cell is a photovoltaic cell that splits water directly into hydrogen and oxygen using only solar illumination.

18. A is correct.

Cu^{2+} is reduced (i.e., gains electrons), while Sn^{2+} is oxidized (i.e., loses electrons).

Salt bridge contains cations (positive ions) and anions (negative ions).

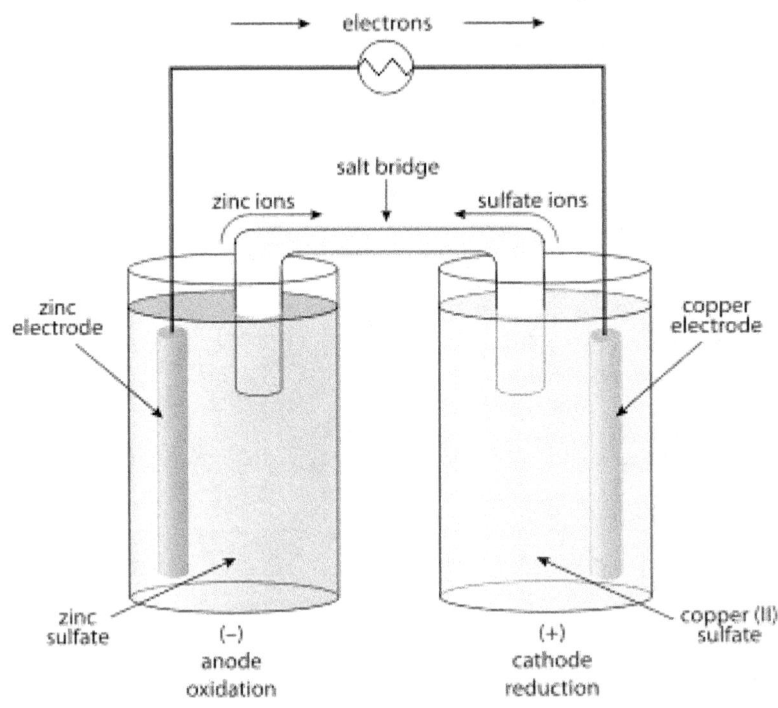

Schematic of a sample Zn–Cu galvanic cell

Anions flow towards the oxidation half-cell because the oxidation product is positively charged, and the anions are required to balance the charges within the cell. The opposite is true for cations; they flow towards the reduction half-cell.

19. B is correct.

Electrochemistry is the branch of physical chemistry that studies chemical reactions at the interface of an ionic conductor (i.e., the electrolyte) and an electrode.

Electric charges move between the electrolyte and the electrode through a series of redox reactions, and chemical energy is converted to electrical energy.

Copyright © Sterling Test Prep.

Notes for active learning

Any duplication (copies, uploads, PDFs) is illegal.

Notes for active learning

Notes for active learning

Copyright © Sterling Test Prep.

CHAPTER 10

Atomic & Electronic Structure

- Origins of Modern Atomic Theory
- Bohr Model and Emission Spectrum of Hydrogen
- Quantum Theory of Atoms
- Quantum Mechanics
- Electronic Configuration
- Atomic Nucleus
- Atomic Number and Atomic Weight
- Nuclear Forces
- Radioactivity
- Decay Types
- Radioactive Decay Rates
- Decay Series
- Mass Deficit, Energy Liberated and Binding Energy
- Mass Spectrometry
- Practice Questions & Detailed Explanations

Page intentionally left blank

Copyright © Sterling Test Prep.

Origins of Modern Atomic Theory

Dalton's model of an atom

John Dalton (1755-1844) proposed the *atomic structure* in the early 1800s. Dalton reintroduced atomic theory to explain chemical reactions.

Dalton's theory centered on five main concepts:

1. All matter is made of indivisible particles of *atoms*.

2. An *element* is made up of identical atoms.

3. Different elements have atoms with different masses.

4. Chemical compounds are made of atoms in specific integer ratios.

5. Atoms are neither created nor destroyed in chemical reactions.

Discovery of the electron

Nobel laureate J. J. Thomson (1856-1940) discovered the *electron* in the late 1800s. By performing cathode ray experiments shown below.

Thomson discovered that the electron is negatively charged by measuring the electron's charge-to-mass ratio and identified electrons as a fundamental particle.

To test the properties of the particles, Thomson placed two oppositely charged electric plates around the cathode ray. The cathode ray was deflected from the negatively charged electric plate and towards the positively charged plate, indicating that the cathode ray was composed of negatively charged particles.

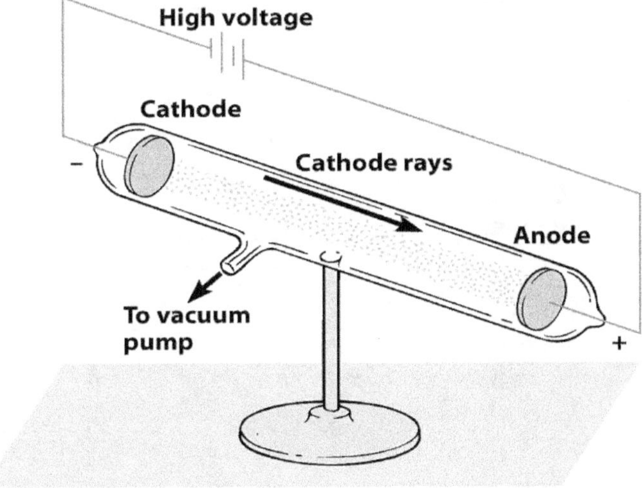

Cathode ray experiment with cathode and anode

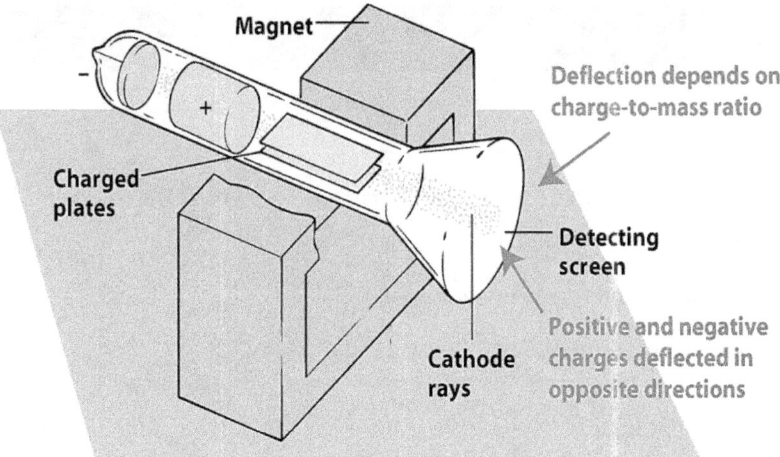

Cathode rays to measure charge-to-mass ratio

Mass of an electron

Around 1906, Nobel laureate Robert Millikan (1868-1953) studied minuscule particles, specifically, charged oil droplets in an electric field.

Millikan observed that the charge on the oil droplets was a multiple of the electron charge. This discovery, along with Thomson's results, was used to calculate the mass of the electron. He observed that the charge changed in steps between one passage of the droplet and the next and were multiples (*quantized*) of a common elementary charge, now known to be 1.602×10^{-19} C.

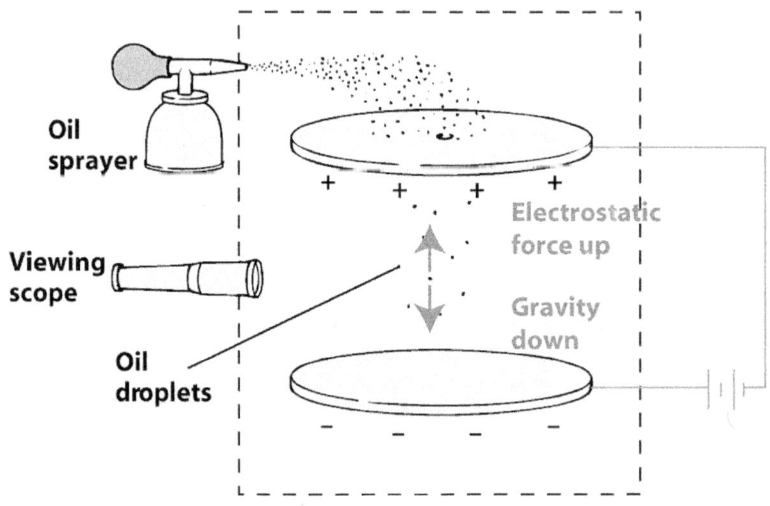

Millikan observed charged oil droplets

Copyright © Sterling Test Prep.

Charge-to-mass ratio

Thomson's result (charge-to-mass) was the quantity:

$$\frac{q}{m} = 1.7584 \times 10^{11} \frac{C}{kg}$$

Millikan's result (q = charge) was:

$$q = 1.60 \times 10^{-19} C$$

Combined, the mass of an electron is:

$$m = \frac{q}{1.7584 \times 10^{11}}$$

$$m = 9.11 \times 10^{-31} kg$$

Thomson's "plum pudding" model

Dalton claimed that atoms were indivisible and that no smaller quantities were involved. However, Thomson's and Millikan's experiments, along with the discovery of the electron, proved Dalton's statement false.

Thomson's atomic model attempted to reconcile the two properties of atoms known at that time: 1) there are electrons, and 2) atoms have no net electric charge. Based on this knowledge, Thomson theorized that there must be an equal amount of positive charge to cancel the electrons' negative charge. He proposed that the atom was spherical, and this positive charge was everywhere with the negatively charged electrons being distributed evenly throughout it and repelling each other.

This model became known as the *"plum pudding"* model with electrons embedded in a blob of positively charged matter like *"raisins in plum pudding."*

While significant at the time as the first model to describe internal atomic structure, Thomson's model became obsolete following Ernest Rutherford's discovery of the atomic nucleus in 1911. Thomson's model attempted to explain the atom's physical and chemical properties by the number and arrangement of its electrons. While he was generally on the right track, his approach was based on classical mechanics and did not account for quantized energy.

Protons

Nobel laureate Ernest Rutherford (1871-1937), in 1911, conducted an experiment that scattered *alpha particles* by bouncing them off gold foil. Most of the gold atoms are in empty space. Most particles passed through without deflection, but a few scattered at large angles, meaning that a small part of the gold atom must be positively charged.

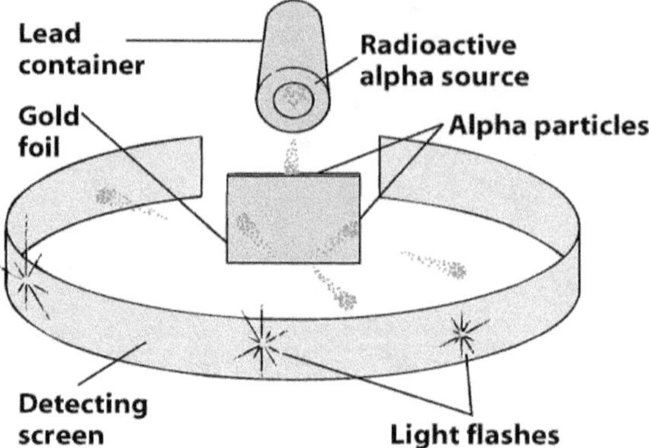

Alpha particles were discovered by being deflected by the dense gold foil

Alpha particles are *positively charged* particles with *two protons and two neutrons* bound together. They are emitted by an atomic nucleus undergoing radioactive decay and identical to a helium atom nucleus.

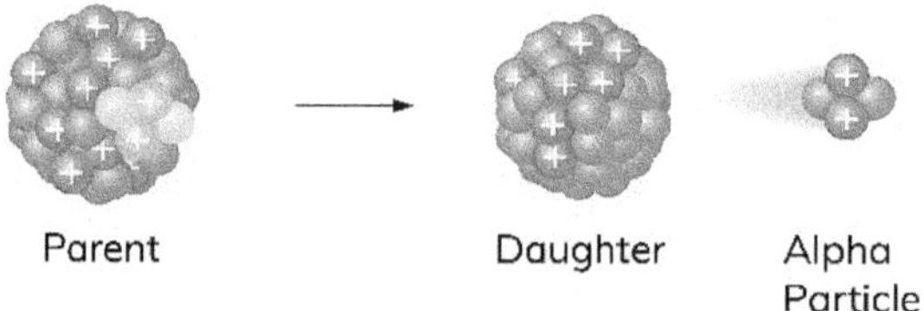

Alpha particles are emitted by radioactive decay of an atomic nucleus

Rutherford concluded that an atom's positive charge resides in a tiny nucleus that contains most of the atom's mass (unlike the spread-out blob of matter containing electrons, as the *plum pudding* model suggested). He named these positively charged subatomic particles *protons*.

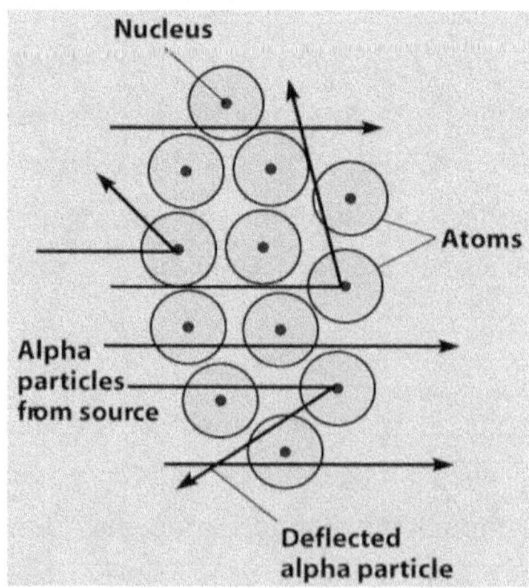

Alpha particle deflection contributed to the current model of the atom's nucleus

Copyright © Sterling Test Prep.

Neutrons

Rutherford discovered the atomic nucleus in 1911 and observed the proton in 1919. However, it seemed there must be something in the nucleus besides protons. For instance, helium was known to have an atomic number of 2 but a mass number of 4.

Some scientists thought there were additional protons in the nucleus and an equal number of electrons to cancel out the additional charge. In 1920, Rutherford proposed that an electron and a proton could combine to form a new, *neutral particle*.

In 1932, Nobel laureate James Chadwick (1891-1974) added the existence of neutral *neutrons* in the nucleus to the atomic theory. Neutrons were discovered later than the other subatomic particles because they have no charge; other methods were needed to identify the neutrons other than those used to discover protons and electrons.

Notes for active learning

Copyright © Sterling Test Prep.

Bohr Model and Emission Spectrum of Hydrogen

Bohr model of atoms

In 1913, Nobel laureate Niels Bohr (1885-1962) proposed a model depicting the atom in which the electron orbits the nucleus as the Earth orbits the Sun.

Electrostatic attraction pulls the electron toward the nucleus.

However, the electron orbits at high enough speeds to prevent it from crashing into the nucleus.

Relationship of the electron orbit around the nucleus leads to several important implications:

1. Electrons only exist *in precise allowed orbits.*

2. Within an orbit, the *electron does not radiate.*

3. Radiation is emitted or absorbed *when changing orbits.*

Three implications of Bohr model

1. Electrons *occupy specific orbitals*, which determine their energy.

Orbits are characterized by the orbital number n, where $n = 1$ equals the ground state and represents the electron's lowest energy level.

The higher the orbital number n of the electron, the higher the energy, and the further the electron is from the nucleus.

2. Electrons *do not radiate* while orbiting the nucleus. As mentioned in earlier chapters, accelerating electric charges emit energy in the form of electromagnetic radiation.

According to classical physics, the electron would radiate away its energy (due to the energy radiated from centripetal acceleration) and crash into the nucleus.

Bohr's model assumes electrons do not radiate energy within energy levels and stay in orbits.

3. Electrons *absorb or release energy when changing energy levels.*

For a ground-state electron to achieve a higher orbital, it must absorb a photon (and thus absorb the photon's energy).

If the electron drops to a lower orbital, then the electron must emit a *photon* equal to the energy difference between orbital levels.

Photons are elementary particles that are a quantum of the electromagnetic field, including electromagnetic radiation such as light and radio waves, and are the force carriers for the electromagnetic force.

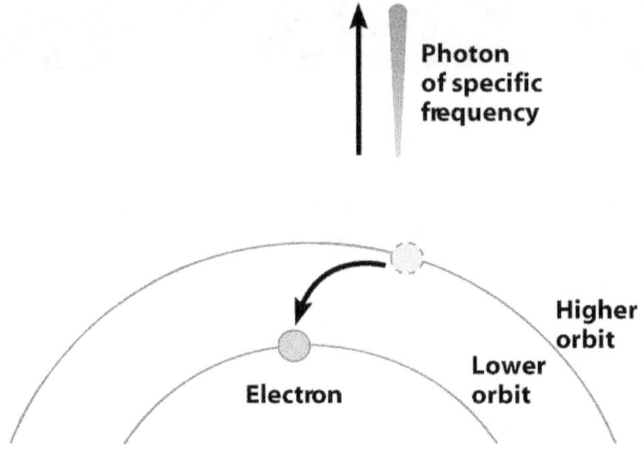

Electrons occupy specific orbitals

Emission line spectrum

An essential application of the Bohr model is identifying elements through their *line spectrum*.

When a high potential difference energizes an element, the electrons are constantly raised to higher energy levels and then dropped back to lower energies.

Emission spectrum of an element is the frequency of electromagnetic radiation emitted due to electrons transitioning from a high-energy state to a lower-energy state. These energies relate the energy differences in the electron orbitals and allow quantization of the specific orbital energies.

Electromagnetic energy of the emitted photons equals the energy difference between the two energy states. Photons (i.e., massless particles) released in the process contain specific energies characterized by frequency.

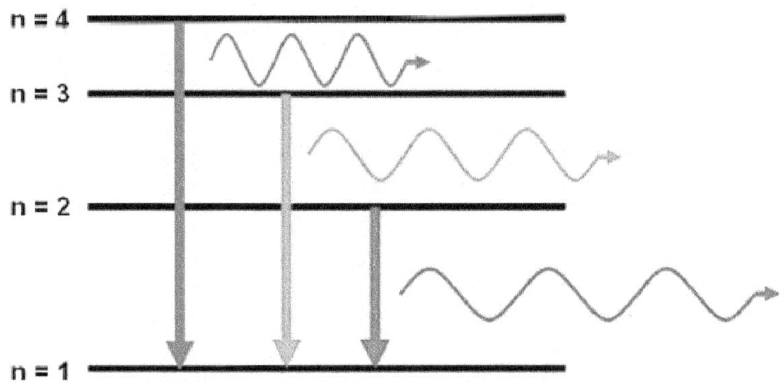

Electromagnetic radiation is emitted due to electrons transitioning from a
high-energy to a lower-energy state

Copyright © Sterling Test Prep.

Rydberg formula

Emission lines form regular patterns and can be represented by a simple equation.

Emission lines are calculated by combining simple whole numbers using the *Rydberg formula*, which governs the wavelength of the emitted or absorbed radiation:

$$\frac{1}{\lambda} = R \left(\frac{1}{n_f^2} - \frac{1}{n_i^2} \right)$$

where λ is the wavelength, n_f is the final energy level, n_i is the initial energy level, and R is the Rydberg constant of $1.0973 \times 10^7\,\mathrm{m}^{-1}$.

Rydberg formula calculates the wavelength of light emitted by an electron moving between the energy levels of an atom. It works well for transitions between energy levels of a hydrogen atom with only one electron.

However, the Rydberg formula gives incorrect results for atoms with multiple electrons because the amount of screening for inner or outer electron transitions varies, and the equation is too simplistic to compensate for the differences.

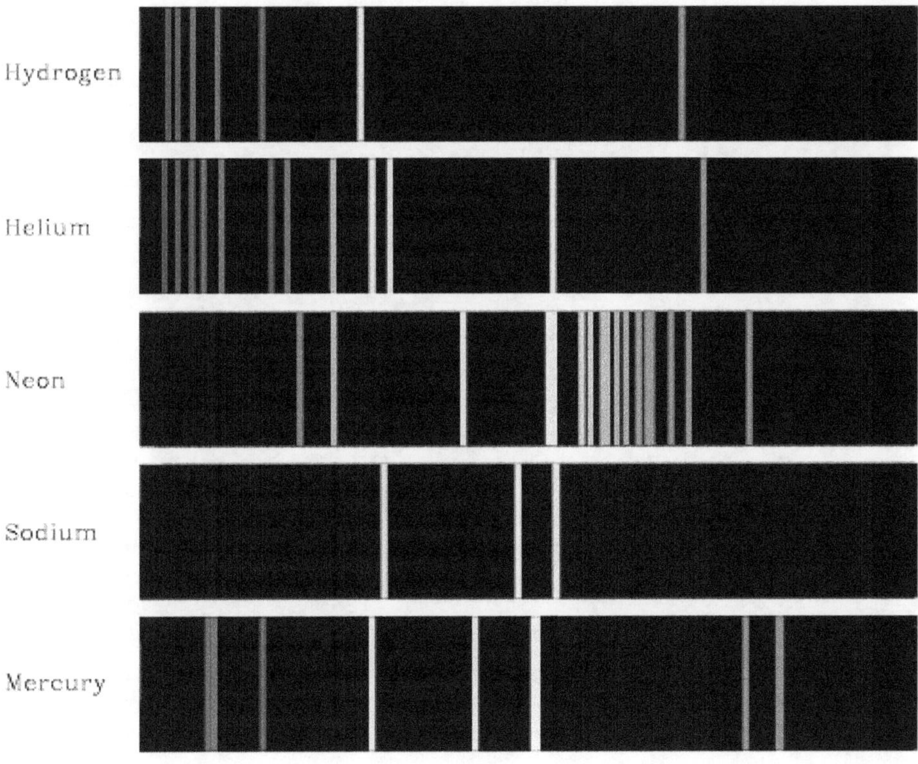

Intensity vs. wavelength emission line spectra for sample elements

Any duplication (copies, uploads, PDFs) is illegal.

Named emission line spectra

Lyman series involve transitions to or from the ground state (n = 1).

Balmer series (in which all lines are in the visible region) corresponds to n = 2.

Paschen series corresponds to n = 3.

Brackett series corresponds to n = 4.

Pfund series corresponds to n = 5.

Humphrey series corresponds to n = 6.

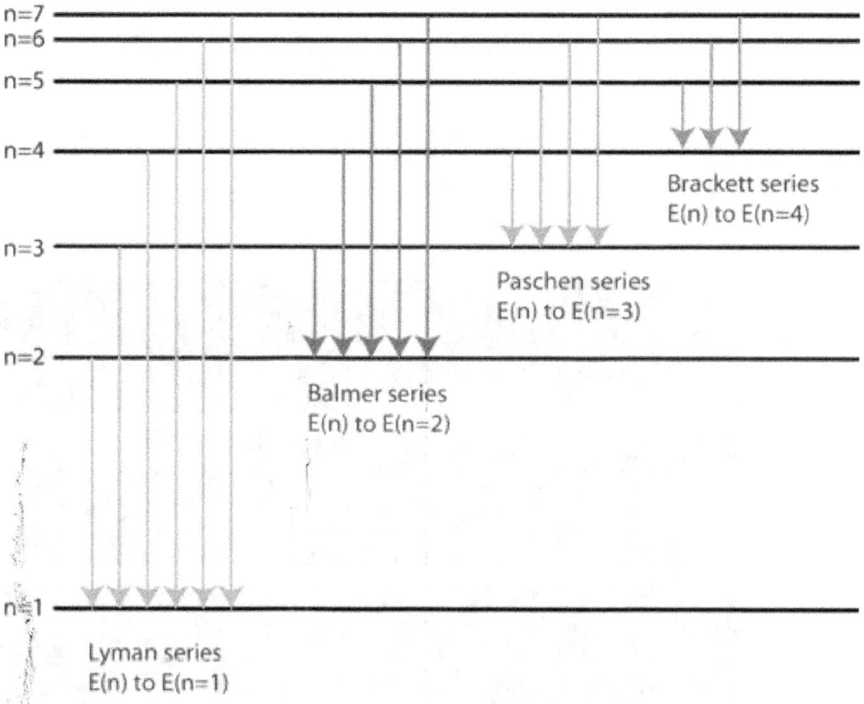

Electron transitions for a hydrogen atom

Further series are unnamed but follow the pattern and equation of the Rydberg equation.

Series are increasingly spread out and occur at increasing wavelengths with lines increasingly faint, corresponding to less rare atomic events. The seventh series of atomic H was demonstrated experimentally at infrared wavelengths in 1972.

Copyright © Sterling Test Prep.

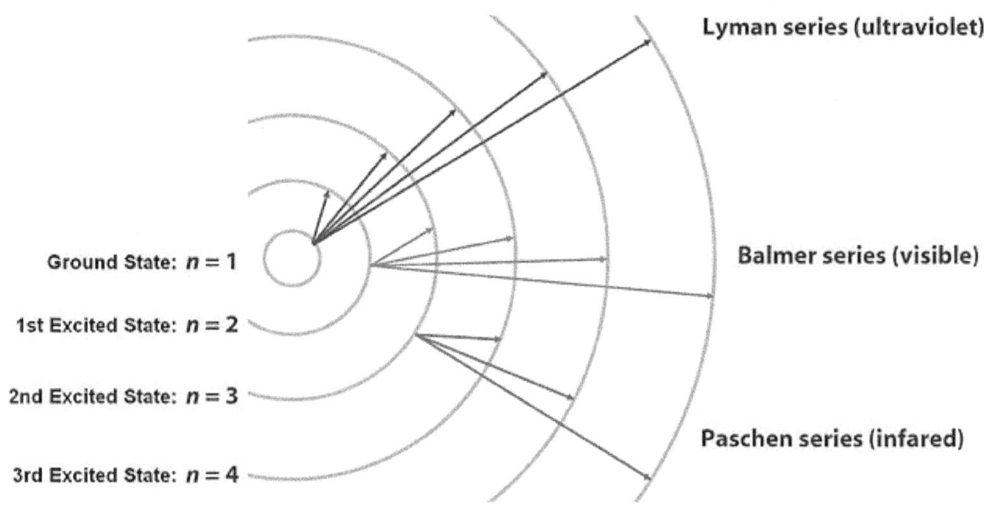

Energy transitions between orbitals for the H atom

Hydrogen line spectra

Emission line spectra of a hydrogen (H) atom have lines representing the changes in quantum energy levels of the atom's single electron.

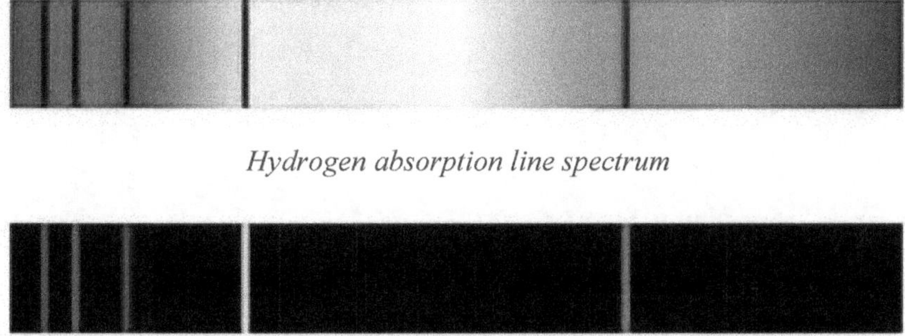

Hydrogen absorption line spectrum

Hydrogen emission line spectrum

Balmer series

In 1885, Johann Balmer (1825-1898) proposed the Balmer series for the orbital transitions of electrons from any orbital back to n = 2. Wavelength = $1/\lambda = R(1/2^2 - 1/n^2)$

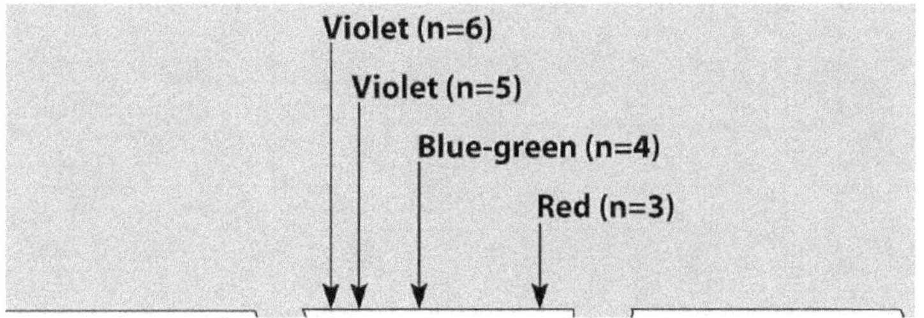

Balmer series to account for electron transitions between orbitals

Any duplication (copies, uploads, PDFs) is illegal.

Notes for active learning

ACS Physical Chemistry: Comprehensive Content

Notes for active learning

Copyright © Sterling Test Prep.

Quantum Theory of Atoms

Atomic energy levels

Quantum theory of the atom describes the series of energy states that an electron can occupy.

Lowest energy state is the *"ground state,"* and higher states are *"excited states."*

The energy of the photon emitted (or absorbed) during a transition between the specific energy levels equals the *difference* in state energies.

Violet light represents the higher energy transitions (n = 5 or 6 to n = 2) because violet has the highest visible spectrum energy (mnemonic of ROY G BIV).

Red light has lower energy than violet because of a lower energy transition (n = 3 to n = 2).

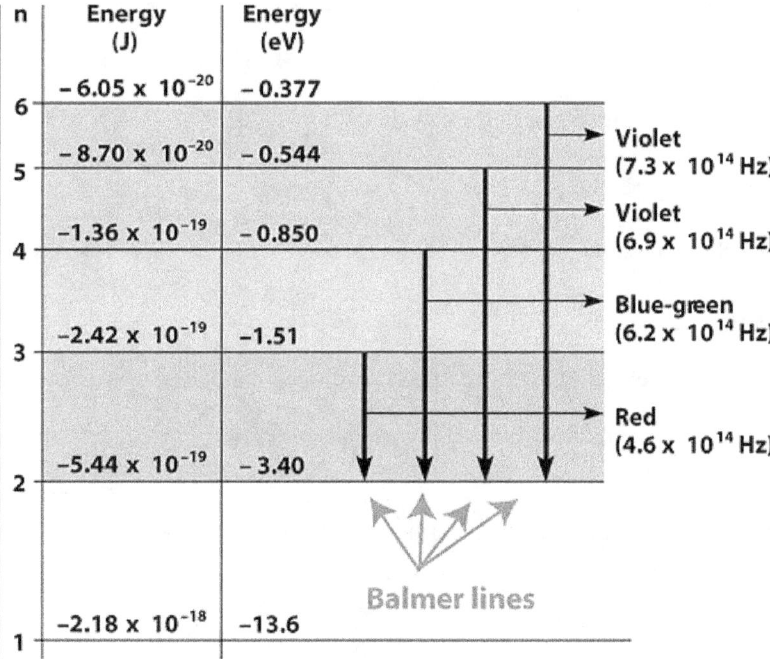

Hydrogen atom and its energy levels

Any duplication (copies, uploads, PDFs) is illegal.

Wave–particle duality

Bohr's theory only holds for the line spectrum of H with only two electrons. When more electrons are added, the theory cannot account for the complications of heavier atoms.

Experiments established the wave-particle duality of light and matter. For example, Young's two-slit experiment (discussed in light wave theory) produced *interference patterns* for photons and electrons.

Wave theory

In 1932, Nobel laureate Louis de Broglie (1892-1987) proposed the existence of *matter waves* regarding the atomic theory.

Matter waves are central to the theory of quantum mechanics as part of wave-particle duality. At scales where measurements have been practical, matter exhibits wave-like behavior. For example, a beam of electrons can be diffracted, like a beam of light or a water wave.

Wavelength is related to *momentum,* and the matter waves in atoms are standing waves.

Louis de Broglie derived the formula for wavelength:

$$\lambda = \frac{h}{mv}$$

where h is Planck's constant 6.63×10^{-34} J·s, λ is particle wavelength, m is mass and v speed.

Wave mechanics

Nobel laureate Erwin Schrödinger (1887-1961) developed the theory of *wave mechanics.*

Wave mechanics treats atoms as three-dimensional systems of waves and incorporates the successful ideas of the Bohr model, along with other foundational principles.

Schrödinger's wave mechanics theory describes the hydrogen atom and other atoms with many electrons, providing a fundamental understanding of chemistry.

Quantum (wave) mechanical theory states that an electron behaves as a standing wave. It describes an electron's possible location in an orbital.

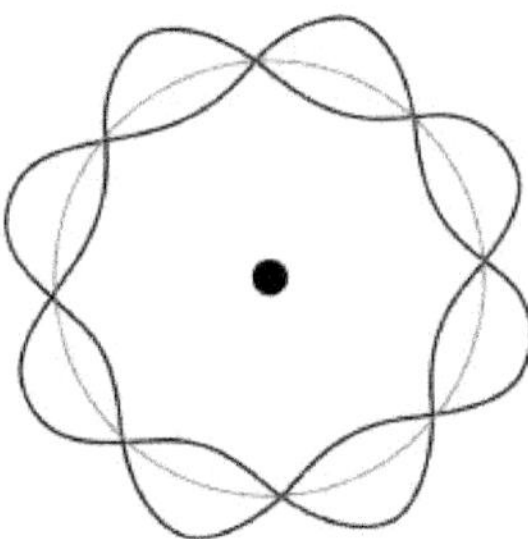

Matter waves around the nucleus with wavelength related to momentum

Copyright © Sterling Test Prep.

Quantum Mechanics

Wave functions and probability distributions

Quantum mechanics model is a visualization of *wave functions* and *probability distributions*. It does not represent an electron's motion but indicates the likelihood of finding an electron within a specific orbital within the atom. It is based on probability and not certainty, unlike previous models of the atom.

Quantum numbers specify electronic quantum states and that an electron is delocalized. They define and describe a particular electron in an atom.

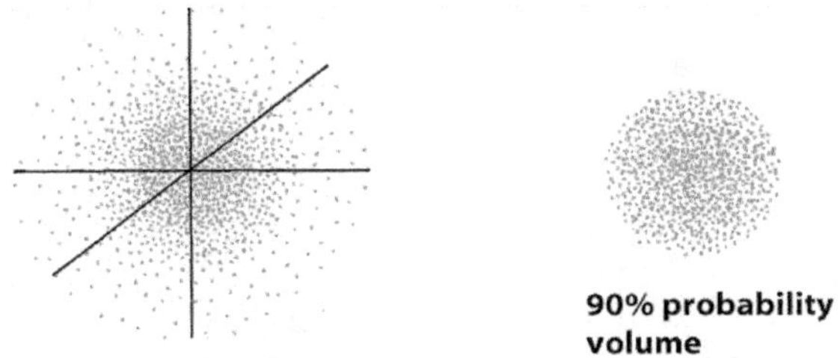

Probability distribution for the location of an orbiting electron

Principal quantum number

Principal quantum number (n) states the energy level (i.e., *shells*) where an electron will most likely be found. n represents the *orbital size* and *average distance from the nucleus*.

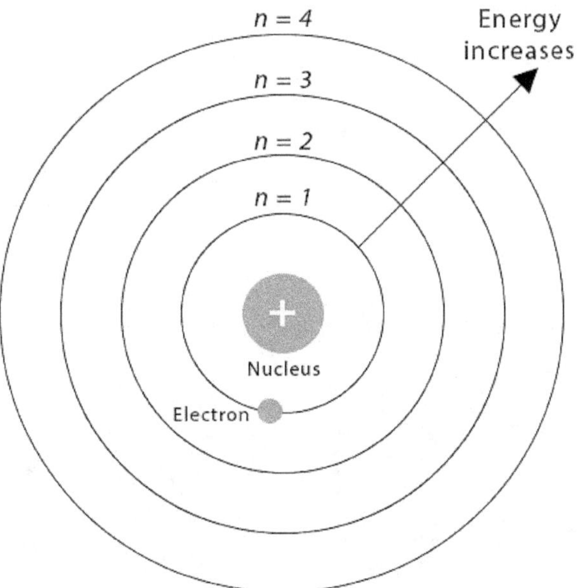

Principal quantum number (n) with relative atomic size and energy

Any duplication (copies, uploads, PDFs) is illegal.

Principal quantum number has positive integer values of $n = 1, 2, 3\ldots$

Electron shells (or principal energy levels) are labeled by principal quantum numbers ($n = 1$, $2, 3\ldots$) but can be labeled alphabetically as $n = K, L, M\ldots$

Higher n shells indicate *larger* orbitals *further* from the nucleus and *higher energy* levels.

Angular momentum quantum number

Angular momentum (or *azimuthal*) *quantum number* is the second quantum number designated with the lowercase letter l. It describes the *shape of the orbital* electron occupies.

Lowest value of l is 0, and the highest, depending on the principal quantum number, is $n - 1$.

Angular momentum quantum number describes the spatial distribution (or shape) of the orbits labeled s, p, d, f, g, h, etc. These are *subshells*.

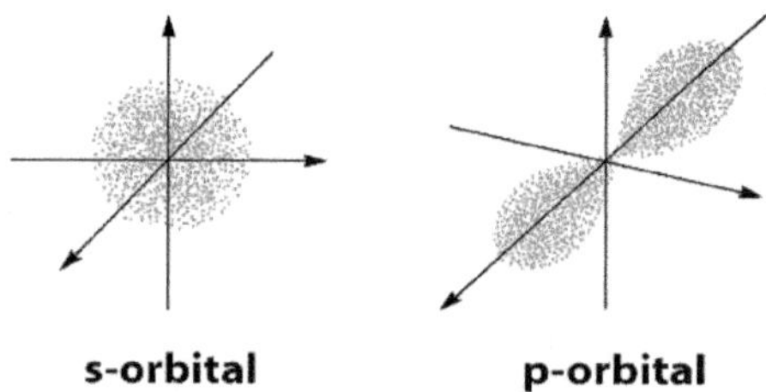

Subshells as spatial distributions of orbitals

Orbital	
s	$l = 0$
p	$l = 1$
d	$l = 2$
f	$l = 3$

Orbitals and associated angular momentum quantum number

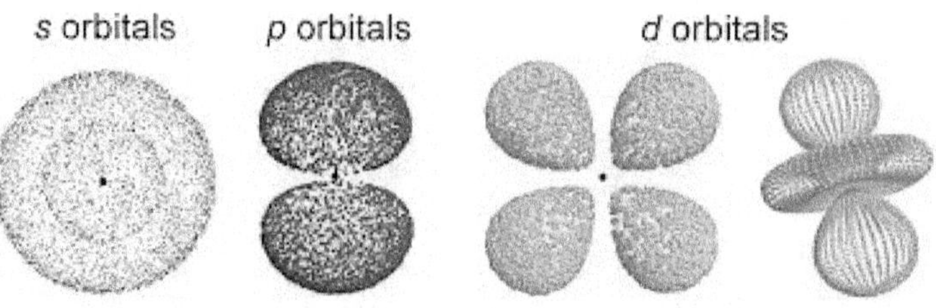

S orbital (l = 0), p orbitals (l = 1) and d orbitals (l = 2)

Copyright © Sterling Test Prep.

Magnetic quantum number

Magnetic quantum number (*m*), the third quantum number, states the spatial orientation of the orbit within the energy (*n*) and shape (*l*). Magnetic quantum number (m_l) divides the subshell into specific orbitals that hold the electron.

Number of orbitals is given by $2l + 1$. m_l has values of $-l$ to l, with increments of one, including zero ($-l,\ ..,\ 0,\ ...,\ +l$).

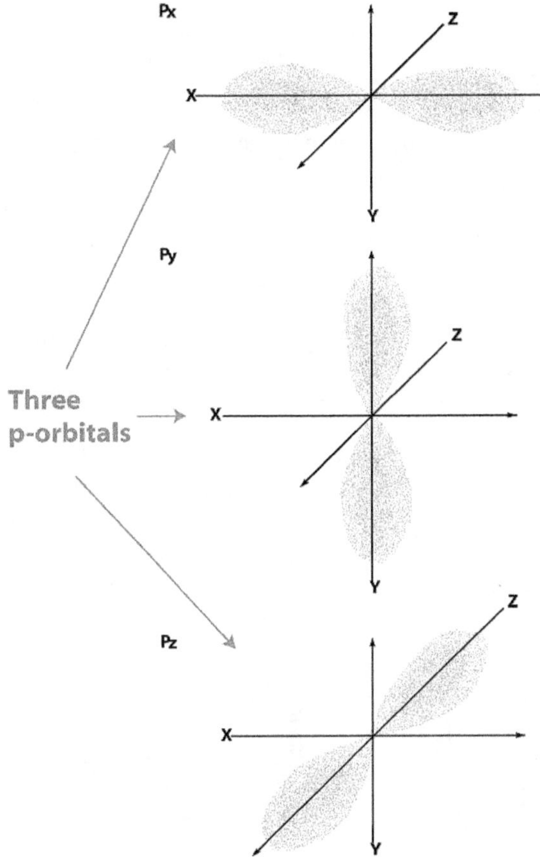

p orbitals orient along the x-, y- and z-axis and can accommodate 6 electrons

m_l range from $-l$ to $+l$, including 0.

- *s* subshell has $l = 0$; the m_l has a single value (0) and one orbital.

  s subshell can hold 2 electrons.

- *p* subshell has $l = 1$; the m_l has values of $-1, 0, +1$.

  p subshell has three orbitals holding up to 6 electrons.

Any duplication (copies, uploads, PDFs) is illegal.   603

- d subshell has $l = 2$; $m_l = -2, -1, 0, +1, +2$.

  d subshell has five orbitals with up to 10 electrons.

- f subshell has $l = 3$; $m_l = -3, -2, -1, 0, +1, +2, +3$.

  f subshell has seven orbitals with up to 14 electrons.

Each succeeding subshell holds four more electrons than its predecessor.

Spin quantum number

Spin quantum number, the fourth quantum number, is either $+\frac{1}{2}$ or $-\frac{1}{2}$ and represents the electron spin orientation. Spin quantum number has only two values because an electron only spins in one of two directions ($+\frac{1}{2}$ or $-\frac{1}{2}$).

By convention, the first electron to occupy an orbital has a spin number of $+\frac{1}{2}$ while the second electron has a spin number of $-\frac{1}{2}$.

In orbital diagrams, the $+\frac{1}{2}$ electrons are represented by upward arrows, and the $-\frac{1}{2}$ electrons are indicated by downward arrows.

Energy level	Number of subshells	Subshells	# of orbitals in subshell	Total number of electrons in a subshell
1	1	s	1	2
2	2	s	1	2
		p	3	6
3	3	s	1	2
		p	3	6
		d	5	10
4	4	s	1	2
		p	3	6
		d	5	10
		f	7	14

Quantum numbers represent the locations of the electrons within the atom

Copyright © Sterling Test Prep.

Electronic Configuration

Pauli Exclusion Principle

In 1925, Nobel laureate Wolfgang Pauli (1900-1958) proposed the *Pauli Exclusion Principle,* which states that no two electrons in the same atom may have the same four quantum numbers (n, l, m_l and m_s).

Maximum number of electrons in an orbital is two, and each electron must possess opposite spins (i.e., $+\frac{1}{2}$ or $-\frac{1}{2}$) because electrons of the same magnetic spin cannot be in the same subshell of an orbital.

Pauli's exclusion principle states that no two electrons in the same atom can have the *same set of quantum numbers*.

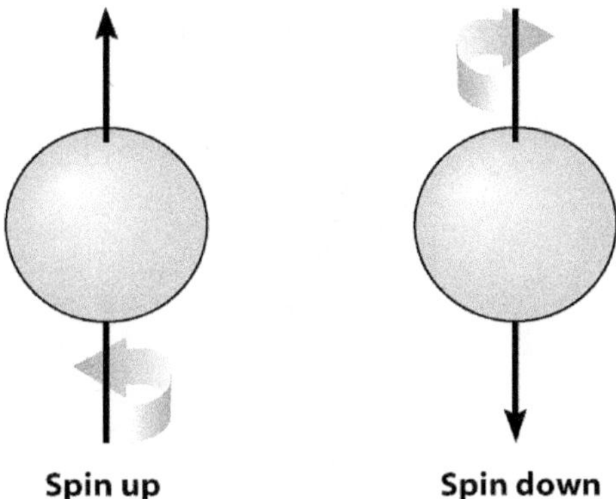

Each orbital contains only two electrons, and the electrons must have opposite spins

Electrons in atomic orbitals

Electron configuration places electrons in atomic orbitals and specifies their quantum state.

Chemical properties of an atom are determined by examining the atom's electronic structure.

When writing electron configurations, following the Pauli exclusion principle is imperative.

Each electron has unique quantum numbers, with a maximum of two electrons per orbital; one spins up, and one spins down.

Electrons fill available orbitals in the order of increasing energy (n).

Each shell only holds a finite number of subshells and the corresponding number of electrons.

Any duplication (copies, uploads, PDFs) is illegal.

Subshells

Electron capacity in each subshell of an orbital is as follows:

s = 2 electrons in 1 subshell

p = 6 electrons in 3 subshells

d = 10 electrons in 5 subshells

f = 14 electrons in 7 subshells

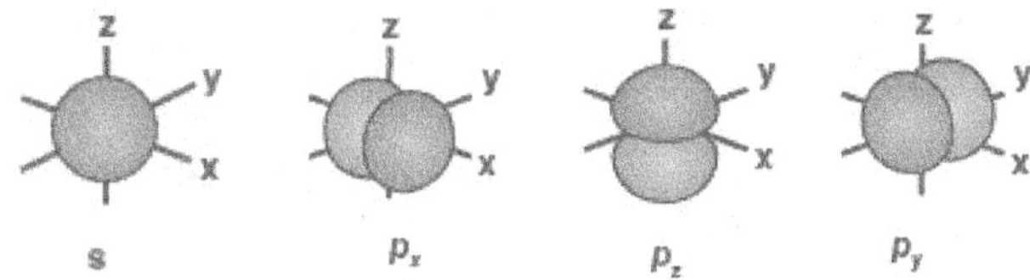

One s orbital (on the left) and three p orbitals (on the right)

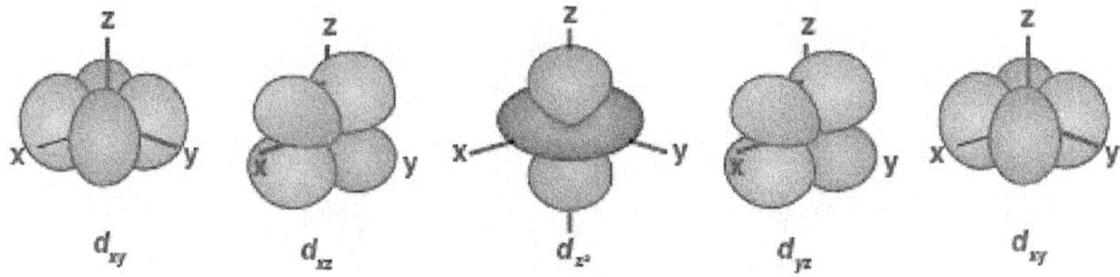

Five d orbitals

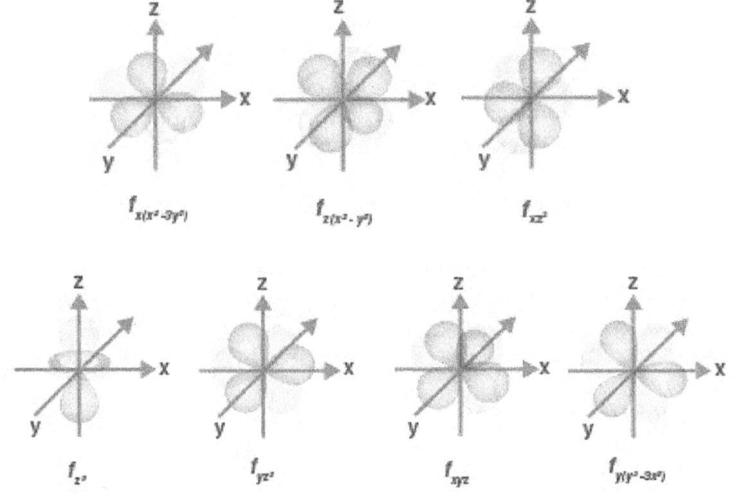

Ten f orbital

Copyright © Sterling Test Prep.

Orbitals hold two electrons

Each orbital contains a maximum of two electrons. It is possible to fill half an orbital, but it must be in the outer (valence) shell. For example, strontium has 38 electrons.

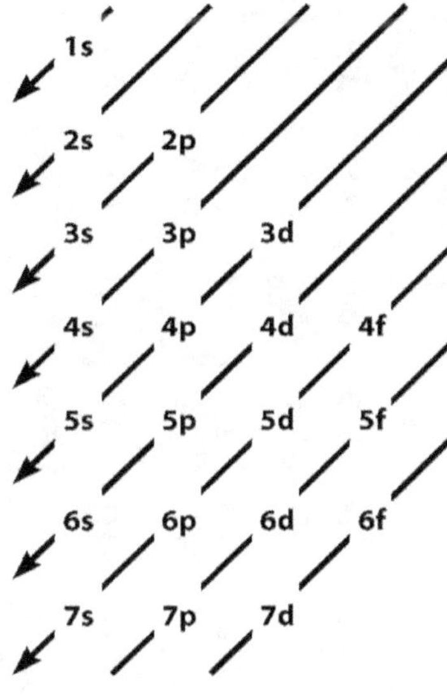

$$1s^2\,2s^2\,2p^6\,3s^2\,3p^6\,4s^2\,3d^{10}\,4p^6\,5s^2$$

Noble gases (VIIIA) have filled valence shells and are inert (or do not react readily).

> Elements with 1 to 3 outer electrons *lose valence electrons* to become positive ions (e.g., metals).

> Elements with 5 to 7 outer electrons tend *to gain electrons* and form negative ions (such as nonmetals).

Metalloids are elements that have a mixture of properties of metals and nonmetals. There is no standard definition of a metalloid or agreement on which elements belong to this category.

> Six commonly recognized metalloids are boron, silicon, germanium, arsenic, antimony, and tellurium.

> Carbon, aluminum, selenium, polonium, and astatine are five less frequently classified metalloid elements.

> Periodic Table shows all eleven metalloid elements in a diagonal in the p-block, extending from boron at the upper left to astatine at the lower right.

Any duplication (copies, uploads, PDFs) is illegal.

Periodic Table of the Elements

Key: Atomic Number — Symbol — Valence — Name — Atomic Mass

Group	1 IA/1A	2 IIA/2A	3 IIIB/3B	4 IVB/4B	5 VB/5B	6 VIB/6B	7 VIIB/7B	8 VIII/8	9 VIII/8	10 VIII	11 IB/1B	12 IIB/2B	13 IIIA/3A	14 IVA/4A	15 VA/5A	16 VIA/6A	17 VIIA/7A	18 VIIIA/8A
1	1 H Hydrogen 1.008 (−1,+1)																	2 He Helium 4.003 (0)
2	3 Li Lithium 6.941 (+1)	4 Be Beryllium 9.012 (+2)											5 B Boron 10.811 (+3)	6 C Carbon 12.011 (+4,+3,+2,−1,−4,−3)	7 N Nitrogen 14.007 (+5,+3,−3)	8 O Oxygen 15.999 (−2)	9 F Fluorine 18.998 (−1)	10 Ne Neon 20.180 (0)
3	11 Na Sodium 22.990 (+1)	12 Mg Magnesium 24.305 (+2)											13 Al Aluminum 26.982 (+3)	14 Si Silicon 28.086 (+4,−4)	15 P Phosphorus 30.974 (+5,+3,−3)	16 S Sulfur 32.066 (+6,+4,+2,−2)	17 Cl Chlorine 35.453 (+7,+5,+3,+1,−1)	18 Ar Argon 39.948 (0)
4	19 K Potassium 39.098 (+1)	20 Ca Calcium 40.078 (+2)	21 Sc Scandium 44.956 (+3)	22 Ti Titanium 47.88 (+4)	23 V Vanadium 50.942 (+5)	24 Cr Chromium 51.996 (+6,+3)	25 Mn Manganese 54.938 (+7,+4,+2)	26 Fe Iron 55.845 (+6,+3,+2)	27 Co Cobalt 58.933 (+3,+2)	28 Ni Nickel 58.693 (+2)	29 Cu Copper 63.546 (+2,+1)	30 Zn Zinc 65.38 (+2)	31 Ga Gallium 69.723 (+3)	32 Ge Germanium 72.631 (+4,+2,−4)	33 As Arsenic 74.922 (+5,+3,−3)	34 Se Selenium 78.971 (+6,+4,+2,−2)	35 Br Bromine 79.904 (+5,+3,+1,−1)	36 Kr Krypton 84.798 (+2,0)
5	37 Rb Rubidium 85.468 (+1)	38 Sr Strontium 87.62 (+2)	39 Y Yttrium 88.906 (+3)	40 Zr Zirconium 91.224 (+4)	41 Nb Niobium 92.906 (+5)	42 Mo Molybdenum 95.95 (+6,+4)	43 Tc Technetium 98.907 (+7,+4)	44 Ru Ruthenium 101.07 (+4,+3)	45 Rh Rhodium 102.906 (+4,+3)	46 Pd Palladium 106.42 (+4,+2)	47 Ag Silver 107.868 (+1)	48 Cd Cadmium 112.414 (+2)	49 In Indium 114.818 (+3)	50 Sn Tin 118.711 (+4,+2)	51 Sb Antimony 121.760 (+5,+3,−3)	52 Te Tellurium 127.6 (+6,+4,+2,−2)	53 I Iodine 126.904 (+7,+5,+1,−1)	54 Xe Xenon 131.294 (+2)
6	55 Cs Cesium 132.905 (+1)	56 Ba Barium 137.328 (+2)	57–71	72 Hf Hafnium 178.49 (+4)	73 Ta Tantalum 180.948 (+5)	74 W Tungsten 183.85 (+6,+4)	75 Re Rhenium 186.207 (+7)	76 Os Osmium 190.23 (+8)	77 Ir Iridium 192.22 (+4,+3)	78 Pt Platinum 195.08 (+4,+2)	79 Au Gold 196.967 (+3,+1)	80 Hg Mercury 200.59 (+2,+1)	81 Tl Thallium 204.383 (+3,+1)	82 Pb Lead 207.2 (+4,+2)	83 Bi Bismuth 208.980 (+3)	84 Po Polonium [208.982] (+4,+2)	85 At Astatine 209.987 (−1)	86 Rn Radon 222.018 (0)
7	87 Fr Francium 223.020 (+1)	88 Ra Radium 226.025 (+2)	89–103	104 Rf Rutherfordium [261]	105 Db Dubnium [262]	106 Sg Seaborgium [266]	107 Bh Bohrium [264]	108 Hs Hassium [269]	109 Mt Meitnerium [278] unknown	110 Ds Darmstadtium [281] unknown	111 Rg Roentgenium [280] unknown	112 Cn Copernicium [285]	113 Nh Nihonium [286] unknown	114 Fl Flerovium [289] unknown	115 Mc Moscovium [289] unknown	116 Lv Livermorium [293] unknown	117 Ts Tennessine [294] unknown	118 Og Oganesson [294] unknown

Lanthanide Series

57 La Lanthanum 138.905 (+3)	58 Ce Cerium 140.116 (+4,+3)	59 Pr Praseodymium 140.908 (+3)	60 Nd Neodymium 144.243 (+3)	61 Pm Promethium 144.913 (+3)	62 Sm Samarium 150.36 (+3)	63 Eu Europium 151.964 (+3,+2)	64 Gd Gadolinium 157.25 (+3)	65 Tb Terbium 158.925 (+3)	66 Dy Dysprosium 162.500 (+3)	67 Ho Holmium 164.930 (+3)	68 Er Erbium 167.259 (+3)	69 Tm Thulium 168.934 (+3)	70 Yb Ytterbium 173.055 (+3)	71 Lu Lutetium 174.967 (+3)

Actinide Series

89 Ac Actinium 227.028 (+3)	90 Th Thorium 232.038 (+4)	91 Pa Protactinium 231.036 (+5)	92 U Uranium 238.029 (+6)	93 Np Neptunium 237.048 (+5)	94 Pu Plutonium 244.064 (+4)	95 Am Americium 243.061 (+3)	96 Cm Curium 247.070 (+3)	97 Bk Berkelium 247.070 (+3)	98 Cf Californium 251.080 (+3)	99 Es Einsteinium [254] (+3)	100 Fm Fermium 257.095 (+3)	101 Md Mendelevium 258.1 (+3)	102 No Nobelium 259.101 (+2)	103 Lr Lawrencium [262] (+3)

Copyright © Sterling Test Prep.

Quantized energy levels for electrons

In 1900, Nobel laureate Max Planck (1858-1947) introduced *quantized energy*.

In 1905, Albert Einstein (1879-1955) discovered that light was made up of quantized photons. Einstein found that higher-frequency photons have more energetic photons. He expressed this with the equation:

$$E = hf$$

where h = Planck's constant = 6.63×10^{-34} J·s, E = photon energy and f = photon frequency.

Distinct emission spectrum lines indicate that electron energy is quantized into discrete energy levels. If electron energy is not quantized, a continuous spectrum would be observed.

Energy of a particular level is calculated using the quantum number and given by the equation:

$$E_n = -\frac{13.6}{n^2}(eV)$$

The equation is negative, so all energies are *negative*.

Electron binding energy

Negative energies mean that energy contributes to the "stability" of the system — the electron's *binding energy*.

More negative (or lower) energy means the compound is more stable in the orbit and more difficult to liberate the electron.

Less negative (or higher) energy means that the compound is less stable in the orbit, and it is easier to remove the electron.

There is no binding energy at the highest energy, 0 eV, so the electron dissociates.

For atoms other than hydrogen, the shape of the energy level curve stays the same. However, the numerator is a constant other than 13.6 eV.

Relationship for atoms other than hydrogen is:

$$E = -\frac{Z^2 R_E}{n^2}$$

where Z is the atomic number.

Higher Z values give more *negative binding energy* (i.e., more stable) because the greater the charge, the more electrostatic attraction.

To compare how tightly bound nuclei are, divide the binding energy by the mass number A to calculate the binding energy per nucleon.

Any duplication (copies, uploads, PDFs) is illegal.

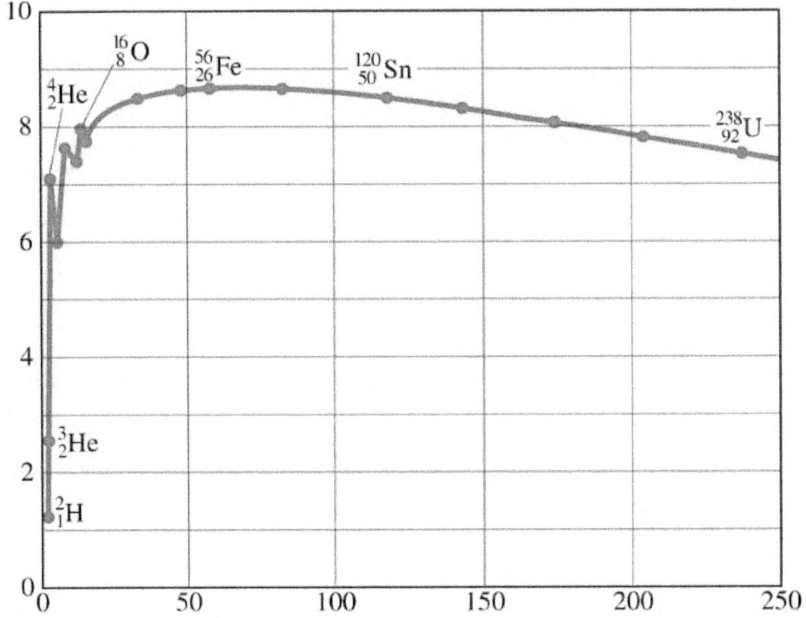

Binding energy per nucleon (MeV) vs. number of nucleons A (or mass number)

Energy emitted or absorbed for energy level transitions

Energy of the emitted or absorbed radiation is:

$$E = hf = hv = h\frac{c}{\lambda}$$

where E is energy, f and v represent frequency, and c is the speed of light.

Energy is *emitted* from transitions to lower energy levels ($n_f < n_i$).

Energy is *absorbed* from transitions to higher energy levels ($n_f > n_i$).

Photon energy is computed using the equation:

$$hf = E_H - E_L$$

Copyright © Sterling Test Prep.

Atomic Nucleus

Nuclides

Nuclei are made of protons and neutrons.

Protons have a positive charge and a mass of:

$$m_p = 1.67262 \times 10^{-27} \text{ kg}$$

Neutrons are electrically neutral and slightly more massive than protons:

$$m_n = 1.67493 \times 10^{-27} \text{ kg}$$

Neutrons and protons are collectively *nucleons* (as they reside in the nucleus).

Nuclides have different nuclei.

Number of protons defines an element and refers to the neutral atom (having no charge); two atoms with the same number of protons are the same element.

Isotopes

Isotope is an atom with the same atomic number as its counterpart on the periodic table but with a different number of neutrons.

Isotopes often have similar chemical properties but different stabilities. Some isotopes decay and give off radiation particles, while others are stable and do not decay.

For many elements, several different isotopes exist in nature.

Natural abundance is the percentage of an element that consists of a particular isotope in nature.

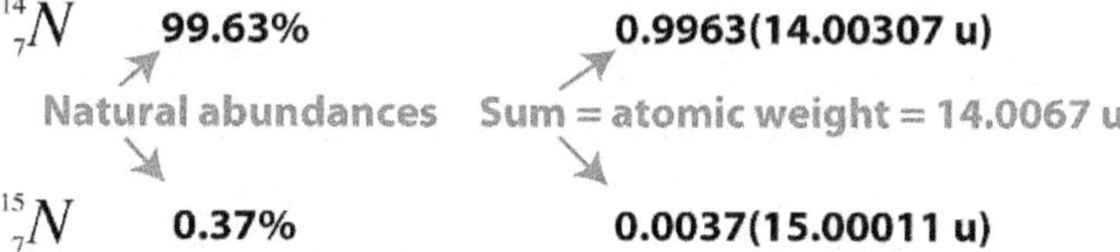

Natural abundance is the percentage of a particular isotope of an element in nature

Any duplication (copies, uploads, PDFs) is illegal.

Atomic particles

Atomic particles			
Name	Mass (amu)	Charge	Location
Proton	1	+1	In the nucleus
Neutron	1	0	In the nucleus
Electron	0	−1	Surrounding the nucleus

Atomic particle, mass, charge, and location of proton, neutron, and electron

Subatomic particles

Subatomic particles are smaller than an atom. According to the Standard Model of particle physics, a subatomic particle can be a:

composite particles are composed of other particles (e.g., a baryon, like a proton or a neutron, composed of three quarks; or a meson, composed of two quarks) or an

elementary particle, which is not composed of other particles (e.g., quarks or electrons, muons, and tau particles).

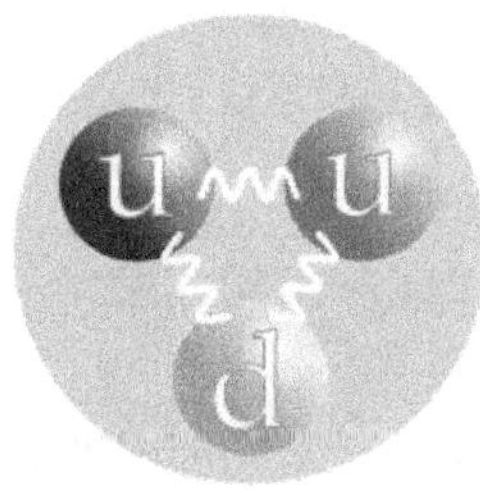

Proton is a composite particle with two up quarks and one down quark

Particle physics and nuclear physics study these particles and how they interact.

Most force-carrying particles like photons or gluons are called *bosons*, and

1) have discrete quanta of energy, do not have rest mass or discrete diameters (other than pure energy wavelength), and

2) unlike particles with rest mass and cannot overlap or combine.

Copyright © Sterling Test Prep.

Quantum field theory

Interactions of particles in quantum field theory are understood as the creation and annihilation of *quanta* of corresponding fundamental interactions, blending particle physics with field theory.

Particle physicists use diverse definitions of a particle, including:

collapsed wave function

quantum excitation of a field

irreducible representation of the Poincare group

an observed phenomenon

Standard Model of Elementary Particles

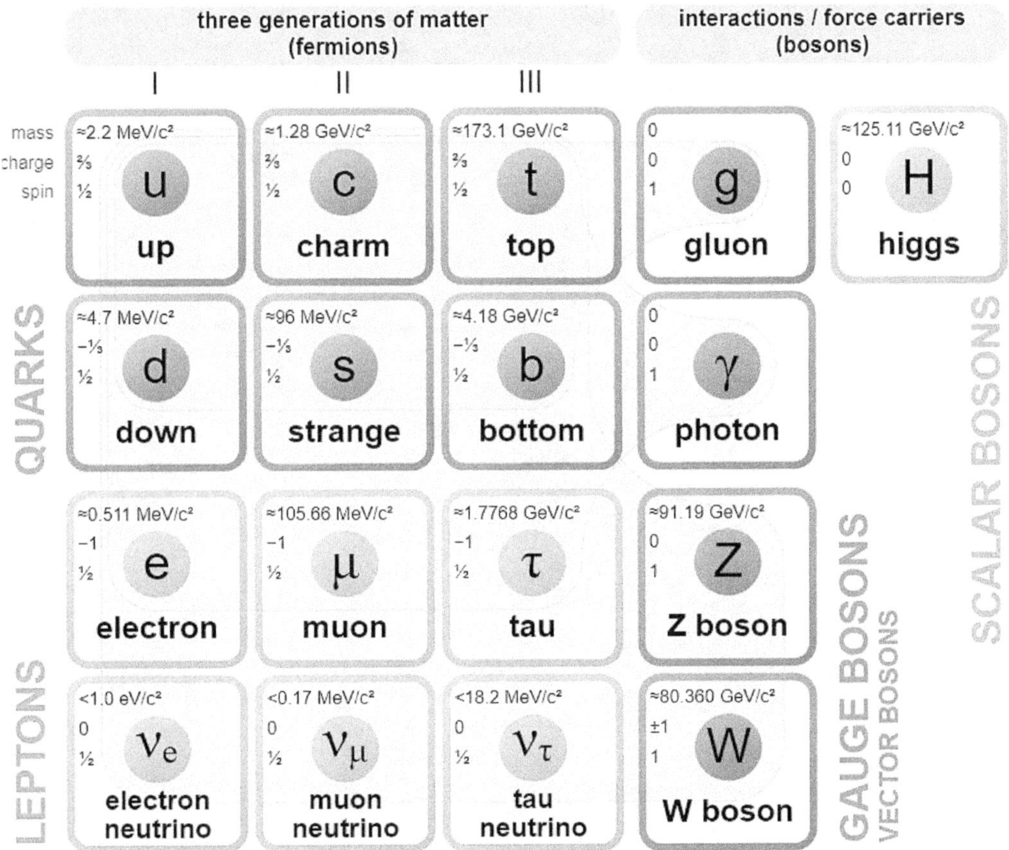

Standard model of elementary particles

Any duplication (copies, uploads, PDFs) is illegal.

Notes for active learning

Copyright © Sterling Test Prep.

Atomic Number and Atomic Weight

Atomic number

Number of protons determines an element's *atomic number* (Z). As of February 2024, 118 elements have been identified.

Number of nucleons is equal to the *atomic mass* number (A).

Atomic mass number is found by adding the number of protons and neutrons.

Neutrons are determined by subtracting the atomic number (Z) from atomic mass number (A).

Expression:

$$\text{The neutron number } N = A - Z$$

Atomic weight and mass

Atomic weight is the weighted average of *atomic mass* for isotopes of a given atom and used for an element.

Atomic mass is used for isotopes.

A and Z are sufficient to specify a nuclide.

Nuclides are symbolized as follows:

$$^{A}_{Z}X$$

X is the chemical symbol for the element; it contains the same information as Z but in a more recognizable form.

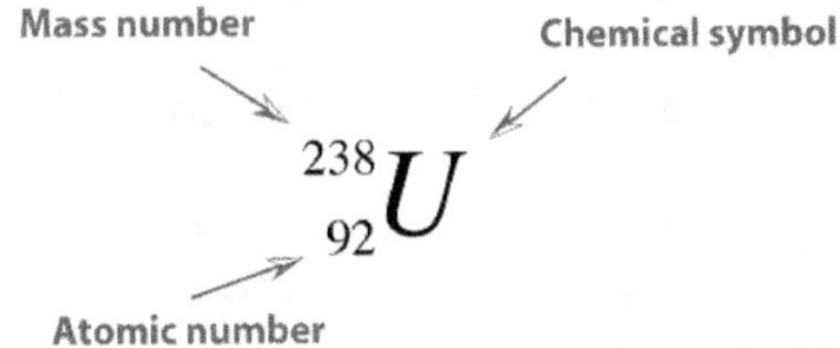

Chemical symbol with atomic number and mass number

Atomic size

Because of wave-particle duality, the size of the nucleus is fuzzy.

Measurements of high-energy electron scattering yield the radius to be:

$$r \approx (1.2 \times 10^{-15} \ m)(A^{\frac{1}{3}})$$

Masses of atoms are measured with reference to the carbon-12 (or ^{12}C) atom, which is assigned a mass of exactly 12 amu (or Daltons).

$$1 \ amu = 1.6605 \times 10^{-27} \ kg$$

$$1.6605 \times 10^{-27} \ kg = 931.5 \ MeV/c^2$$

$$1 \ amu = 931.5 \ MeV/c^2$$

From the table, the electron is considerably less massive than a nucleon.

Rest mass

Rest mass (or *invariant mass*) is the mass of a body as measured when the body is at rest relative to an observer.

Electron mass (or *mass of an electron* and *electron rest mass*) is a fundamental constant defined as the mass of a stationary electron and denoted by m_e.

Electron mass is the invariant mass of an electron.

Rest mass is an *inherent property* of the body.

Rest energy of a particle is given by:

$$E_0 = m_0 c^2$$

where m_0 is the particle's rest mass and c is the speed of light.

Object	kg	Atomic mass u	MeV/c^2
Electron	9.1094×10^{-31}	0.00054858	0.51100
Proton	1.67262×10^{-27}	1.007276	938.27
$^{1}_{1}H$ atom	1.67353×10^{-27}	1.007825	938.78
Neutron	1.67493×10^{-27}	1.008665	939.57

Rest mass (in kilograms, unified atomic mass units, and MeV/c^2

Copyright © Sterling Test Prep.

Nuclear Forces

Strong nuclear and electromagnetic forces

Two forces in the nucleus are the *strong nuclear* and *electromagnetic forces*.

Strong nuclear force binds the nucleons together, contributing to the *binding energy*.

Electromagnetic force is due to the electrostatic repulsion in the group of positively charged protons in the nucleus (i.e., like-charges repel).

Nuclei stay together because the *strong nuclear force exceeds the electromagnetic repulsion*.

Strong force is *short-ranged*, less than 10^{-15} m.

Proton–proton chain reaction

Repulsion between proton particles that it overcomes is the *proton-proton Coulomb repulsion*.

$$^1_1H + {}^1_1H \rightarrow {}^2_1H + {}^0_1e$$

$$^2_1H + {}^2_1H \rightarrow {}^3_2He + {}^1_0n$$

$$^3_2He + {}^3_2He \rightarrow {}^4_2He + 2{}^1_1H$$

Coulomb's Law

Coulomb's Law (F) relates the repulsion force on two charged particles depending on their charge (q) and distance (r) from each other.

The electrostatic force between two electric charges can be quantified by Coulomb's law, according to which the force between two electrically charged point particles is directly proportional to the product of the magnitude of the two charges inversely proportional to the square of the distance.

Force (F) is significant when the distance is small.

$$F_{Coulomb} = \frac{kq_1q_2}{r^2}$$

Any duplication (copies, uploads, PDFs) is illegal.

Notes for active learning

Copyright © Sterling Test Prep.

Radioactivity

Natural radioactivity

Natural radioactivity is a spontaneous emission of particles or energy from an unstable nucleus.

At the end of the 19th century, minerals were found to darken a photographic plate even in the absence of light. This phenomenon is radioactivity.

When an atom is unstable, it will undergo *radioactive decay*.

Maximum stability for a nucleon number is 2, 8, 20, 28, 50, 82, or 126.

The higher the binding energy per nucleon, the more stable the nucleus.

Massive nuclei require extra neutrons to overcome Coulomb repulsion of protons to be stable.

Specific neutron-to-proton ratios

A band of stability depends on the ratio of protons to neutrons, which makes an atom stable.

Coulomb force is long-range; therefore, extra neutrons are needed for stability in high-Z nuclei:

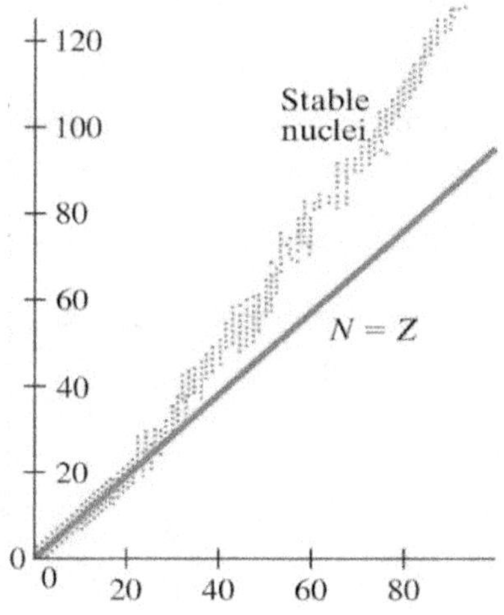

Neutron number (N = A − Z) vs. proton number (Z)

Pairs of protons and pairs of neutrons add stability to the atom.

Conversely, an odd number of protons and neutrons makes a less stable atom.

Specific neutron-to-proton ratios are desired for added stability: 1:1 for isotopes with up to 20 protons and (1+1):1 for increasingly heavy isotopes.

Any duplication (copies, uploads, PDFs) is illegal.

Unstable nuclei undergo decay

Unstable nuclei undergo decay; the *weak nuclear force* governs many such decays.

Three types of radioactive decay are *alpha*, *beta*, and *gamma decay*. These are *ionizing radiation* because they ionize material when passing through.

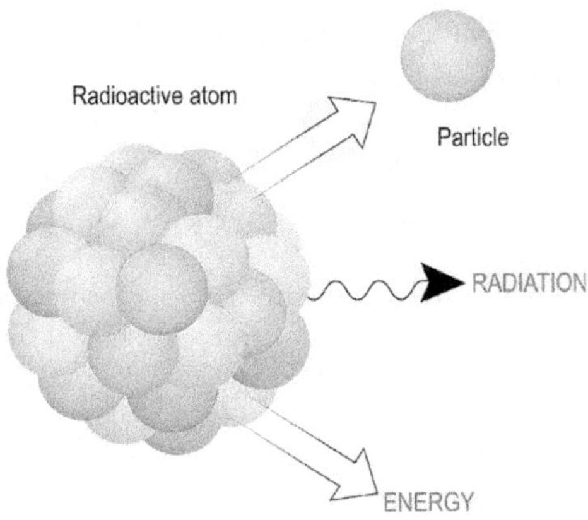

Radioactive decay creates new particles while releasing energy

Alpha emission is the expulsion of a helium nucleus, is the least penetrating, and can be stopped by a piece of paper.

Beta emission is the expulsion of an electron. Beta emission is more penetrating than alpha decay and can be stopped by 1 cm of aluminum.

Gamma decay is an emission of a high-energy photon (electromagnetic radiation).

Gamma decay is the most penetrating emission and requires a thick, dense shield (e.g., 13.5 feet of water, 6.5 feet of concrete, or about 1.3 feet of lead) to block transmission.

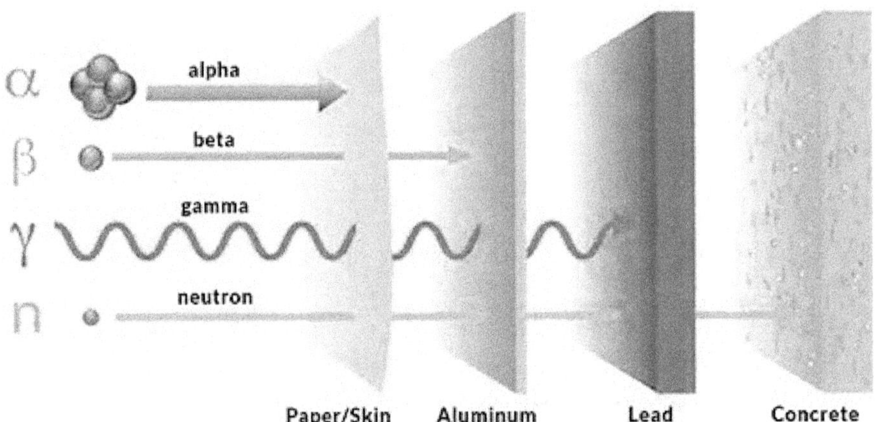

Radioactive particles penetrating paper, aluminum, and lead by alpha,
beta, gamma particles, and neutrons

Copyright © Sterling Test Prep.

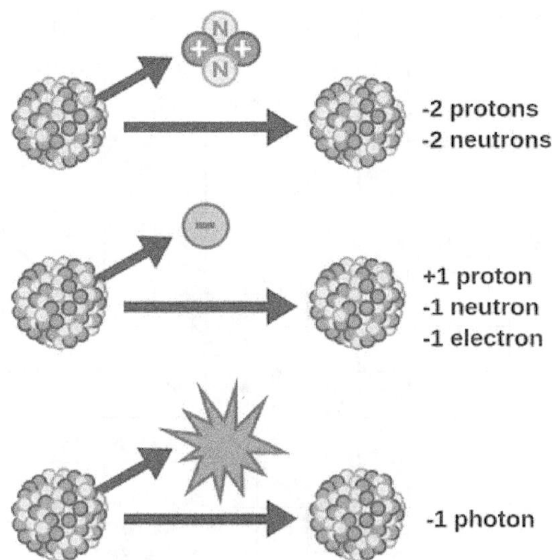

Alpha (top), beta (middle) and gamma (bottom) decay with emitted particles

Unstable condition	Type of decay	Emitted	Product nucleus
More than 83 protons	Alpha emission	$^4_2\alpha$ (^4_2He)	Lost 2 protons and 2 neutrons
Neutron-to-proton ratio too large	Beta emission	$^0_{-1}\beta$ ($^0_{-1}e$)	Gained 1 proton, no mass change
Excited nucleus	Gamma emission	$^0_0\gamma$	No change
Neutron-to-proton ratio too small	Other emissions	^0_1e	Lost 1 proton, no mass change

Radioactive decay with emitted particles and product nucleus

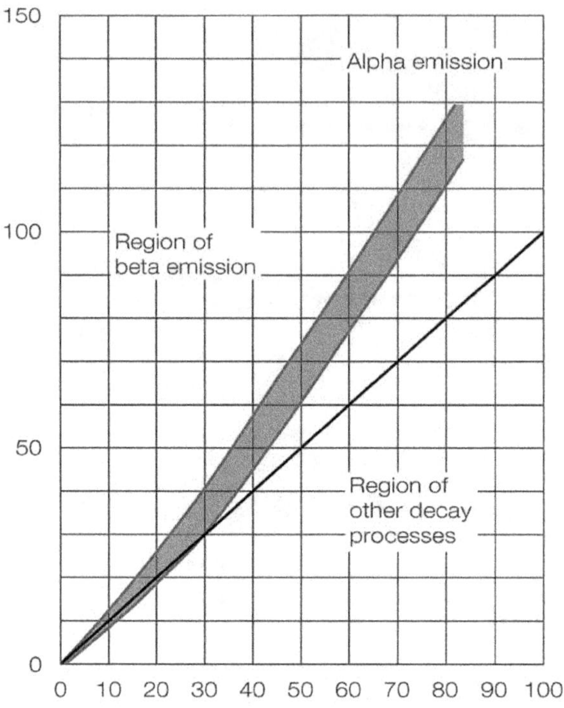

Number of neutrons vs. protons for alpha emission, beta emission, and other decay processes

Magnetic field deflections

Alpha and beta rays are bent in *opposite directions* in a magnetic field, while a magnetic field does *not* bend gamma rays.

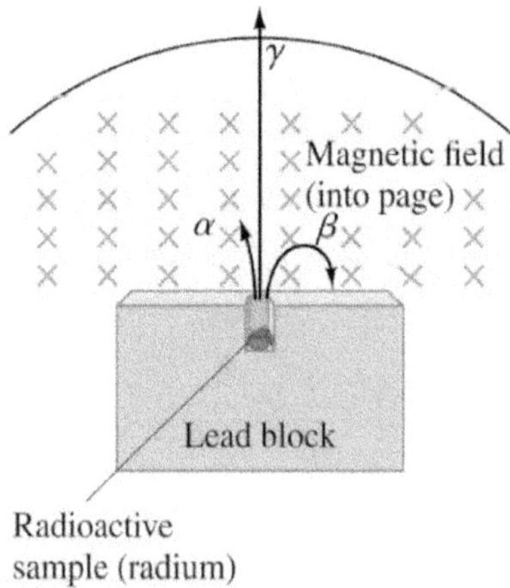

Magnetic field bends charged particles

Copyright © Sterling Test Prep.

Decay Types

Conservation of mass and charge

Conservation of mass dictates that the total atomic weight before the decay must equal the total atomic weight after.

Conservation of charge dictates that the total atomic number before the decay must equal the total atomic number after.

Radioactive decay supports the conservation that the total number of nucleons cannot change.

Do not get thrown off by unfamiliar particles. Incorporate these numbers into calculations if they have a weight and a charge. Identifying decay products is just math.

Atomic number (bottom number) determines what element it is.

Alpha decay

For example, the alpha decay of radium-227 (^{227}Ra) into radon-222 (^{222}Rn):

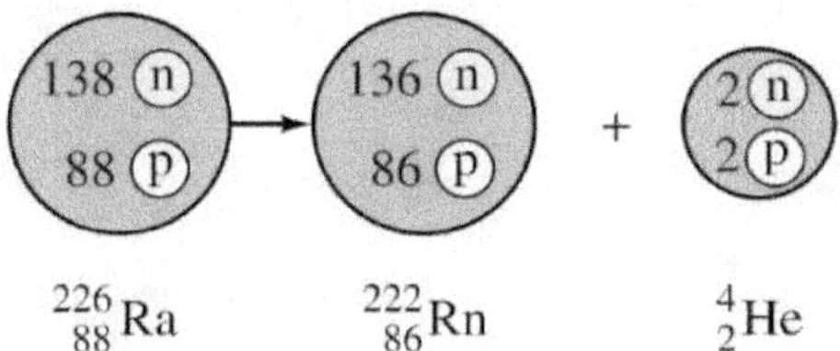

Alpha decay of radium-227 (^{227}Ra) into radon-222 (^{222}Rn)

Alpha decay can be written as:

$$^{A}_{Z}N \rightarrow {}^{A-4}_{Z-2}N' + {}^{4}_{2}He$$

Alpha decay occurs when the strong nuclear force cannot hold a large nucleus together.

Mass of the parent nucleus is greater than the sum of the masses of the daughter nucleus and the alpha particle; this difference is the *disintegration energy*.

Alpha decay is more likely than other nuclear disintegration because the alpha particle is stable.

Some smoke detectors use alpha radiation—the presence of smoke absorbs the alpha rays and keeps them from striking the collector plate.

Any duplication (copies, uploads, PDFs) is illegal.

Beta decay

Beta-decay occurs when a nucleus emits an e^{-1}. For example, the decay of carbon-14 (^{14}C):

$$^{14}_{6}C \rightarrow {}^{14}_{7}N + e^- + \text{neutrino}$$

The nucleus has 14 nucleons, but it has one more proton and one fewer neutron. This decay is an example of an interaction that proceeds via the weak nuclear force.

An electron in beta decay is not orbital; it is *created* in the decay.

Fundamental process is a neutron decaying to a proton, electron, and neutrino:

$$n \rightarrow p + e^- + neutrino$$

Neutrinos

Neutrinos were discovered through energy and momentum conservation analysis in beta decay—it could not be a two-particle decay.

Neutrinos are notoriously difficult to detect because they interact weakly, and direct evidence for their existence was unavailable for some time.

Neutrino symbol is the Greek letter nu (v); the beta decay of carbon-14 is written as (the bar over the neutrino means that it is an antineutrino):

$$^{14}_{6}C \rightarrow {}^{14}_{7}N + e^- + \overline{v}$$

Beta decay can occur when the nucleus emits a positron rather than an electron:

$$^{19}_{10}Ne \rightarrow {}^{19}_{9}F + e^+ + \overline{v}$$

Nucleus can capture one of its inner electrons:

$$^{7}_{4}Be + e^- \rightarrow {}^{7}_{3}Li + v$$

Gamma decay

Gamma rays are high-energy photons. They are emitted when a nucleus decays from an excited state to a lower state, just as electrons returning to a lower state emit photons.

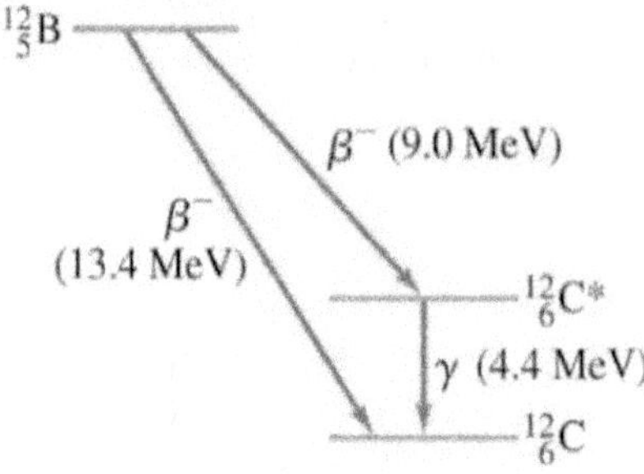

Gamma rays are high-energy photons emitted when a nucleus decays from an excited state

Copyright © Sterling Test Prep.

Radioactive Decay Rates

Half-life

Half-life is the time required for ½ a radioactive sample to decay.

For example, a 1 kg sample of an unstable isotope with a one-day half-life:

       After 1 day - 500 g remain

       After 2 days - 250 g remain

       After 3 days - 125 g remain

Isotope	Half-life	Mode of decay
$^{3}_{1}H$ (tritium)	12.26 years	Beta
$^{14}_{6}C$	5,730 years	Beta
$^{90}_{38}Sr$	28 years	Beta
$^{131}_{53}I$	8 days	Beta
$^{133}_{54}Xe$	5.27 days	Beta
$^{238}_{92}U$	4.51×10^{9} years	Alpha
$^{242}_{94}Pu$	3.79×10^{5} years	Alpha
$^{240}_{94}Pu$	6,760 years	Alpha
$^{239}_{94}Pu$	24,360 years	Alpha
$^{40}_{19}K$	1.3×10^{9} years	Alpha

Half-life of sample radioactive isotopes

Any duplication (copies, uploads, PDFs) is illegal.

Exponential decay

Half-life graphs are shown with decreasing exponential:

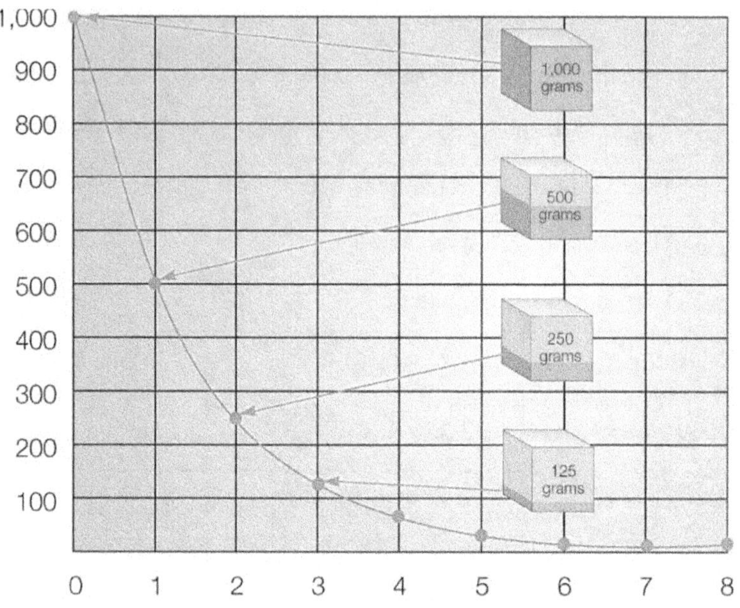

Mass (g) vs. time (days) with decreasing exponential half-life

Decay rate as a function of time

Nuclear decay is a random process; the decay of one nucleus does not influence another.

Decay rate is *unaffected* by temperature, pressure, volume, or other environmental factor.

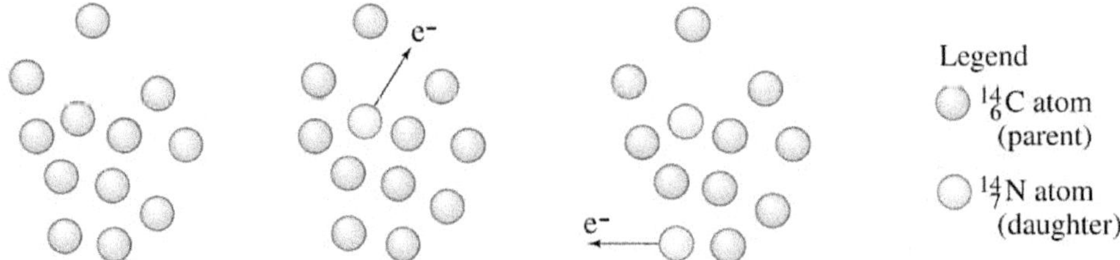

^{14}C with 6 protons, the parent atom decays into ^{14}N with 7 protons daughter atom

Number of decays in a short time interval is proportional to the number of nuclei and time:

$$\Delta N = -\lambda N \Delta t$$

Here, λ is a constant characteristic of that nuclide, the decay constant.

The number of decays equation uses calculus for N as a function of time:

$$N = N_0 e^{-\lambda t}$$

Copyright © Sterling Test Prep.

Plotting decay rates

Number of parental nuclei plotted:

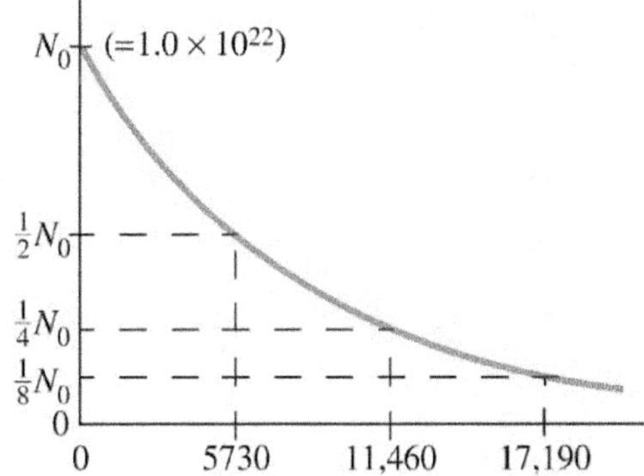

Number of parental nuclei (N) vs. time (years)

Decays per second plotted:

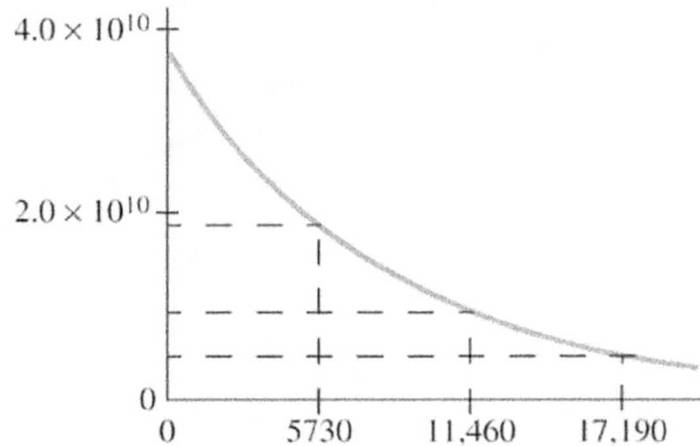

Decays per second (ΔN/Δt) vs. time (years)

Half-life calculations

Half-life is related to the decay constant:

$$T_{\frac{1}{2}} = \frac{\ln 2}{\lambda} = \frac{0.693}{\lambda}$$

It can also be written:

$$N_t = N_{t=0} \cdot \left(\frac{1}{2}\right)^{\#half-lives} = N_{t=0} \cdot \left(\frac{1}{2}\right)^{\frac{t}{half-life}}$$

where $N_{t=0}$ is amount of original starting material, N_t is remaining material, and t is time.

The above is the formal half-life equation, but it may be easier to multiply rather than divide. Therefore, a more user-friendly equation is:

$$N_{t=0} = N_t \cdot 2^{\#half-lives} = N_t \cdot 2^{\frac{t}{half-life}}$$

Any duplication (copies, uploads, PDFs) is illegal.

Semi-log plots

Semi-log plots convert exponential curves into *straight lines*.

- Something that *curves up* becomes a *straight line with a positive slope*.

- Something that *curves down* becomes a *straight line with a negative slope*.

- For exponential decay, a semi-log plot graphs the *log of amount vs. time*.

- For exponential decay, a semi-log plot is a straight line with a *negative slope*.

- Semi-log plot intercepts the *x*-axis, where the original *y*-value is 1.

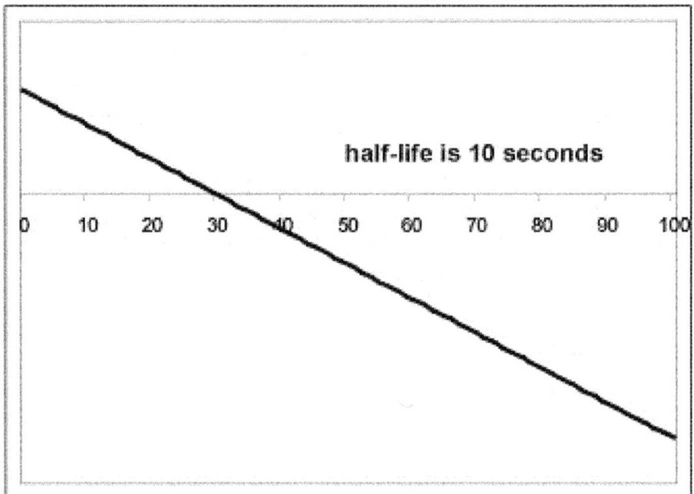

*Exponential decay in a semi-log plot of the amount of material (logarithmic
scale) vs. time in seconds (plotted in non-logarithmic scale)*

Copyright © Sterling Test Prep.

Decay Series

Novel isotopes

Decay series occurs when one radioactive isotope decays to another radioactive isotope, which decays to another, and so on.

Decay series allows for the creation of nuclei that otherwise would not exist in nature.

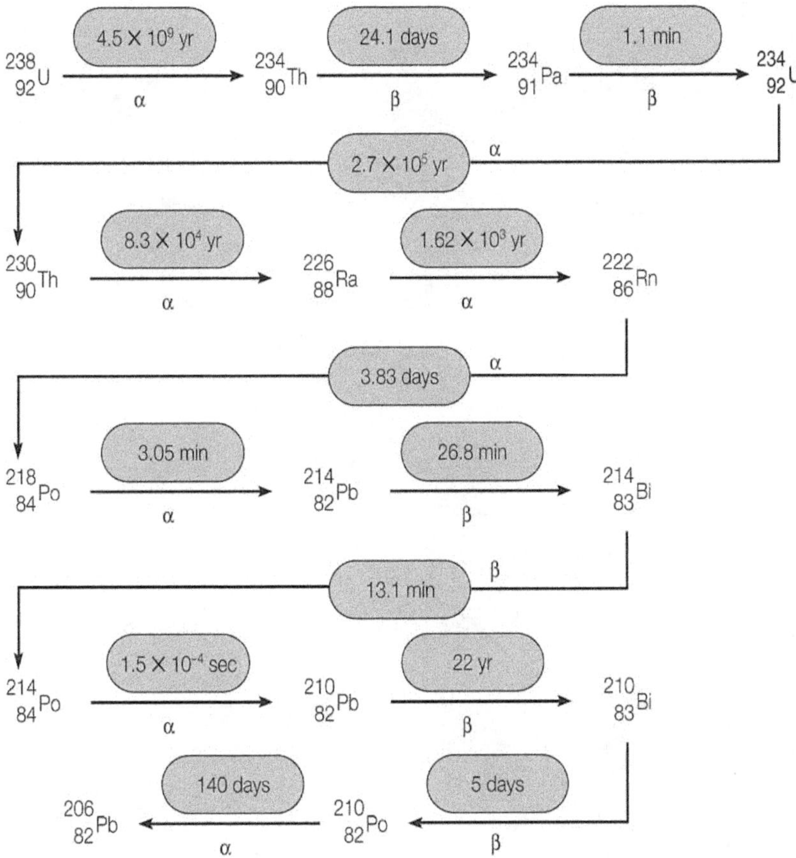

Decay series occurs when one radioactive substance decays into another

Radioactive dating with carbon

Radioactive dating analyzes the fraction of carbon in organic material that is carbon-14. The ratio of ^{14}C to ^{12}C in the atmosphere has been roughly constant over thousands of years.

A living plant or tree constantly exchanges carbon with the atmosphere and will have the same carbon ratio in its tissue. When the plant dies, this exchange stops.

^{14}Carbon has a half-life of about 5,730 years; it decays and becomes a smaller fraction of the plant tissue's total carbon. This fraction can be measured, and the age of the tissue is deduced.

Any duplication (copies, uploads, PDFs) is illegal.

Objects older than about 60,000 years cannot be dated this way because there is little ^{14}carbon remaining. Other isotopes are helpful for geologic time-scale dating.

Uranium-238 (^{238}U) has a half-life of 4.5×10^9 years and has been used to date the oldest rocks on Earth as about 4 billion years old.

Radioactive decay timelines

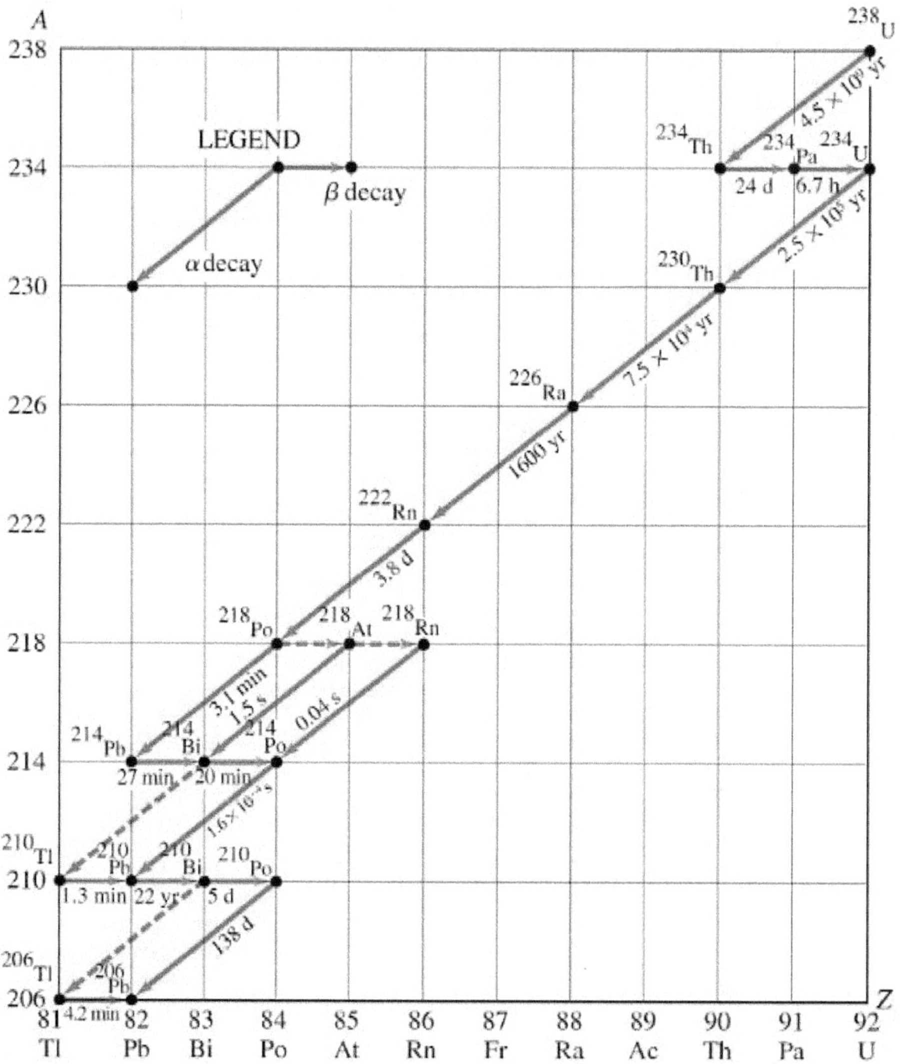

Elements undergo alpha and beta decay with corresponding years noted

Copyright © Sterling Test Prep.

Energy barriers

Energy is released when nuclei decay by alpha emission, but nuclei do not decay immediately.

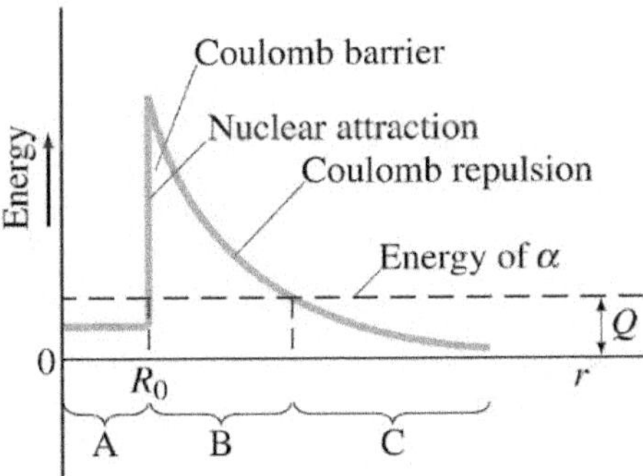

Energy barriers for nuclei decay by alpha emissions

Although energy is released in the decay, there is an energy barrier. The alpha particle escapes through a quantum mechanical phenomenon of *tunneling*.

Heisenberg uncertainty principle, energy conservation differences can be violated temporarily:

$$Q = M_p c^2 - (M_D + m_\alpha)c^2$$

The higher the energy barrier, the less time the alpha particle has to get through it, and the less likely it is to happen; this accounts for the extensive variation in half-lives for alpha decay.

Notes for active learning

Copyright © Sterling Test Prep.

Mass Deficit, Energy Liberated and Binding Energy

Mass deficit

Mass deficit is the difference between the masses of the *reactants and products.*

Nuclear energy is an interconversion of mass and energy. Nuclear energy is connected to the origins of the universe and the life cycles of stars.

Big Bang theory states that incredibly hot, dense, primordial plasma cooled, creating protons and neutrons. Continued cooling leads to hydrogen atoms, which collapse gravitationally into 1st generation stars.

Nuclei and binding energy

Gravitational attraction is the ultimate source of nuclear energy.

Binding energy is required to break a nucleus into individual protons and neutrons.

Ratio of *binding energy* to the *nucleon number* indicates how stable a nucleus is.

Iron-56 is the most stable nucleus:

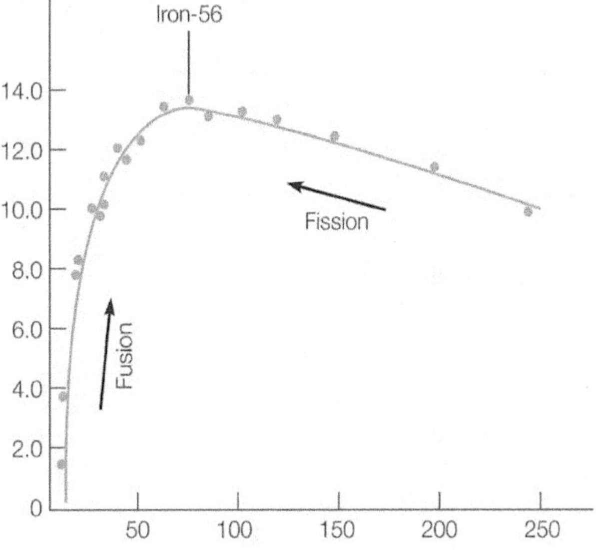

Binding energy per nucleon (10^{-13} J) vs. mass number

Any duplication (copies, uploads, PDFs) is illegal.

Conserving charge and mass number

Balanced equations present nuclear reactions where the charge is conserved, as is mass number.

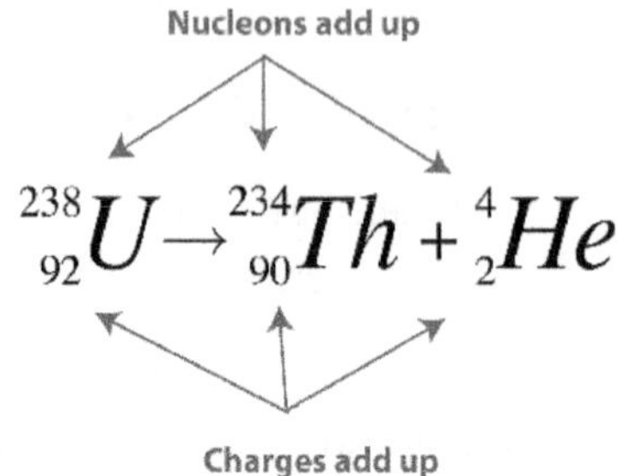

$$_{92}^{238}U \rightarrow {}_{90}^{234}Th + {}_{2}^{4}He$$

Mass number and charge are conserved

Classifying nuclear particles

Particle	Symbol	Mass number	Charge
Proton	$_{1}^{1}H$ (or $_{1}^{1}p$)	1	1+
Electron	$_{-1}^{0}e$ (or $_{-1}^{0}\beta$)	0	1−
Neutron	$_{0}^{1}n$	1	0
Gamma Photon	$_{0}^{0}\gamma$	0	0

Names, symbols, and properties of particles in nuclear equations

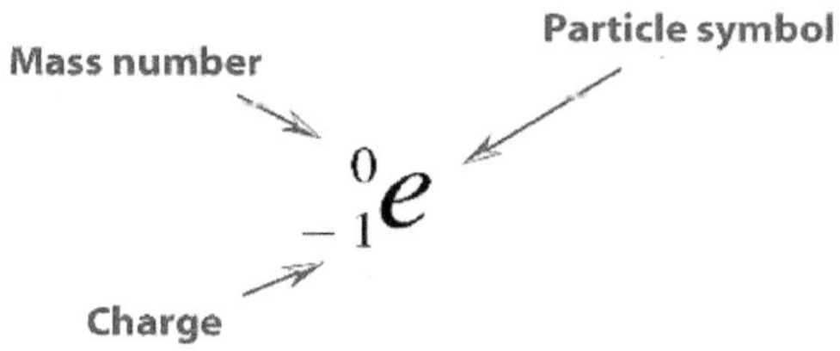

Particles labeled with mass number, charge, and symbol

Copyright © Sterling Test Prep.

Mass converts to energy

The difference in mass before and after a reaction is the *mass deficit* (or *mass defect*). If the total mass before the reaction differs after the reaction, then the difference in mass is energy.

Energy is liberated when mass is lost during a reaction because mass and energy are conserved. Total mass and energy before a reaction equals the total mass and energy afterward.

Energy that makes up for the mass deficit is calculated by:

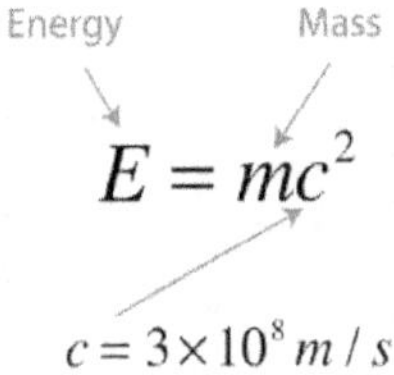

$$E = mc^2$$

$$c = 3 \times 10^8 \, m/s$$

Energy-mass deficit relationship

Mass of the nucleons is higher than the mass of the atom. Mass has become energy (e.g., radiation or kinetic energy) released during the formation of the nucleus. The difference between the mass of the constituents and the nucleus is the *binding energy* of the nucleus.

Energy liberated equals binding energy

Energy liberated equals the *binding energy*, which is used when converting ΔM into its equivalent in energy (ΔMc^2).

$$M_{nucleons} = M_{atom} + \text{binding energy}/c^2$$

Binding energy commonly refers to nuclear binding energy (energy binding the nucleons).

Binding energy per nucleon is strongest for Iron (^{56}Fe) and weakest for Deuterium (the 2-nucleon isotope of hydrogen).

Electron binding energy is less commonly used because it is called *ionization energy*.

$$m_{nucleons} - m_{atom} = \text{mass deficit (also mass defect)} = \Delta m$$

$$\Delta m = 4.0330 \, u - 4.0026 \, u = 0.0304 \, u = 28.3 \, Mev$$

Reactant masses

$$2(1.0078u) + 2(1.0087u) = 4.0330u$$

$$2\,^1_1H + 2\,^1_0n \rightarrow \,^4_2He$$

4.0026u

Mass defect = binding energy

Helium mass

Mass deficit equals binding energy

Any duplication (copies, uploads, PDFs) is illegal.

Notes for active learning

Copyright © Sterling Test Prep.

Mass Spectrometry

Identifying chemical composition

Mass spectrometry is a technique that identifies the chemicals in a substance.

Magnetic fields deflect charged particles.

Mass spectrometers convert atoms into ions, increasing their acceleration so they all have the same kinetic energy and deflecting them with a magnetic field.

It creates ions by removing electrons from the atom—even if an atom typically forms a negative ion, a mass spectrometer usually works with positive ions.

Particles are deflected, and their mass can be calculated because the deflection force is known. The curve of their deflection path separates atoms; lighter ions are deflected more.

The higher the positive charge on an ion, the more the magnetic field deflects it.

Mass spectrometer

In a typical mass spectrometer procedure, a sample (i.e., solid, liquid, or gaseous) is ionized by bombarding it with a beam of electrons. This may cause sample molecules to break into positively charged fragments or become positively charged without fragmenting.

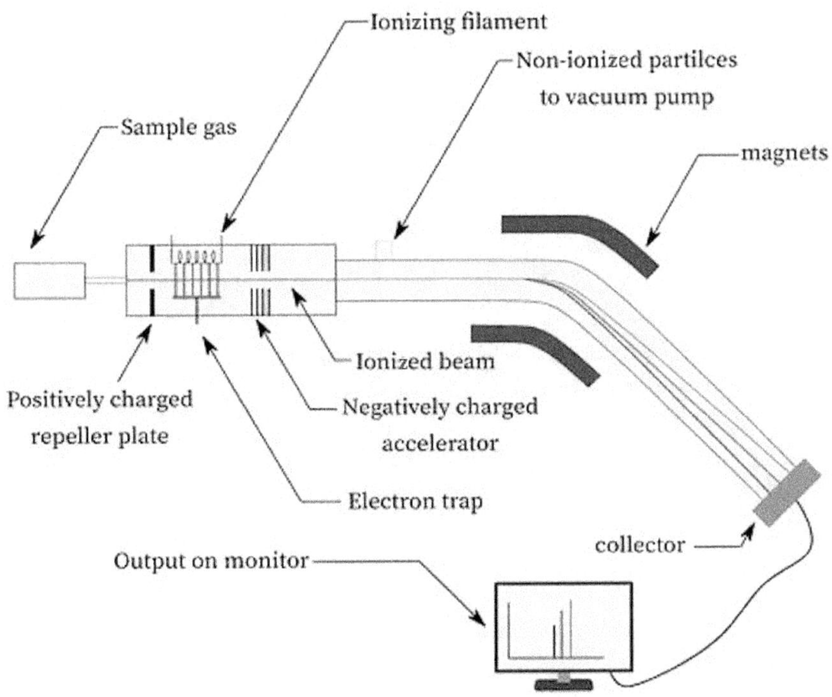

Mass spectrometers measure the mass-to-charge ratio of a sample

Mass-to-charge ratio

Mass spectrometers measure the *mass-to-charge ratio* of a sample and the number of ions in the gas phase, producing a *mass spectrum.*

A mass spectrum is produced, resembling a bar graph (shown below). The mass spectrum gives the concentration of the atom, organized by their mass-to-charge ratio.

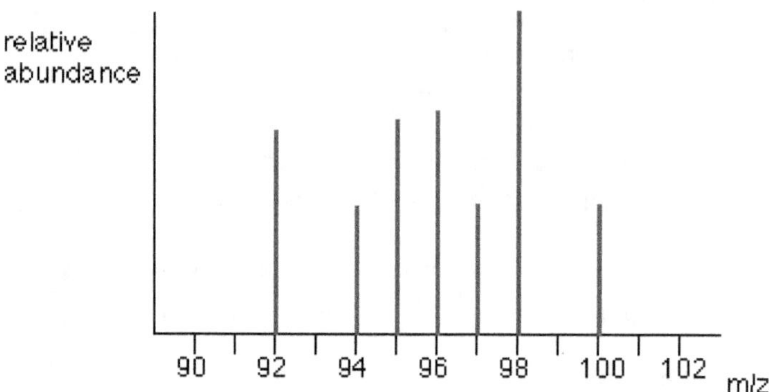

Mass spectrum plotting relative abundance against the mass-charge ratio

Copyright © Sterling Test Prep.

Practice Questions

1. In a nuclear reaction, the mass of the products is less than the mass of the reactants. Why is this not observed in a chemical reaction?

    A. In chemical reactions, the mass is held constant by the nucleus

    B. In chemical reactions, the mass deficit is balanced by a mass surplus

    C. The mass deficit in chemical reactions is too small to be observed

    D. The mass does not convert to energy in chemical reactions

2. The isotope $^{13}_{7}N$ decays by positron emission to what isotope?

    A. $^{14}_{6}C$ **C.** $^{13}_{6}C$

    B. $^{11}_{7}N$ **D.** $^{12}_{6}C$

3. The attraction of the nucleus on the outermost electron in an atom tends to:

    A. decrease from right to left and bottom to top on the periodic table

    B. decrease from left to right and bottom to top on the periodic table

    C. decrease from left to right and top to bottom of the periodic table

    D. decrease from right to left and top to bottom on the periodic table

4. Which of the statements best describes the strong nuclear force?

    A. The strength of the force increases with distance

    B. The force is powerful and is effective over a large range of distances

    C. The electrical force is stronger than the nuclear force

    D. The force is powerful but is effective only within a short range of distances

5. Natural line broadening can be understood regarding the:

    A. Schrodinger wave equation   **C.** de Broglie wavelength

    B. Pauli exclusion principle   **D.** uncertainty principle

6. Which characteristics describe the mass, charge, and location of a proton, respectively?

    A. approximate mass 5×10^{-4} amu; charge +2; inside nucleus

    B. approximate mass 1 amu; charge 0; inside nucleus

    C. approximate mass 1 amu; charge +1; inside nucleus

    D. approximate mass 5×10^{-4} amu; charge −1; outside nucleus

Any duplication (copies, uploads, PDFs) is illegal.

7. Which of the following subshell notations for electron occupancy is NOT possible?

A. $4f^{11}$ **B.** $2p^1$ **C.** $5s^3$ **D.** $4p^5$

8. The main reason that there is a limit to the size of a stable nucleus is the:

A. weakness of the electrostatic force

B. weakness of the gravitational force

C. short-range effect of the strong nuclear force

D. limited range of the gravitational force

9. Elements combine in fixed mass ratios to form compounds. This fixed mass ratio requires the following elements:

A. have unambiguous atomic numbers

B. are always chemically active

C. are composed of continuous matter without subunits

D. are composed of discrete subunits called atoms

10. Which of the following compounds does NOT exist?

A. H^-, because H forms only positive ions

B. PbO_2, because the charge on a Pb ion is only +2

C. SF_6, because F does not have an empty d orbital to form an expanded octet

D. OCl_6, because O does not have d orbitals to form an expanded octet

E. $^-CH_3$, because C lacks d orbitals

11. Paramagnetism, the ability to be pulled into a magnetic field, is demonstrated by:

A. any substance containing unpaired electrons

B. nonmetal elements that have unpaired p orbital electrons

C. transition elements that have unpaired d orbital electrons

D. nonmetal elements that have paired p orbital electrons

12. Ionizing radiation is:

A. a neutron that has acquired a charge, resulting in the formation of an ion

B. high-energy radiation that removes electrons from atoms or molecules

C. radiation that only interacts with ions

D. equivalent to a proton

Copyright © Sterling Test Prep.

13. What nucleus results when ^{55}Ni decays by positron emission?

   A. ^{55}Ca   **C.** ^{55}Co

   B. ^{55}Ni   **D.** ^{55}Fe

14. Which element has the electron configuration $1s^2 2s^2 2p^6 3s^2 3p^6 4s^2 3d^{10} 4p^6 5s^2 4d^1$?

   A. Y   **B.** La   **C.** Si   **D.** Sc

15. What is its temperature if a star has a peak intensity at 580 nm? (Use Wien's displacement constant $b = 2.9 \times 10^{-3}$ K·m)

   A. 5,000 °C   **C.** 5,000 °F

   B. 5,000 K   **D.** 2,000 K

Any duplication (copies, uploads, PDFs) is illegal.

Notes for active learning

Notes for active learning

Copyright © Sterling Test Prep.

Detailed Explanations

1. D is correct.

Mass is conserved in a chemical reaction; no particles are created or destroyed. Atoms are rearranged to form products from reactants.

In nuclear reactions, the mass difference is due to mass converting into energy.

2. C is correct.

Positron emission occurs during β^+.

The decay equation:

$$^{13}_{7}\text{N} \rightarrow\ ^{13}_{6}\text{C} + e^+ + \nu_e$$

A proton converts into a neutron, and a positron is ejected with an electron neutrino. The daughter nuclide is ^{13}C.

3. D is correct.

The nucleus's attraction on the outermost electrons determines the *ionization energy*, increasing right and up on the periodic table.

4. D is correct.

The strong nuclear force is the strongest of the four fundamental forces. However, it only acts within a small distance range (about the nucleus's diameter).

5. D is correct.

Natural line broadening extends a spectral line over a range of frequencies. This occurs partly due to uncertainty principle, which relates the time an atom is excited to emitted photon energy.

$$\Delta E \Delta t > \hbar\,/\,2$$

where $\hbar$ is the symbol for the reduced Planck's constant

Because energy is related to frequency by:

$$E = hf$$

where h is Planck's constant

Range of frequencies observed (broadening) is due to the uncertainty in energy outlined by the uncertainty principle.

6. C is correct.

Protons are the positively charged particles located inside the nucleus of an atom.

Like neutrons (also in the nucleus), protons have a mass of approximately 1 amu.

Protons have a +1 charge, while neutrons have a charge of 0 (i.e., neutral).

7. C is correct.

Each orbital can hold two electrons.

Maximum number of electrons in each shell:

 The *s* subshell has 1 spherical orbital and can accommodate 2 electrons

 The *p* subshell has 3 dumbbell-shaped orbitals and can accommodate 6 electrons

 The *d* subshell has 5 lobe-shaped orbitals and can accommodate 10 electrons

 The *f* subshell has 7 orbitals and can accommodate 14 electrons

Therefore, the *s* subshell can accommodate only 2 electrons.

8. C is correct.

Nuclei with atomic numbers greater than 83 are inherently unstable and thus radioactive. This limit in size is because the strong nuclear force has a short range, and as the nucleus gets larger, the strong nuclear force cannot overcome Coulomb repulsion from protons within the nucleus.

9. D is correct.

Its molar mass can measure the mass of an element or compound.

Molar mass relates the mass of the element or compound to a discrete number of subunits (atoms for elements, molecules for compounds).

10. D is correct.

O contains only *s* and *p* orbitals and cannot form 6 bonds due to the lack of *d* orbitals.

Hydrogen forms a negative ion (i.e., hydride) in NaH (sodium hydride), $LiAlH_4$ (lithium aluminum hydride), and $NaBH_4$ (sodium borohydride).

Each of these hydrides (i.e., H^-) is a strong base with a pK_a greater than 32.

Lead (Pb), although not a transition metal, does have two common valences: +2 and +4.

Fluorine in SF_6 does not have an expanded octet because each F is bonded only to S. The sulfur has an expanded octet, but it is in the 3rd row of the periodic table and has empty *d* orbitals.

Copyright © Sterling Test Prep.

11. A is correct.

Paramagnetic elements are substances that move into a magnetic field and have one or more unpaired electrons.

Most transition metals and their compounds in oxidation states involving incomplete inner electron subshells are paramagnetic.

An applied magnetic field repels diamagnetic elements because they create an induced magnetic field in a direction opposite to an externally applied magnetic field.

12. B is correct.

Ionizing radiation can be high-energy charged particles (alpha, beta) or electromagnetic waves (X-rays, gamma rays).

All are ionizing because they possess enough energy to remove and ionize electrons from atoms (or molecules).

13. C is correct.

Nuclear reaction:

$$^{55}_{28}\text{Ni} \rightarrow {}^{55}_{27}\text{Co} + e^+ + v_e$$

where Co = product, e^+ = positron and v_e = electron neutrino

In positron emission (β^+ decay), a proton in the nucleus converts to a neutron while releasing a positron and an electron neutrino.

The atomic number decreases by one, but the mass number stays constant.

14. A is correct.

Identify an element using the periodic table by using its atomic number.

The *atomic number* equals the number of protons or electrons.

The total number of electrons is determined by summing the electrons in the provided electron configuration:

$$2 + 2 + 6 + 2 + 6 + 2 + 10 + 6 + 2 + 1 = 39$$

Element #39 in the periodic table is yttrium (Y).

Any duplication (copies, uploads, PDFs) is illegal.

15. B is correct.

Use Wien's Displacement Law:

$$\lambda_{max} = b \,/\, T$$

$$T = (2.9 \times 10^{-3}\,\text{K·m}) \,/\, (580 \times 10^{-9}\,\text{m})$$

$$T = 5{,}000\,\text{K}$$

Copyright © Sterling Test Prep.

Notes for active learning

Notes for active learning

Copyright © Sterling Test Prep.

APPENDIX

Copyright © Sterling Test Prep.

Periodic Table of the Elements

Key to the element boxes:

Atomic Number · Valence · **Symbol** · Name · Atomic Mass

Each cell below is listed as: Symbol · Atomic Number · Atomic Mass.

1 IA / 1A	2 IIA / 2A	3 IIIB / 3B	4 IVB / 4B	5 VB / 5B	6 VIB / 6B	7 VIIB / 7B	8 VIII / 8	9 VIII / 8	10	11 IB / 1B	12 IIB / 2B	13 IIIA / 3A	14 IVA / 4A	15 VA / 5A	16 VIA / 6A	17 VIIA / 7A	18 VIIIA / 8A
H 1 1.008																	He 2 4.003
Li 3 6.941	Be 4 9.012											B 5 10.811	C 6 12.011	N 7 14.007	O 8 15.999	F 9 18.998	Ne 10 20.180
Na 11 22.990	Mg 12 24.305											Al 13 26.982	Si 14 28.086	P 15 30.974	S 16 32.066	Cl 17 35.453	Ar 18 39.948
K 19 39.098	Ca 20 40.078	Sc 21 44.956	Ti 22 47.88	V 23 50.942	Cr 24 51.996	Mn 25 54.938	Fe 26 55.845	Co 27 58.933	Ni 28 58.693	Cu 29 63.546	Zn 30 65.38	Ga 31 69.723	Ge 32 72.631	As 33 74.922	Se 34 78.971	Br 35 79.904	Kr 36 84.798
Rb 37 85.468	Sr 38 87.62	Y 39 88.906	Zr 40 91.224	Nb 41 92.906	Mo 42 95.95	Tc 43 98.907	Ru 44 101.07	Rh 45 102.906	Pd 46 106.42	Ag 47 107.868	Cd 48 112.414	In 49 114.818	Sn 50 118.711	Sb 51 121.760	Te 52 127.6	I 53 126.904	Xe 54 131.294
Cs 55 132.905	Ba 56 137.328	57-71	Hf 72 178.49	Ta 73 180.948	W 74 183.85	Re 75 186.207	Os 76 190.23	Ir 77 192.22	Pt 78 195.08	Au 79 196.967	Hg 80 200.59	Tl 81 204.383	Pb 82 207.2	Bi 83 208.980	Po 84 [208.982]	At 85 209.987	Rn 86 222.018
Fr 87 223.020	Ra 88 226.025	89-103	Rf 104 [261]	Db 105 [262]	Sg 106 [266]	Bh 107 [264]	Hs 108 [269]	Mt 109 [278]	Ds 110 [281]	Rg 111 [280]	Cn 112 [285]	Nh 113 [286]	Fl 114 [289]	Mc 115 [289]	Lv 116 [293]	Ts 117 [294]	Og 118 [294]

Lanthanide Series

57	58	59	60	61	62	63	64	65	66	67	68	69	70	71
La 138.905	Ce 140.116	Pr 140.908	Nd 144.243	Pm 144.913	Sm 150.36	Eu 151.964	Gd 157.25	Tb 158.925	Dy 162.500	Ho 164.930	Er 167.259	Tm 168.934	Yb 173.055	Lu 174.967

Actinide Series

89	90	91	92	93	94	95	96	97	98	99	100	101	102	103
Ac 227.028	Th 232.038	Pa 231.036	U 238.029	Np 237.048	Pu 244.064	Am 243.061	Cm 247.070	Bk 247.070	Cf 251.080	Es [254]	Fm 257.095	Md 258.1	No 259.101	Lr [262]

Any duplication (copies, uploads, PDFs) is illegal.

Notes for active learning

Copyright © Sterling Test Prep.

Common Equations

Throughout the test the following symbols have the definitions specified unless otherwise noted.

L, mL	=	liter(s), milliliter(s)	mm Hg	=	millimeters of mercury
g	=	gram(s)	J, kJ	=	joule(s), kilojoule(s)
nm	=	nanometer(s)	V	=	volt(s)
atm	=	atmosphere(s)	mol	=	mole(s)

ATOMIC STRUCTURE

$$E = h\nu$$
$$c = \lambda\nu$$

E = energy
ν = frequency
λ = wavelength

Planck's constant, $h = 6.626 \times 10^{-34}$ J s

Speed of light, $c = 2.998 \times 10^8$ m s^{-1}

Avogadro's number $= 6.022 \times 10^{23}$ mol^{-1}

Electron charge, $e = -1.602 \times 10^{-19}$ coulomb

EQUILIBRIUM

$$K_c = \frac{[C]^c[D]^d}{[A]^a[B]^b} \text{, where } a\,A + b\,B \rightleftarrows c\,C + d\,D$$

$$K_p = \frac{(P_C)^c(P_D)^d}{(P_A)^a(P_B)^b}$$

$$K_a = \frac{[H^+][A^-]}{[HA]}$$

$$K_b = \frac{[OH^-][HB^+]}{[B]}$$

$$K_w = [H^+][OH^-] = 1.0 \times 10^{-14} \text{ at } 25°C$$
$$= K_a \times K_b$$

$$pH = -\log[H^+], \quad pOH = -\log[OH^-]$$

$$14 = pH + pOH$$

$$pH = pK_a + \log\frac{[A^-]}{[HA]}$$

$$pK_a = -\log K_a, \quad pK_b = -\log K_b$$

Equilibrium Constants

K_c (molar concentrations)
K_p (gas pressures)
K_a (weak acid)
K_b (weak base)
K_w (water)

KINETICS

$$\ln[A]_t - \ln[A]_0 = -kt$$

$$\frac{1}{[A]_t} - \frac{1}{[A]_0} = kt$$

$$t_{1/2} = \frac{0.693}{k}$$

k = rate constant
t = time
$t_{1/2}$ = half-life

Any duplication (copies, uploads, PDFs) is illegal.

GASES, LIQUIDS, AND SOLUTIONS

$$PV = nRT$$

$$P_A = P_{total} \times X_A, \text{ where } X_A = \frac{\text{moles A}}{\text{total moles}}$$

$$P_{total} = P_A + P_B + P_C + \ldots$$

$$n = \frac{m}{M}$$

$$K = {}^\circ C + 273$$

$$D = \frac{m}{V}$$

$$KE \text{ per molecule} = \frac{1}{2}mv^2$$

Molarity, M = moles of solute per liter of solution

$$A = abc$$

P = pressure
V = volume
T = temperature
n = number of moles
m = mass
M = molar mass
D = density
KE = kinetic energy
v = velocity
A = absorbance
a = molar absorptivity
b = path length
c = concentration

Gas constant, R = 8.314 J mol^{-1} K^{-1}
$\qquad$ = 0.08206 L atm mol^{-1} K^{-1}
$\qquad$ = 62.36 L torr mol^{-1} K^{-1}
1 atm = 760 mm Hg
$\qquad$ = 760 torr
STP = $0.00\,^\circ$C and 10^5 Pa

THERMOCHEMISTRY/ ELECTROCHEMISTRY

$$q = mc\Delta T$$

$$\Delta S^\circ = \sum S^\circ \text{ products} - \sum S^\circ \text{ reactants}$$

$$\Delta H^\circ = \sum \Delta H_f^\circ \text{ products} - \sum \Delta H_f^\circ \text{ reactants}$$

$$\Delta G^\circ = \sum \Delta G_f^\circ \text{ products} - \sum \Delta G_f^\circ \text{ reactants}$$

$$\Delta G^\circ = \Delta H^\circ - T\Delta S^\circ$$

$$= -RT \ln K$$

$$= -n F E^\circ$$

$$I = \frac{q}{t}$$

q = heat
m = mass
c = specific heat capacity
T = temperature
S° = standard entropy
H° = standard enthalpy
G° = standard free energy
n = number of moles
E° = standard reduction potential
I = current (amperes)
q = charge (coulombs)
t = time (seconds)

Faraday's constant, F = 96,485 coulombs per mole of electrons

$$1 \text{ volt} = \frac{1 \text{ joule}}{1 \text{ coulomb}}$$

Copyright © Sterling Test Prep.

Solubility Rules for Ionic Compounds

Use the following rules to *predict the solubility of ionic compounds* in water. These rules generally relate to the compound's lattice and hydration energies of the individual ions.

One of the factors determining the *lattice energy* is the magnitudes of charges on the ions (i.e., the greater the magnitude of the charges, the greater the lattice energy). Simplified, the greater the lattice energy, the lower the solubility.

The rules focus on the magnitudes of the charges and ignore other influences on the lattice energies, such as hydration energy. In addition, the rules disregard the effect of temperature.

The solubility rules are listed in order of *decreasing importance*. For example, if the first rule applies, disregard the subsequent rules. Similarly, if the second rule is applicable, disregard the third rule. Thus, only use the third rule when rules one and two are not applicable.

The following ions are nearly always soluble: $C_2H_3O_2^-$, ClO_3^-, ClO_4^-. K^+, Na^+, NH_4^+, and NO_3^-.

General principle: Hydrides (H^-) decompose in water to yield H_2 and ^-OH. Many metal oxides react with water to produce hydroxides (^-OH).There are exceptions to the solubility rules besides those listed. However, the ions normally encountered follow these rules.

Rule 1: compounds containing a +1 or −1 ion are typically soluble

Examples include:	$AlCl_3$, $FeCl_2$, K_2SO_4, $NaBr$, Rb_3PO_4
Exceptions (Insoluble):	OH^- (other than with cations that produce strong bases)
	Hg_2^{2+} (other than with NO_3^-, $C_2H_3O_2^-$, ClO_3^-, and ClO_4^-)
	Pb^{2+}, Hg^{2+} with most −1 ions
	IB metals (column 11)

Rule 2: compounds containing a +3 or −3 or higher ion are typically NOT soluble

Examples include:	$AlPO_4$, $Ca_3(PO_4)_2$, $TiO2$
Exceptions (Soluble):	cations combined with sulfate or dichromate

Rule 3: compounds containing −2 ion are typically NOT soluble

Examples include:	$CaCO_3$, FeS, ZnC_2O_4
Exceptions (Soluble):	dichromates and sulfates
	(insoluble if combined with Ba, Ca, Hg, Pb, and Sr)

Solubility examples for ionic compounds

AgBr	Insoluble	Rule 1	Ag^+
AgCl	Insoluble	Rule 1	$Ag^+ + Cl^-$
AgIO$_3$	Insoluble	Rule 1	$Ag^+ + IO_3^-$
AgOH	Insoluble	Rule 1	weak base
Ag$_2$S	Insoluble	Rule 1	Ag^+
Hg$_2$Cl$_2$	Insoluble	Rule 1	Hg_2^{2+}
PbCl$_2$	Insoluble	Rule 1	Pb^{2+}
AgNO$_3$	Soluble	Rule 1	$NO_3^- + Ag^+$
Ba(NO$_3$)$_2$	Soluble	Rule 1	NO_3^-
Ba(OH)$_2$	Soluble	Rule 1	strong base
NaCl	Soluble	Rule 1	$Na^+ + Cl^-$
NaClO$_3$	Soluble	Rule 1	$Na^+ + ClO_3^-$
NaOH	Soluble	Rule 1	strong base
Na$_3$PO$_4$	Soluble	Rule 1	Na^+
(NH$_4$)$_2$CO$_3$	Soluble	Rule 1	NH_4^+
NH$_4$IO$_3$	Soluble	Rule 1	$NH_4^+ + IO_3^-$
FeAsO$_4$	Insoluble	Rule 2	$Fe^{3+} + AsO_4^{3-}$
Fe$_2$(SO$_4$)$_3$	Soluble	Rule 2	SO_4^{2-}
BaCO$_3$	Insoluble	Rule 3	CO_3^{2-}
BaSO$_4$	Insoluble	Rule 3	$Ba^{2-} + SO_4^{2-}$
MgSO$_4$	Insoluble	Rule 3	SO_4^{2-}
PbCrO$_4$	Insoluble	Rule 3	CrO_4^{2-}
ZnS	Insoluble	Rule 3	S^{2-}
Ba(IO$_3$)$_2$	Insoluble	exception	

Copyright © Sterling Test Prep.

ACS Physical Chemistry Preparation and Test-Taking Strategies

Test preparation strategies

The best way to do well in ACS Physical Chemistry is to be good at chemistry. There is no way around knowing the subject; proper preparation is key to success. Prepare to answer with confidence as many questions as possible.

Study in advance

Assuming you completed all necessary classwork and did fairly well at it, devote 2 to 3 months to studying. The information is manageable by studying regularly before the test.

Cramming is not a successful tactic. However, do not study too far in advance. Studying more than four months ahead is not advised and may result in fatigue and poor knowledge retention.

Develop a realistic study and practice schedule

Cramming eight hours a day is unfeasible and leads to burnout, which is detrimental to performance.

Commit to a realistic study and practice schedule.

Remove distractions

During this preparation period, temporarily eliminate distractions.

However, balance is critical, and it is crucial not to neglect physical well-being and social or family life.

Prepare with full intensity but do not jeopardize your health or emotional well-being.

Develop an understanding over memorization

When studying, devote time to each topic.

After a study session, write a short concept outline to clarify relationships and increase knowledge retention.

Make flashcards

Avoid commercial flashcards because making cards help build and retain knowledge.

Consider using self-made flashcards to develop knowledge retention and quiz what you know.

Find a study partner

Occasionally studying with a friend to prepare for the test can motivate and provide accountability.

Explaining concepts to another improves and fine-tunes your understanding, integrates knowledge, bolsters competence, and identifies deficiencies in comprehension.

Take practice tests

Do not take practice tests too early.

First, develop a comprehensive understanding of concepts.

In the last weeks, use practice tests to fine-tune your final preparation. If you are not scoring well on practice tests, you want time to improve without undue stress.

Alternate studying and practicing to increase knowledge retention and identify areas for study.

Taking practice tests accustoms you to the challenges of test-taking.

Test day strategies

Be well-rested and eat the right foods

Get a full night's sleep before the test for proper mental and physical capacity. If you are up late the night before, you will have difficulty concentrating and focusing on the test.

Avoid foods and drinks that lead to drowsiness (carbohydrates and protein).

Avoid drinks high in sugar, causing glucose to spike and crash.

Pack in advance

Check what you are allowed to bring to the test. Pay attention to the required check-in items (e.g., confirmation, and identification).

Pack the day before to avoid the stress of not frantically looking for things on test day.

Arrive at the testing center early

Starting right is an advantage. Allow time to check in and remain calm before the test begins.

Plan your route to the center correctly without additional challenges and unnecessary stress. Map and test the route to the center in advance and determine needed parking.

If you are unfamiliar with the test location, visit before the test day to practice and avoid travel errors.

 Copyright © Sterling Test Prep.

Maintain a positive attitude

Avoid falling into a mental spiral of negative emotions. Too much worry leads to underperformance.

If you become anxious, chances are higher for lower performance in preparation and during the test.

Inner peace helps during preparation and the high-stakes test.

To do well on the test requires logical, systematic, and analytical thinking, so relax and remain calm.

Focus on progress

Do not be concerned with other test-takers.

Someone proceeding rapidly through the exam may be rushing or guessing on questions.

Take breaks

Do not skip the available timed breaks. Refreshing breaks help you finish strong.

Eat a light snack to replenish your energy. Your mind and body will appreciate the available breaks.

The best approach to any test is *not* to keep your head down the whole session.

While there is no time to waste, breathe deeply for a few seconds between questions.

Momentarily clear your thoughts to relax your mind and muscles.

Time management strategies

Timing

Besides good preparation, time management is the critical strategy for the exam.

Timed practice is not the objective at this stage.

While practicing, note how many questions you would have completed in the allotted time.

Average time per question

In advance, determine the average time allotted for each question.

Use two different approaches depending on which preparation phase you are working.

During the first preparation phase, acquire, fortify, and refine your knowledge.

During final practice, use the time needed to develop analytical thought processes related to specific questions.

Work systematically

Note your comprehension compared to the correct answers to learn and identify conceptual weaknesses.

Do not overlook explanations to questions as a source of content, analysis, and interdependent relationships.

During the second preparation phase, do not spend more than the average allotted time on each question when taking practice tests.

Pace your response time to develop a consistent pace and complete the test within the allotted time.

If you are time-constrained during the final practice phase, work more efficiently, or your score will suffer.

Focus on the easy questions and skip the unfamiliar

Easy or difficult questions are worth the same points. Score more points for three quickly answered questions than one hard-earned victory.

Answer the familiar and easy questions to maximize points if time runs out.

Identify strengths and weaknesses

Skip unfamiliar questions in the first round, as challenging questions require more than average time.

In the second review, questions you cannot approach systematically or lack fundamental knowledge will not be answered through analysis.

Use the elimination and educated guessing strategy to select an answer and move on to another question.

Do not overinvest in any question

Some questions consume more time than the average and make you think of investing more time.

Stop thinking that investing more time in challenging questions is productive.

Do not get entangled with questions while losing track of time.

The test is timed, so do not spend too much time on any questions.

Copyright © Sterling Test Prep.

Look at every question on the exam

It is unfortunate not to earn points for a question you could have quickly answered because you did not see it.

If you are in the first half of the test and spending more than the average on a question, select the best option, note the question number, and move on.

Do not rush through the remaining questions, causing you to miss more answers.

If time allows, return to marked questions and take a fresh look.

However, do not change the original answer unless you have a reason to change it.

University studies on students show that hastily changing answers often replace correct and incorrect answers.

Multiple-choice questions

Answer all questions

How many questions are correct, not how much work went into selecting the answers matters. An educated guess earns the same points as an answer known with confidence.

On the test, you need to think and analyze information quickly.

The skill of analyzing information quickly cannot be gained from a college course, prep course, or textbook.

Working efficiently and effectively is a skill developed through focused effort and applied practice.

Strategic approach

Strategies, approaches, and perspectives for answering multiple-choice questions help maximize points.

Many strategies seem like common sense. However, these helpful approaches might be overlooked under the pressure of a timed test.

While no strategy replaces comprehensive preparation, apply probability for success on unfamiliar questions.

Understand the question

Know what the question is asking before selecting an answer.

It is surprising when students do not read (and reread) carefully and rush to select the wrong answer.

The test-makers anticipate hasty mistakes, and many enticing answers include specious choices.

A successful student reads and understands the question precisely before looking at the answers.

Focus on the answer

Separate the vital information from distracters and understand the design and thrust of the question.

Answer the question and not merely pick a factually accurate statement or answer a misconstrued question.

Rephrasing the question helps articulate precisely what the correct response requires.

When rephrasing, do not change the question's meaning; assume it is direct and to the point as written.

After selecting the answer, review the question and verify that the choice selected answers the question.

Answer the question before looking at choices

This valuable strategy is applicable if the question asks for generalized factual details.

Form a thought response first, then look for the choice that matches your preordained answer.

Select the predetermined statement as it is likely correct.

Factually correct, but wrong

Questions often have incorrect choices that are factually correct but do not answer the question.

Predetermine the answer and do not choose merely a factually correct statement.

Verify that the choice answers the question.

Do not fall for the familiar

When in doubt, it is comforting to choose what is familiar. If you recognize a term or concept, you may be tempted to pick that choice impetuously.

However, do not go with familiar answers merely because they are familiar.

Before selecting the answer and how it relates to the question, think about it.

Experiments questions

Determine the purpose, methods, variables, and controls of the experiment.

Understanding the information presented helps answer the question.

　　　　　　　　　　　　　　　　Copyright © Sterling Test Prep.

With multiple experiments, understand variations of the same experiment by focusing on the differences.

For example, focus on changes between the first and second experiments, second and third, and first and third.

Direct comparison between experiments helps organize and apply the information to the answer.

Words of caution

The words *"all," "none,"* and *"except"* require attention. Be alert with questions containing these words, as the answer may not be apparent on the first read of the question.

Double-check the question

After selecting an answer, return to the question to ensure the selected choice answers the question as asked.

Fill the answers carefully

Many mistakes happen when filling in answers.

Filling answers correctly is simple but crucial.

Note the question number and enter answers accordingly. If you skip a question, skip it on the answer sheet.

Quantitative multiple-choice questions

Know the equations

Many exam questions require scientific equations, so understand when to use each.

As you work with this book, learn to apply formulas and equations, and use them in many questions.

Manipulate the formulas

Know how to rearrange the formulas.

Many questions require manipulating equations to calculate the correct answer.

Familiarity includes manipulating the terms, understanding relationships, and isolating variables.

Any duplication (copies, uploads, PDFs) is illegal.

Estimating

Estimating helps choose the answer quickly for quantitative questions with a sense of the order of magnitude.

Often, estimation enables the correct answer to be identified quickly compared to the time needed for calculations.

Estimating is especially applicable to questions where the answer choices have different orders of magnitude.

It saves time to estimate instead of computing the solution.

Evaluate the units

For quantitative problems, analyze the units to build relationships between the question and the correct answer.

Understand what value is sought and eliminate wrong choices with improper units.

Make visual notes

Write, draw, or graph the question to determine the information provided, the objective, and the concept tested.

A chart or table often makes the solution apparent, even if a question does not require a graphic answer.

Elimination strategies

If the correct answer is not immediately apparent, use the process of elimination.

Use the strategy of educated guessing by eliminating one or two answers.

Usually, at least one answer choice is easily identified as wrong. Eliminating one choice increases the odds of selecting the correct one.

Process of elimination

Eliminate choices:

- Use proportionality for quantitative questions to eliminate choices too high or too low.

- Eliminate answers that are *almost right* or *half right*.

   Consider *half right* as *wrong* since distractor choices are purposely included.

Copyright © Sterling Test Prep.

- If two answers are direct opposites, the correct answer is likely one of them.

  However, note if they are direct opposites or another reason to consider them correct. Therefore, eliminate the other choices and narrow the search for the correct one.

- For numerical questions, eliminate the smallest and largest numbers (unless for a reason).

Roman numeral questions

Roman numeral questions present several statements and ask which is/are correct.

These questions are tricky for most test-takers because they have more than one potentially correct statement.

Roman numeral questions are often included in combinations with more than one answer.

Eliminating a wrong Roman numeral statement eliminates choices that include it.

Educated guessing

Correct ways to guess

Do not assume you must get every question right; this will add unnecessary stress during the exam.

You will (likely) need to guess for some questions.

Answer as many questions correctly as possible without wasting time.

For challenging questions, random guessing does not help. Use educated guessing after elimination.

Playing the odds

Guessing is a form of "partial credit" because while you might not be sure of the correct answer, you have the relevant knowledge to identify some wrong choices.

There is a 25% chance of correctly guessing random responses since questions have four choices. Therefore, the odds are guessing 1 question correctly to 3 incorrectly.

Guessing after elimination of answers

After eliminating one answer as wrong, you have a 33% chance of a lucky guess. Therefore, your odds move from 1 question right to 2 questions wrong.

While this may not seem like a dramatic increase, it can make an appreciable difference in your score.

Any duplication (copies, uploads, PDFs) is illegal.

Confidently eliminating two wrong choices increases the chances of guessing correctly to 50%!

When using elimination:

- Do not rely on gut feelings alone to answer questions quickly.

  Understand and recognize the difference between *knowing* and *gut feeling* about the answer.

  Gut feelings should sparingly be used after the process of elimination.

- Do not fall for answers that sound "clever," and choose "bizarre" answers.

  Choose them only with a reason to believe they may be correct.

Eliminating Roman numeral choices

A workable strategy for Roman numeral questions is to guess the wrong statement.

For example: A. I only

B. III only

C. I and II only

D. I and III only

Notice that statement II does not have an answer dedicated to it. This indicates that statement II is likely wrong and eliminates choice C, narrowing your search to three choices.

However, if you are confident that statement II is the answer, do not apply this strategy.

                                    Copyright © Sterling Test Prep.

If you benefited from this book, please leave a review on Amazon so others can learn from your input. Reviews help us understand our customers' needs and experiences while keeping our commitment to quality.

Customer Satisfaction Guarantee

Your feedback is important because we strive to provide the highest quality educational materials. Email us comments or suggestions.

info@sterling-prep.com

We reply to emails – check your spam folder

ACS Physical Chemistry Practice Questions provides high-yield practice questions covering topics tested on ACS. Develop the ability to apply your knowledge and quickly choose the correct answer to increase your test score.

ACS Organic Chemistry Practice Questions provides high-yield practice questions covering topics tested on ACS Organic Chemistry exam. Develop the ability to apply your knowledge and quickly choose the correct answer to increase your test score.

Visit our Amazon

www.ingramcontent.com/pod-product-compliance
Lightning Source LLC
Chambersburg PA
CBHW081933160726
47999CB00008B/2368